MW01506118

NOISE MATTERS

For three who stood with me at the temple of the moon
and the one who saw the flaming feathers
and danced around the orchids

NOISE MATTERS

THE EVOLUTION OF COMMUNICATION

R. HAVEN WILEY

Harvard University Press

Cambridge, Massachusetts   London, England

2015

Copyright © 2015 by the President and Fellows of Harvard College
All rights reserved
Printed in the United States of America

First printing

LIBRARY OF CONGRESS CATALOGING-IN-PUBLICATION DATA
Wiley, R. Haven.
Noise matters : the evolution of communication / R. Haven Wiley.
pages   cm
Includes bibliographical references and index.
ISBN 978-0-674-74412-7 (alk. paper)
1. Auditory adaptation.   2. Noise.   3. Hearing—Physiological
aspects.   4. Auditory perception.   5. Hearing levels.   I. Title.
QP465.W55   2015
612.8′5—dc23   2014042000

CONTENTS

FIGURES

NOISE MATTERS

INTRODUCTION

THIS BOOK DESCRIBES a discovery. It is that communication between living organisms cannot avoid noise. Noise is not just annoying. It is inescapable. It is not just a conditional aspect of communication, but an essential element. And this new vista opens an ocean of possibilities.

So this book is not an introductory briefing about well-explored territory. It is not the estuarial musings after a lifetime meandering in the details of research. It is not a comprehensive report on a carefully cultivated specialty. It might contain elements of all of these, but the intent is to reconnoiter new terrain.

Although the influence of noise on communication might seem intuitively obvious, it has taken me four decades of research on animal communication to see the possibilities. No doubt I might have reached this point sooner, but it is also true that the problems turned out to be more complex than they seemed at first. Each step along the way, often following contributions of others, raised questions that were important to answer before taking the next step—or even before discerning the way ahead.

Part of the complexity of communication in noise comes from the nature of communication itself. Because it is a relationship between two parties—signalers and receivers, as I shall call them—when both of these parties are living organisms, both can evolve. The joint evolution of two sets of organisms in relation to each other is more complicated than the evolution of one set in relation to a more or less fixed feature of the environment. For instance, understanding how a population of animals or plants adapts to a particular environment, although not without its intricacies, lacks the complexity of joint evolution by interacting parties. Such joint evolution or coevolution has attracted attention for some time. Understanding how predators and their prey evolve jointly (or parasites and their hosts, or males and females, or parents and young) has presented challenges addressed by many evolutionary biologists. Many of these interactions include special cases of communication. Adding noise to these interactions complicates their evolution further.

The complication increases as a result of a trade-off that any receiver faces in the presence of noise. Almost two decades ago I proposed a way to understand how receivers should evolve to optimize their responses when constrained by this trade-off. My proposal, however, assumed that the relevant signals had fixed properties. It was clear that this proposal could not provide a complete understanding of the evolution of communication in noise, because signalers—and consequently their signals— evolve at the same time as receivers. My proposal in fact made it clear that the optimal properties of receivers depend on the properties of signals and vice versa: the optimal properties of signals depend on the properties of receivers.

This reciprocal dependence of course suggests that receivers and signals might evolve complementary features. Indeed, a correspondence between signals and sensory receptors has long been recognized by ethologists and neurophysiologists. As a rule, sensory organs respond especially well to the signals each species uses for communication. Once again, however, the addition of noise complicates matters. Signalers as well as receivers face inevitable trade-offs in optimizing their behavior.

These trade-offs for the participants in noisy communication change everything. Perfection in communication is unattainable. Instead, both parties evolve to a joint equilibrium, one at which both receivers and signalers do the best they can provided the other does so also. I eventually developed a way to explore this kind of joint evolution. This book formalizes my intuition of an equilibrium and traces its manifold consequences.

Because this project aims to explore a new way to think about communication, the argument must sway more than the usual skeptics. And to do so despite the suggestion near the end that skepticism has an important place in a noisy world. The central objective is nevertheless to persuade you, dear reader, that noise is indeed inescapable. Noise does not just annoy, it changes everything. Noise makes a difference. This demonstration occupies the central part of the book. The earlier chapters set the stage. The later ones broaden the perspective.

Part I of this book starts with an introduction to communication, information, and noise. Subsequent chapters consider how acoustic signals are produced and received so as to convey information and minimize noise. Next I consider how acoustic signals change during transmission from signaler to receiver and how animals have adapted to minimize this form of noise. Attention is also directed to other forms of noise, including the pervasive noise generated by humans in many environments today. Much less studied is the noise that results from properties of the signaler's and receiver's own nervous systems. Although the focus remains on sound and acoustic communication, the discussion reveals that all modalities of communication share the same general features of communication in noise. The stage is thus set for pursuing an analysis of how we should expect signalers and receivers to evolve in the presence of noise.

Part II presents this analysis. It requires, first of all, a more precise understanding of the properties of signals, receivers, and signalers—and of course noise. Because the issue is the evolution of communication, it is also essential to be clear about the properties of natural selection and the interaction of genes and environment in the development of behavior, even potentially complex behavior such as communication. Several mathematical chapters then provide a way to address the evolution of receivers and signalers—both separately and jointly. An evaluation of this exercise leads finally to an emphasis on two conclusions. First, signalers and receivers are not expected to evolve to escape noise. Ideal communication is not attainable. Second, honesty in communication is the norm, but deception by signalers and manipulation by receivers can never be excluded.

Part III of this book indicates how these consequences of noisy communication affect our understanding of animal behavior. As I proceeded over the decades, the broad relevance of communication in noise became progressively apparent. Because much social behavior depends on communication, noise alters possibilities for the evolution of social behavior. For instance, it changes the arguments about the evolution of honesty in communication. Sexual selection of males and females acquires a definite direction as a result of noise. And noise sets limits for the evolution of cooperation and thus complexity in societies. The applications extend beyond behavior. For instance, the principles of noisy communication apply to all of the inter- and intracellular signaling under investigation by the enterprise of molecular biology.

Part IV goes even farther. As the arguments in the preceding sections developed over the decades, I could see that the principles of noisy communication do indeed reach far. Not only do these principles apply to human communication—which is of course an extension of communication by nonhuman animals—but they also apply to forms of communication designed by humans, including all forms of electronic communication, analog or digital. The optimization of communication in noise encounters similar problems of diminishing returns and thus presents similar problems in every instance.

Moreover, the evolution of noisy communication has implications that extend to philosophical problems. Because the joint evolution of signalers and receivers cannot escape from noise, noise is an inevitable feature of communication. This noise isolates each receiver in its own partially subjective world. As a result, problems in the philosophy of language appear in a new light.

Furthermore, any organism, including any person, perceiving the external world is a receiver responding to signals. Perception has all the properties of reception, and perception in noise raises problems of optimization similar to those for reception of signals in noise. The evolution of perception thus puts limits on how we can think about acquiring knowledge of the world around us.

In the end, science itself is revealed as noisy perception of the external world. Even this book must become the subject of its own conclusions. First, it results from my perceptions, with the limitations that noise imposes, as developed in the course of this book, and, second, it attempts to communicate these perceptions to you, the reader, again with the limitations that noise imposes. Truly, noise matters.

From start to finish, the objective of this book is thus not only to indicate what noise is and how it affects the evolution of communication but also to explore the unexpected consequences of noise in communication. For such wide-ranging applications and implications there are many possible audiences, each with its particular focus. After considering options, I rejected the possibility of limiting the book to any one focus and thus excluding others. I might never have a chance to return to the neglected possibilities.

Recognizing that a decision to include everyone risks pleasing no one, I have tried at least to clear a channel for the central argument. From this channel each chapter takes the form of a meander in the general flow. Any reader wishing to shortcut across any one or more of these loops should find the current on the other side (although any shortcut requires acceptance of some intervening conclusions without discussion). Some of these excursions violate a principle for presentations to a diverse audience, for they include equations—even calculus—and complex plots. Those readers trained in mathematics and physics can no doubt float through these places faster than I was initially able to and also see directions that I missed. My goal in these passages, however, is to keep the way clear so everybody can rejoin on the other side. Readers are thus invited to explore, sometimes skipping sections and sometimes returning to divarications previously bypassed, but always finding the way ahead.

Bibliographic notes at the back of the book provide documentation for topics discussed in the text and guides for further reading. In hopes of encouraging readers to foray beyond their usual range of expertise, the chapters themselves are cleared of all citations (but include occasional pointers for coordination with the Bibliographic Notes). My hope is that everyone might explore the broadening perspectives, but anyone can dig where interested. I beg for some forbearance from readers who realize that I too have wandered far from my expertise, and for their sensitivity in considering why noise matters.

PART ONE

NOISE AND WAYS TO REDUCE IT

REALIZING THE WAYS in which animals manage their lives in a noisy world is a first step in understanding how communication evolves. Noise proves to be a pervasive theme for understanding communication. It takes many forms, as Chapter 1 explains. It also introduces a continuing, although not exclusive, focus of this book on acoustic communication. Sound is particularly easy to study, and thus acoustic communication provides many examples for discussion. Chapters 2 and 3 examine production and perception of sounds as ways to deal with noisy communication. In Chapter 4, the changes in sounds during transmission through natural environments suggest ways that animals might adapt to minimize these changes. Attenuation and degradation of sounds during transmission increase noise, especially at long range, and long-range communication is a specialty of many animals. Chapters 5 and 6 are then in a position to consider more complex ways that signalers and receivers might counteract noise in communication. All of these means for dealing with noise take us a long way in understanding the evolution of communication in noise. Nevertheless, the chapters in Part II will reveal some fundamental issues that remain.

1

NOISE AND SIGNALS INTRODUCED

NOISE AS ERRORS BY RECEIVERS

Noise, as we usually think of it, is sound that has no interest for us yet makes sounds of interest hard to hear. As a result of noise, we make mistakes. We sometimes misinterpret what we hear or miss important information. However intuitively sensible, this informal understanding of noise has several shortcomings. First, what counts as noise changes from person to person and from situation to situation. It seems entirely subjective. Furthermore, our intuition about noise focuses on sound, but other sensory modalities have similar issues.

To an objective observer, the only sure indication of noise is the receiver's mistakes. By determining which features of the situation result in mistakes, we can determine the properties of the relevant noise. By focusing on the receiver's mistakes, we not only avoid the reliance on subjectivity but can easily extend our studies to any modality. The first person to realize that receivers' errors are the key to understanding noise in communication was Claude Shannon in his seminal 1948 paper "The Mathematical Theory of Communication." He diagrams noise (see Figure 1.1) as an additional source that becomes mixed with the

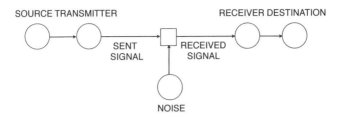

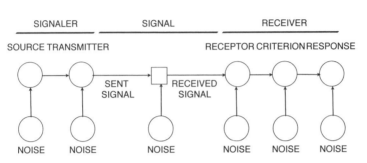

FIGURE 1.1. Diagrams of noisy communication. On top is an adaptation of Shannon's diagram (1948), which indicates that noise alters a signal during transmission from a signaler to a receiver. He nevertheless describes an independent observer studying noise by recording errors by the receiver. A diagram of noisy communication should thus include not only alterations during transmission but alterations (errors) by the receiver and also alterations (errors) by the signaler. The bottom diagram shows this comprehensive view of noisy communication. It also includes the components of a signaler and a receiver, described in more detail in Part II.

source of information. To measure noise he proposes "an observer (or auxiliary device) who can see both what is sent and what is recovered (with errors due to noise). This observer notes the errors in the recovered message and [can send data to the receiver] to correct the errors." His measure of noise is the amount of information necessary to correct the receiver's errors.

Shannon's insight that noise is measured by a receiver's errors makes it clear that we cannot separate the notion of noise from receivers' errors. The causes of these errors are manifold. Not all of them originate in the receiver itself. They include extraneous sources of energy that mix with signals and mask them or otherwise confuse receivers. In addition, noise results from any degradation of signals that makes them less certainly detected or discriminated. Noise also results from internal errors of signalers as well as those of receivers. In all cases, though, the consequence of noise is error by receivers.

AN AVIAN EXAMPLE

Consider a songbird, such as the Hooded Warbler that my students and I studied for more than a decade in North Carolina. During spring and summer, male Hooded Warblers produce loud songs within their territories, areas some 10 hectares (25 acres) in size that are defended aggressively against other males of the same species. Our experiments, described in more detail in later chapters, showed that these songs are signals that males use in order to advertise their territories and that other males use to recognize the presence and identity of rivals or neighbors. It is possible that they are also signals that females use in making their choices of where and with whom to mate. A male Hooded Warbler sings its ten or so patterns over and over. Each pattern varies only slightly in successive performances, so in this respect they are like human songs with a recognizable melody. Each of these patterns has features shared by the songs of all Hooded Warblers, but each also has some features peculiar to the individual singer. By the time the sound of one of these songs has traveled 50 and even as much as 200 meters through a forest to a potential listener, it has changed considerably. Reverberations as a result of reflections from solid surfaces, such as trunks and foliage of trees, and attenuation, as a result of scattering from foliage and absorption by molecules in the atmosphere, in addition to spreading, alter almost every feature of the song. The song has been transformed. Comparisons of a song recorded close to a singing male and at a distance through a forest show that most of the details vanish in transmission.

In addition to signals, which are those patterns of energy of interest to a receiver, there is always energy of no interest to a receiver but nevertheless affecting its sensory receptors. A warbler listening to singing males also hears other sound. Some extraneous sound comes from nearly "white" noise (with a wide spectrum of frequencies) from rustling of nearby foliage or from "pink" noise (with low frequencies predominating) of traffic on highways up to two or more kilometers away. Rain and running water produce noise in a wide spectrum of frequencies and patterns too. In rain many birds simply stop singing (and presumably listening too).

A listening Hooded Warbler also has to deal with many other species signaling at the same time. Of species in the same forest where we worked, the Red-eyed Vireo routinely produces a song that so closely resembles a Hooded Warbler's song at a distance that I myself never managed to avoid momentary false alarms when listening for distant warblers. It remains an unanswered question whether or not Hooded Warblers are similarly misled. It is also possible that some of the songs of these warblers are deceptive. One reason neighboring males share similar song patterns could be that inexperienced young males mimic established older males in order to improve their chances of attracting females or discouraging intruders. For some potential listeners, these deceptive signals might result in erroneous responses.

Finally, there is variation in the physiology of the cells in the warblers' nervous systems. Singing males do not, and presumably cannot, produce exact copies of any one song pattern. Male birds are remarkably good at replicating their song patterns, but careful analysis of their songs shows that they are not perfect. Listeners are also not perfect. Our experiments with playbacks of Hooded Warblers' songs revealed that responses by listening warblers are obviously less than perfectly predictable, even when variation in the signal and in the situation are standardized.

The example of Hooded Warblers introduces all of the sources of noise in communication. The first category of noise is extraneous energy that affects the sensory receptors of receivers. Often this includes random energy in the environment, diffuse sources of energy that are predictable only as statistical ensembles. On some occasions, however, this background energy includes particular patterns, such as signals of other species, that resemble signals of interest to a particular receiver. Furthermore, some patterns of irrelevant energy not only resemble signals of interest by chance but have so evolved to deceive a receiver. A second category of noise arises from degradation of signals during transmission from the sender to the receiver. This degradation usually includes random changes, so the receiver encounters signals partially altered in unpredictable ways from their original structure. These two sources of noise are environmental. A third category of noise arises within the sig-naler or the receiver. Both parties' nervous systems have an element of randomness, so some noise in communication must result from the participants' internal errors in producing or responding to signals. Shannon recognized this source of noise also (although it is omitted from his diagram). He described a noisy channel as one in which "the signal is perturbed by noise during transmission or at one or the other of the terminals." There are thus internal sources of noise as well as external ones.

The first category of noise comprises three progressively more spe-cific forms of extraneous energy affecting a receiver's sensory receptors. First is energy from inanimate sources. More specific is that from other species' signals. Finally there is energy from signals of other individuals of the same species but nevertheless not relevant for a particular receiver. The last two categories result from deterioration of a signal's properties and associations, either during transmission between signaler and re-ceiver or during internal processing by signaler or receiver. All result in the characteristic feature of noise—errors in receivers' responses to signals.

SIGNALS AS THE MEDIUM FOR INFORMATION

The example of the warblers also emphasizes another important feature of communication—information. Shannon's paper, in addition to pro-viding a clear proposal for measuring noise, did the same for measuring information. He proposed that the amount of information in a series of signals depends on the probabilities of those signals. In his widely known equation, the amount of information (H) equals –1 times the grand sum of the probability of occurrence of each signal times the logarithm of its probability: $H = -\Sigma\, p_i \log_2 p_i$, with p_i = probability of occurrence of signal i. Notice that the logarithms are taken with a base of 2 (not the usual 10 or e). Shannon and Warren Weaver in their well-known 1963 book de-scribe how this formula is not only the simplest one that satisfies our intuitive idea of an "amount of information" but also has a straightfor-ward interpretation. It is intuitive because a series of the same signal repeated without alternatives ($p_i = 1.0$) conveys no information, and a series of n signals in which each signal occurs with $p_i = 1/n$ conveys the most information possible for a series of that length. With logarithms to the base 2, H becomes the average number of binary (yes-or-no) questions necessary to guess the next signal in a series. The more guesses necessary, on average, the more information is conveyed by the arrival of the next signal. A signal conveys a lot of information when it settles a lot of questions. The upshot is that signals with greater unpredictability encode more information. Information is the negative of predictability.

What Shannon and Weaver do not emphasize is what constitutes a signal. Yet it is clear from Shannon's discussion of noise and information that a signal must be energy or matter that a receiver can discriminate from other sources of energy or matter. In other words, a signal must have identifying features or properties. In could be a single feature in a restricted place or time, such as a spot of matter that reflects light with the wavelengths of red or a sound of limited duration with a particular wavelength.

Wavelength is directly related to the velocity of a wave and inversely related to its frequency (at least for light and sound as we normally experience them): wavelength equals velocity divided by frequency. Frequency (and hence wavelength) determine the color of light (visible electromagnetic radiation) and the pitch of sound (waves of pressure). So the dominant pitch of a middle A on a piano has a frequency of 440 Hz (hertz, the standard international unit for cycles per second). The velocity of sound in air on the surface of the earth is about 340 meters per second. So the wavelength of a middle A is slightly more than 3/4 meter. Hitting a middle A on a piano produces a signal that we can discriminate from a signal produced by hitting a middle C or any other note.

Suppose we wanted to know what tune a pianist would play next (let's assume the pianist is a beginner and only picks out tunes with one finger, one note at a time). The first note this pianist hits provides some information about the tune that is coming. There are 88 keys on a standard piano these days, so the information from the first key might be $H = -88 (1/88) \log_2 (1/88) = 6.46$ bits (short for binary digits because we are using logarithms to the base 2) or the average number of yes-or-no questions we would need to guess this key. Actually, the bits of information would be considerably fewer, because tunes usually start with a note near the middle of the keyboard, so the actual probability of different starting notes is much higher than 1/88.

To revert to the previous example, notice that our Hooded Warblers face a similar problem. What species' (and perhaps what individual's) song do I (speaking for a warbler) hear? The first note provides some information, which in principle is measurable, as just described. The first note is usually not enough, however, to identify the species and individual, any more than a single note can identify a human tune. Signals usually do not consist of a single feature at a particular time or place, but instead a combination of features arranged in time and space. In other words, signals are patterns of features in time and space. They might be a pattern of tones in time (such as our beginning pianist might play or a bird might sing). It could have extraordinary complexity, such as a romantic symphony or some birds' songs. It could consist of multiple frequencies (tones) at a time, or multiple simultaneous frequencies varying in time. The pattern could be a spatial arrangement of colors. Simultaneous colors (frequencies of light waves) might vary in space to produce varying hues in complex spatial arrangements. A signal could be a dollop of molecules emitted into the air or water or deposited on a twig or the ground. The dollop might contain several kinds of molecules.

In all cases a signal is the predictable temporal and spatial association of its features. The predictable association of features makes a pattern. To the extent that they are predictably associated, the different features of a signal do not themselves convey information. They instead serve to differentiate one signal from other possible signals and from irrelevant features of the environment, the background against which a receiver perceives the signal. To respond to a signal, receivers must detect it (determine that it has occurred) and usually must discriminate it from others (distinguish it from other possible signals that might occur). Signals in the form of patterns of energy or matter serve for both detection and discrimination.

Another issue that Shannon and Weaver do not discuss is what kind of information a signal conveys (as opposed to how much information it conveys). To tell the truth, their discussion of information often seems diametrically opposed to any commonsense understanding of information. Most people think of information as *about* something, and their primary concern is the *reliability* of information. The quantity of information, as discussed by Shannon and Weaver, and its quality, as conceived by most people, constitute two distinct aspects of information. On the one hand, the quantity of information a signal conveys, as Shannon explains, is related to how unexpected it is. A signal that can be predicted with high assurance cannot provide much information for a receiver, regardless of what the information is about. On the other hand, the quality of information conveyed by a signal is related to which events or situations are associated with its occurrence.

For instance, signals correlated with the presence of a potential predator or a potential mate might be similarly unexpected, and thus might convey similar amounts of information, but nevertheless might evoke contrasting responses from a receiver—at least from one that maximizes its survival and reproduction. Shannon's formula is the appropriate measure for the amount of information conveyed by signals; the statistical correlation or association between signals and other events or situations is the appropriate measure for the quality of information. The stronger the statistical correlation or association, the greater the reliability of the information provided by a signal about a particular situation. Shannon's formula and statistical correlations thus measure respectively how much information and what information signals convey—or, more precisely, the quantity and the reliability of information. These aspects of information in signals are, as statisticians would say, orthogonal in their influence on the utility of communication. A lot of information of low reliability is just as nearly useless as a little information of high reliability. As with so many other things, information is only fully understood by measures of both its quantity and its quality.

With this introduction to noise and signals, the following chapters consider how animals (including humans) can counteract the consequences of noise (errors) in communication. To do so, it is necessary to consider how they produce signals, how they perceive them, how signals themselves can change during transmission from a signaler to a receiver, and how their features influence how easily a receiver can detect and discriminate them. Birdsong and the human voice provide a contrast in balancing noise and information. Noise varies in accordance with the biological and physical properties of environments. As a result, animals have different ways to cope with noise in the environments they inhabit. Sound continues to provide the focus in the following chapters, although the principles extend to other modalities as well.

2

PRODUCING ACOUSTIC SIGNALS IN NOISE

CONCENTRATED ENERGY

THE EXAMPLE of male Hooded Warblers singing within their territories introduces an obvious source of noise in communication and an obvious way to counteract it. Much of their communication occurs at long range (often 100 meters or more) through dense forests, conditions that would make it very difficult for two people to communicate by voice. To compensate for the resulting noise in communication, these warblers like many other birds, produce songs that are remarkably loud. Furthermore, at any instant during a song this intensity is concentrated in a single frequency of sound (the feature of sound that we perceive as pitch). A louder signal stands out more from whatever irrelevant sounds are present at the same time. Moreover, since these irrelevant sounds often include many different frequencies at any moment, concentrating the intensity of a signal in a single frequency makes it stand out even more. Producing intense signals of concentrated energy is one of the fundamental ways to counteract noise in communication.

Although the example of the warblers has focused on acoustic signals, the same principle applies to other modalities as well. Visual signals that concentrate the maximal intensity in a single wavelength of light have the same advantage as intense sounds of a single frequency. It is customary to characterize light by wavelength (the property that we perceive as color), but, because the speed of light is constant, the wavelength of light is inversely proportional to the frequency of light, just as the wavelength of sound is inversely proportional to its frequency when the speed of sound is constant. Intense light at a single frequency (wavelength) makes the throat of a hummingbird (at an appropriate angle with the sun) stand out among other colors in the environment, just as intense sound at single frequencies makes the song of a warbler stand out among other sounds. The same applies to touch and odors. Concentration of a particular molecule makes an odor stand out among other odors.

Any of these modalities could illustrate ways that animals have evolved to produce intense and concentrated signals. Sound has the advantage that it is relatively easy to record, analyze, and manipulate. It is also the modality I have studied most thoroughly. This chapter thus focuses on how sounds become signals that counteract noise. Producing intense, concentrated sounds faces some challenges. Intensity requires oscillations in pressure of relatively large amplitude. Concentration requires limitation to a single (or narrow band of) frequencies at any time.

PRODUCING SOUND

Sound, to review some basics, is a wave of pressure in a medium such as the atmosphere or water. Like other waves it consists of a fluctuation propagating outward from its source. The features of a sound wave depend in part on how fast it propagates. Consider pulses of pressure produced at a constant rate. It is not too difficult to imagine that they spread farther apart if they move more rapidly through the medium; each pulse moves farther before the next one starts. In other words, a sound of a particular frequency (rate of producing pulses) has a wavelength (distance between pulses) that is proportional to the speed of the pulses in the medium. The speed of sound in water is more than four times the speed of sound in air. Consequently, a sound of a particular frequency in water has a wavelength that is more than four times longer than a sound of the same frequency in air.

Producing sound requires some mechanism to generate these fluctuations of pressure. Often a vibrating object alternately pushes and pulls on the medium to create small changes in pressure, which then propagate outward as sound. Loudspeakers, drums, and sound boards of musical instruments are examples. Alternatively, sound can result from a mechanism that interrupts the flow of air or water to create intermittent pushes in the medium. The vocal folds of many animals are examples, as are the reeds of wind instruments or the lips of musicians playing brass instruments. In addition, sound can result from vortices in the medium. Whistles, for instance, result from such vortices when exhaled air is forced through an orifice, such as human lips when whistling or the nostrils of a bull elk. The frequency and intensity of the resulting sound is also affected by resonating structures or cavities adjoining the source of vibrations.

For instance, the sound produced by the vibrating reed of a clarinet is constrained by resonance in the tube of the clarinet. A tube open at one end, which a clarinet resembles, has a fundamental resonance at a wavelength equal to approximately four times the length of the tube (see Figure 2.1). At the closed end of the tube the movement of air molecules falls to zero, because the rigid barrier there is essentially inflexible, while at the open end the movement is maximal, because the scope for movement there is greatest. Consequently, the length of a clarinet, from the mouthpiece to the first open finger hole, corresponds to one quarter of a wavelength of sound. It thus resonates with a fundamental frequency corresponding to a wavelength four times the effective length of the clarinet. In addition, there are successive minor resonances at all odd multiples of the fundamental (one, three, and five times the fundamental, and so forth). The mouthpiece of a clarinet is not strictly analogous to the closed end of a tube, because a narrow opening rapidly closes and opens as the reed vibrates. For the sound wave inside the clarinet, the narrow opening makes little difference. On the other hand, the resonating sound wave can constrain the vibration of the reed to match the resonant frequencies. As a result, the energy of the musician's exhalation (aside from some frictional loss) is focused on these frequencies of sound.

THE HUMAN VOICE

The human larynx is superficially similar, because sound is generated by the vibration of the vocal folds (chords), which interrupt the flow of exhaled air much as the reed of a clarinet interrupts the flow of the musician's exhalation. Furthermore, the sound produced by the vocal folds is modified by the resonances of the cavities of the throat (pharynx), nose, and mouth. Together these spaces are the vocal tract. This space is much more complex in shape than a tube (see Figure 2.2), and it has correspondingly more complex patterns of resonance and higher harmonics, including resonances that might not be related harmonically (in integer multiples). These resonances, called formants by linguists, change as a person changes the shape of the vocal tract by altering the positions of lips, tongue, palate, and jaw. In this way we produce the various vowels of human speech, each of which is a continuous sound with different combinations of resonant frequencies (formants).

The complex resonances of the human vocal tract do not constrain the vibrations of the vocal folds, at least not as strongly as the resonances of a clarinet constrain the vibrations of the reed. This independence of the larynx and pharynx is produced by a partial constriction of the "false vocal folds," two folds just downstream from the vocal folds themselves, and by a partial constriction near the epiglottis, at the junction of the trachea and pharynx (throat). These constrictions create an impedance mismatch between the laryngeal and pharyngeal spaces. *Impedance* is a mathematical term for resistance to an alternating flow, such as the minute movements of molecules back and forth as a wave of pressure passes through air. An impedance mismatch between two compartments reduces the transfer of energy between them. It also tends to isolate the resonances of the compartments. A speaking human can thus adjust the resonances of its vocal tract without major effects on the vibration of the vocal folds.

The result is often called the source-filter model of human speech. The vocal tract acts like an acoustic filter. Vibration of the vocal folds produces a series of rapid pulses, each characterized by a wide range of frequencies. Those frequencies that match a resonant frequency of the vocal tract are preferentially passed onward, and the rest partly dissipate

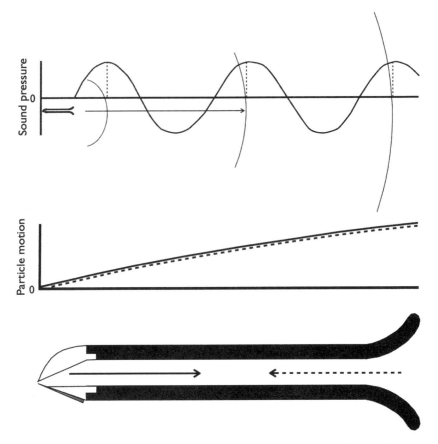

FIGURE 2.1. Clarinet playing its lowest note (all holes closed). This schematic shows a clarinet (much shortened and simplified) as a solid tube with a mouthpiece and a stiff reed at the left end and a horn at the right. The tube has a natural resonant frequency that depends on its length. This frequency is higher when the tube is shorter (as when the fingering changes to open some of the lower holes). The diagram shows what produces this resonance. When a musician blows on the mouthpiece and makes the reed vibrate, small puffs of air enter the clarinet at rapid intervals. Each puff starts a wave traveling at the speed of sound down the tube (indicated by the *solid arrow*). Near the horn most of the sound is reflected back toward the mouthpiece—an impedance mismatch, such as that between the large cross-section of air outside and the much smaller cross-section inside, produces reflection (indicated by the *dashed arrow*). The reflected wave returns toward the mouthpiece and is reflected there again, then again at the horn, and so forth. If a wave has just the right length (about four times the length of the tube or odd integer divisions of this length, 1/3, 1/5, and so forth) it reaches its minimal pressure but maximal movement of air molecules near the horn (at that point nothing opposes movement of molecules back and forth, as shown by the plot of the magnitude of particle motion).

as heat in the pharynx and mouth. What issues from our mouths is predominantly the resonant frequencies of the cavities of the vocal tract, the vowels of speech (in addition there are sounds generated by the lips, tongue, and throat themselves, the consonants of speech). The human voice is thus the result of a broad spectrum of frequencies produced by the vocal folds from which the nonresonant frequencies are partly subtracted. The vocal tract behaves like a passive acoustic filter. Like all passive filters, the human vocal tract reduces the total energy in a complex wave. The dissipation of frequencies other than the resonances attenuates the potential intensity of the human voice.

The human voice thus results from combining a source and a filter, which operate approximately independently. In other words, the prop-

---

The wave reaches the mouthpiece one-quarter of a wavelength later (where there is minimal movement of air molecules but maximal pressure). Thus at the mouthpiece the wave is ready for a boost by the next puff of air. Successive puffs of air augment the energy in the repeatedly reflecting sound—but only for these particular wavelengths and their corresponding frequencies. To think about resonance, remember pushing a child on a swing. The time the swing takes to move forward and back is determined by the length of the suspending ropes and the mass of the child (it is after all a pendulum like those studied by Galileo). If you push with just the right periodicity, your pushes augment the energy of the swing. If you push at any other periodicity, your pushes counteract the natural motion of the swing, and eventually they bring the motion of the swing to a halt. In an analogous way, puffs of air at the right periodicity (frequency) excite resonance in the tube of a clarinet; other periodicities are quenched and dissipate as heat. For the clarinet (or a swing) the situation is even more interesting because the returning sound wave itself influences the movement of the reed (or pushes) by exerting a backpressure for any frequency except the resonant frequencies (the returning wave of any other frequency does not return exactly to zero motion at the start). Consequently, the resonance in the tube of a clarinet entrains the vibrations of the reed. Vibrating reed and tube work together to augment the energy of a particular frequency, the resonant frequency, and its odd integer multiples. The human voice is also produced by a vibrating valve (vocal folds instead of a reed) and resonating cavities. Unlike the clarinet, the human vocal tract has several such cavities. Furthermore, in normal speech, these resonating cavities are partially isolated from vibrations of the vocal folds and thus do not entrain them. This crucial difference allows us to utter human vowels, as explained in this chapter. Some energy escapes from the horn of the clarinet, as it does from a person's mouth, to produce the pressure wave we hear as sound (*upper diagram*, which roughly indicates the wavelength of the lowest frequency of a B-flat clarinet, although the pressure wave would only predominate at distances beyond one wavelength from the horn).

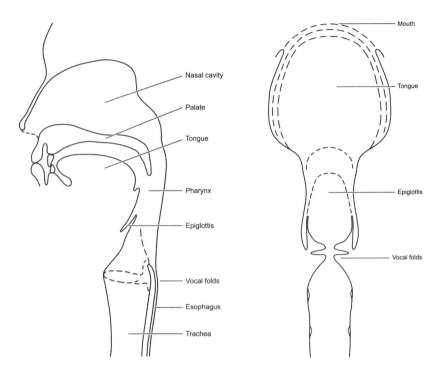

Mouth

Nasal cavity

Palate

Tongue

Tongue

Pharynx

Epiglottis

Epiglottis

Vocal folds

Vocal folds

Esophagus

Trachea

FIGURE 2.2. Human vocal tract. On the *left* is the human vocal tract in side view from the nose and lips through the throat (pharynx) to the vocal folds and the trachea. Vibrations of the vocal folds produce sound with a broad spectrum of frequencies, but resonances in the pharynx and the oral and nasal cavities determine which frequencies are most prominent in the sound issuing from the mouth. Notice the epiglottis, the flap that closes the trachea during swallowing. The space between the epiglottis and the vocal folds (the epiglottal cavity) partly isolates the vocal folds from the resonances of the oral cavity. As a result these resonances can change independently of the vocal folds to produce the formants of human vowels. When singing, the tension on the vocal folds is adjusted by movements of the cartilages that support them to produce different pitches (frequencies) of sound. Highly trained singers can expand the epiglottal cavity and vocal tract to allow the vocal folds to vibrate in unison with one of the resonances of the oral cavity and thus greatly increase the intensity of sound produced. The epiglottal cavity, in one of the great design flaws of human anatomy, creates a risk of choking when hastily swallowing an incompletely masticated lump of food. This side view does not reveal the complexity of the pharyngeal and oral cavities that produce the formants. On the *right* is the human vocal tract in rear view (rotating in the mouth to a view from above) on the same scale as the side view on the left. Below (and behind) the mouth and tongue are the pharynx with the epiglottis and the flanking pyriform cavities. Then come the false vocal (ventricular) folds separated from the vocal folds themselves by the wide but shallow ventricular cavity. The dimensions of these spaces determine the resonances of the vocal tract. Adapted from illustrations in many anatomical textbooks and Titze (2000), as well as dissections of my own.

erties and behavior of the larynx and pharynx do not influence each other much. The overall output approximates a linear combination of the effects of the source and the filter, without a large multiplicative (nonlinear) interaction term. This source-filter model works well to explain the frequencies of formants in normal speech. Singing, however, is a different case.

People sing different pitches primarily by adjusting the tension of their vocal folds. This tension affects the resonant frequencies of the folds and thus the pitch of singing. Some singers also manage to alter the conformation of the vocal tract in order to concentrate energy in the frequencies of the vocal folds. This trick is not easy, but it can yield a substantial increase in the intensity of the broadcast sound. Highly trained singers, especially sopranos and tenors in operas and orchestral concerts (not singers relying on microphones), practice diligently to lengthen the epiglottal tube and to open (widen) the vocal tract. These changes provide a better transition between the impedances of the vocal folds and the throat and thus allow more of the power from the vocal folds to be transferred to the vocal tract and then radiated from the lips. It also allows a transfer of energy the other way, so the vibration of the vocal folds becomes more closely coupled with the resonances of the adjoining spaces. As a result the voice of these singers comes to resemble a clarinet, in some basic features, as the vocal folds come to vibrate more nearly in unison with the resonances imposed by the vocal tract. The resulting sound has lost much less energy than has ordinary speech. On the other hand, the resonances can no longer be easily adjusted independently of the vibration of the vocal folds. The resulting sounds of vowels are simplified. As a consequence we find ourselves impressed by the volumes of sound produced by great performers of opera but disappointed by their pronunciation! Likewise, one cannot talk through a clarinet (see the Bibliographic Notes for a bit more about this point). The human voice thus has two options (and some intermediates): weak volume and complex enunciation, as in normal speech, or strong volume and weak enunciation, as a result of rigorously trained contortion of the vocal tract. Intensity and enunciation are thus inextricably, physically linked features of human sound production.

Human vocalization thus must work around a trade-off between simple intensity and complex modulation. The weak volume of normal human speech is linked with our exceptional ability to modulate the spectral composition of sounds, an ability perhaps unmatched by any other organism. The rapidity with which we can alter combinations of frequencies allows us to encode lots of information in our speech in short amounts of time. Information lies in patterns of energy, and rapid variation of patterns of frequency can encode a lot of information.

Humans, however, pay for this ability to produce the complex modulation of vowels by uncoupling our vocal folds from our pharynx. The constricted epiglottal space allows marvels of spectral modulation at the expense of volume. We are thus adapted, perhaps supremely among all animals, to produce rapid spectral modulation of our vocalizations for communication at relatively short range. Only electronic amplifiers have partially overcome our disadvantage in long-range communication by voice.

Furthermore, we have paid a price for our vocal agility. The epiglottal space can trap a piece of food just the right size and choke us. Mammals with much shorter epiglottal spaces face less risk of choking (but lack the vocal gymnastics of humans). The human advantage of speech, however, has allowed our parents to warn us to chew with our mouths closed (and thus not to speak while swallowing)!

Mammals other than humans have vocal folds not unlike humans in their function. They, however, lack the highly modifiable vocal cavities that can generate vowel sounds. On the other hand, some mammals can produce impressively loud sounds. It would be fascinating to know more about how wolves produce the tonal sounds of their baleful howls. Some coupling of resonances in the pharynx with the vibrations of the larynx seems likely, much as in trained human singers. Other mammals have resonating cavities that are apparently also acoustically coupled with their larynges (see Figure 2.3). Some primates in rainforests have expanded spaces in the cartilages adjacent to their larynges or, in the case of gibbons, pouches in their throats. There is a report that puncturing these sacs in an African rainforest monkey reduces the intensity of its calls. Of course, many frogs also have balloon-like pouches that arise from their

FIGURE 2.3. A roaring male Mantled Howling Monkey reveals his swollen neck, a result of enlarged cartilages in its throat. The resulting cavity presumably serves to augment the low frequencies in its roar, which carry for kilometers through many tropical Central and South American forests—at least those where the monkeys have not been extirpated. Chapter 6 describes how Jay Whitehead showed that these monkeys can track the movements of neighboring groups by assessing the degradation of their calls over distance. He also provided this photo.

throats or the sides of their heads. Frogs have vocal folds that might again have some acoustic coupling with resonances of the air spaces of their lungs and throat pouches. These structures all offer possibilities for increasing the intensity and concentration of vocalizations and hence their range. Recent study of how these structures influence vocalizations has tended to confirm this point.

AVIAN SONG

Hooded Warblers, as already noted, produce impressively loud sounds with energy concentrated primarily in one frequency at a time. Many other birds match this intensity and concentration of sound for their songs and some of their other vocalizations, and so do other organisms. How do they achieve intensity and concentration of acoustic signals?

A bird produces sound primarily with its syrinx, located near the base of the trachea, rather than with its larynx, near the apex of the trachea, as in mammals and frogs (see Figure 2.4). The syrinx varies in its exact location and arrangement among different groups of birds, but it always includes two exceedingly thin membranes, called tympaniform membranes, in the wall of the trachea or each bronchus. Fine muscles control the tension of these membranes just as similar muscles control the tension of vocal folds in the larynx in many other animals. In some birds, such as doves, these membranes vibrate in unison with the frequencies of sound produced by the syrinx.

In order for the tympaniform membranes in the walls of the trachea or bronchi to vibrate, there must be air spaces on both sides of them. In other words, there must be air outside the bronchi as well as inside them. Birds have air sacs throughout their bodies, even in the centers of their bones. All of these air sacs are interconnected with each other and with the birds' lungs. A large air sac occupies part of the abdomen and a smaller one the apex of the thorax, roughly between the clavicles, where it wraps snuggly around the bifurcation of the trachea. The tympaniform membranes of the syrinx thus are backed by this interclavicular air sac (see Figure 2.4).

Hooded Warblers belong to the group of birds known as songbirds, those in the order Passeriformes. Even more specifically, they are in the suborder Passeri, the "true" or oscine songbirds, which include about half of all known species of birds, some 5,000 species. In the oscine songbirds there are tympaniform membranes, as in other birds, but sound is controlled by a valve at the anterior end of each bronchus. This valve consists of two fleshy protuberances, called the medial and lateral labia. The latter can be pushed into the lumen of the bronchus by rotation of a specialized bronchial ring (see Figure 2.3). Sound is thus produced by air from the lungs passing through a valvular structure, somewhat like the vocal folds of a mammalian larynx, except that the syringeal valve is in the bronchi. Furthermore, recent evidence suggests that the syringeal labia, although they vibrate, do not completely close between vibrations. Because the position and tension of all of these components can

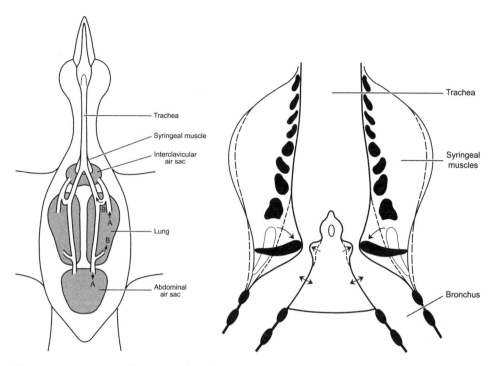

FIGURE 2.4. Avian vocal tract. On the *left,* most of the respiratory system of a bird is viewed from below, from the bill and tongue to the abdominal air sac. The trachea passes down the neck to the syrinx located at the origin of the two bronchi where the trachea divides. This bifurcation, along with the syrinx, is embedded in the cervical air sac, just anterior to the lungs in the bird's thorax. Slender evaginations from the major air sacs extend to every corner of a bird's body, including throughout their bones. A bird's respiratory system is much more efficient than a mammal's in extracting oxygen from the atmosphere because air passes in one direction through the lungs (not in and then out the same way, as in mammals). In the course of two cycles of inhalation and exhalation, air passes down the trachea and each bronchus directly into the abdominal air sac, as indicated by arrows, then forward through the lungs into the interclavicular air sac, and finally from there out each bronchus and the trachea. In the syrinx the thin membranes in the walls of each bronchus (or in some species at the base of the trachea) can vibrate because this air sac envelopes the syrinx. On the *right* is a magnified diagram of the syrinx itself. The cartilaginous rings of the trachea and each bronchus appear in cross-section as ovals. Notice that one of the first bronchial rings (in songbirds) is highly modified, so the minute intrinsic muscles of the syrinx (shown as swellings on either side) can tilt these rings into each bronchus (either one or both at the same time). In this position they produce a valve (syringeal labia) that constricts air flow from the lungs. As a result, the labia and the internal tympaniform membranes vibrate. The sounds produced in this way resonate in the trachea and a cavity formed by the pharynx, mouth, and upper esophagus. Because there is no arrangement to isolate the syrinx acoustically from the rest of these spaces, the resonance presumably entrains the vibration of the syringeal labia and the internal tympaniform membranes, much as the resonance in the cylinder of a clarinet entrains the vibration of the reed. As a result a singing bird can produce remarkably intense sounds for its size. Adapted from illustrations in Brackenbury (1982), Riede and Goller (2010), and Suthers and Zollinger (2008).

be adjusted separately for each bronchus, by means of the tiny syringeal muscles, each bronchus can produce sounds separately. In some species the two bronchi produce sounds simultaneously, in effect a self-duet. Some of the haunting phrases of the Wood Thrush provide good examples. Most songbirds use only one bronchus at a time often in rapid alternation. This surprising result was discovered by Roderick Suthers, Franz Goller, and Carolyn Pytte (1999) by measuring air flow with tiny thermistors inserted into the bronchi of singing birds. Various songbirds often alternate the active bronchus from note to note in a single song, and sometimes even within a single note. In some species the two bronchi seem specialized for lower and higher frequencies and thus together provide a wider register for their songs. The silenced bronchus at any instant is completely closed by occlusion of the two labia of the bronchial valve. The other bronchus remains partly open, and adjustments of the position and tension of the labia presumably control the intricate changes of frequency even in a single note. Exactly how the labia interact with the internal tympaniform membrane is a puzzle for the future.

The challenge for our present purposes, though, is to understand how songbirds produce loud, tonal songs, with a single, intense frequency at any instant. Just as in mammals, resonance of sounds in the air space between the source of sound (a bird's syrinx or a mammal's larynx) and the opening of the mouth influences the sound emitted. In birds this vocal tract includes the trachea as well as the pharynx and mouth. One line of evidence that resonance in the vocal tract affects sounds produced by songbirds comes from a strong correlation between the frequency of sound and movements of the bill. When producing higher frequencies, the bill is opened more, which shortens the effective length of the resonating "tube" between the syrinx and the mouth. Nevertheless, it is not clear that the movements of a bird's bill precisely tune its vocal tract to match the frequencies in its songs. The length of the vocal tract does not change enough as a result of movements of the bill to resonate with the range of frequencies in most birds' songs. Instead it seems likely that the movements of the bill contribute to a more major change in the configuration of the vocal tract. At least in some oscine songbirds, as discovered again by Suthers's team, major adjustments occur in the volumes of the mouth, throat, and upper esophagus, which together constitute an adjustable oropharyngeal-esophageal cavity (OEC). Calculations indicate that changes in the resonance of this cavity can track the fundamental frequency of syringeal sounds. It is still unclear how the thin tympaniform membranes adjacent to the syringeal labia might contribute to production of loud, tonal sounds.

Resonance of a bird's vocal tract was first revealed by an experiment by Steve Nowicki. He coaxed birds to sing in a modified atmosphere, with helium replacing most of the nitrogen, so that the resultant mixture of gases was less dense than air. This change altered the speed of sound and hence the wavelength and the corresponding resonant frequencies in the vocal tract of the bird. As a result, the frequencies in the birds' songs shifted upward, a sure sign that resonance in the bird's airspace affects the production of sound.

One interpretation of these experiments is that the vocal tract acts as a passive filter to remove syringeal frequencies other than those that resonate in the trachea and OEC, much as the human vocal tract acts as a filter to produce normal human vowels by selectively removing frequencies from the laryngeal sound. Experiments have confirmed that the cooing sounds of doves result from a wide spectrum of sound in the syrinx and a passive filter in the trachea and esophageal cavities. There were indications from another study that the same might also apply to songbirds.

An alternative interpretation, however, is that resonance of the vocal tract is coupled with the vibration of the labial valve in the syrinx. This coupling would result in a mechanism somewhat like the vocal tract of a trained human singer. In this case, a helium mixture would raise the frequencies produced by a songbird not by filtering but by changing the frequencies of vibration at the source—the syrinx. For humans, normal speech in helium produces higher vowels—the "Donald Duck effect" that many of us have experienced after inhaling air from a helium balloon. Nevertheless, the fundamental frequency of the normal voice, the vibration of the vocal folds, does not change in a helium mixture. Helium only changes the resonance of the oral cavities and thus the frequencies of the formants of vowels. The effect of helium on the normal human voice is thus clear evidence that the vocal folds normally vibrate nearly independently of the resonances of the cavities of the vocal tract and that the vocal tract acts as a passive filter to remove the energy in nonresonant frequencies. It is not clear whether a bird's syrinx also remains unaffected by helium mixtures, as expected for independent source and filter, or whether the syrinx vibrates in unison with the vocal cavities.

The membranes and labia of the syringes of songbirds are capable of especially intricate control. These birds have as many as 12 pairs of minute muscles that adjust the components of the syrinx on these membranes in multiple directions. It is perhaps the most complex sound-producing mechanism of any living organism. Furthermore the resonating air space potentially includes the bird's entire system of air sacs, which extend throughout its body. It thus seems probable, as Nowicki and Peter Marler have suggested, that Passeriformes achieve or even exceed the great power and tonal concentration of trained singers. At least the oscine passerines, the true songbirds, have the added advantage of great versatility in modulating the frequency of sound.

In addition to the power produced by the source of a sound and its coupled resonant spaces, the power actually conveyed to the atmosphere depends on the efficiency with which sound radiates from the organism. The way to increase this efficiency is to provide a gradual transition from the acoustic impedance near the source to the impedance of the surrounding atmosphere. As discussed previously, acoustic impedance depends on, among other things, the dimensions of the space in which sound travels, so a structure that flares from the narrower dimensions of the source to the wider dimensions of the surrounding space provides just such an impedance-matching device. The flaring ends of many orchestral instruments, such as a trumpet, have this effect. Mole crickets build burrows in which they produce stridulations underground. Nevertheless, these sounds are transmitted to listeners through the air above ground. The flaring dimensions of their burrows serve as impedance-matching devices for their stridulations. The bills, mouths, and throats of birds presumably also provide some impedance matching for the radiation of their songs.

Flaring horns are not the only possibilities for matching impedances between source and medium. Balloons can also serve this purpose. Many birds have large resonating structures associated with sound production. The expandable esophageal sacs of doves (as well as some grouse, bustards, and primates) are in direct contact with the trachea and thus must

FIGURE 2.5. A displaying male Greater Sage-Grouse has bare patches of skin, which bulge grotesquely as a result of an enormous underlying air sac. In this and other grouse, the sacs are enlargements of the esophagus inflated by air diverted from the trachea. The males' displays are followed by extravagant belches! The display of the male sage-grouse involves inflating these sacs, then compressing the air in them (by contracting musculature in the wall of the esophagus and in the overlying skin), and finally suddenly releasing this compressed air into two expanding balloons of relaxed sac and skin. The maximal expansion is over in an instant. The result is an abrupt pop, audible at dawn for a kilometer or more across the sagebrush prairies of western North America.

vibrate in unison with them. These large sacs serve to as a transition from the impedance of the internal air spaces to that of the external atmosphere and thus improve the radiation of sound. Other birds have extraordinarily elongated trachea, which loop within the sternum. One wonders if the sternum does not become an impedance-matching sounding board, somewhat like that of a piano.

The inflated sacs in the necks of many birds must help them to produce low-frequency sounds, as a result of the large resonating cavities they provide, but they would also improve the radiation of these sounds (see Figure 2.5). The anatomy of these sacs is extremely diverse. Some are balloon-like expansions of the esophagus and others of the trachea. All of these structures could improve the radiation of sound from its source and thus increase the intensity of a bird's vocalizations, as similar structures in mammals seem to do.

Although how it is accomplished is not yet completely understood, a male Hooded Warbler barely a half an ounce (15 grams) in mass produces a song with an intensity of 80–90 decibels (in relation to the minimal intensity for human hearing) measured one meter from its bill. These intensities are typical for the long-distance songs of songbirds. The Screaming Piha of South American rainforests (a suboscine songbird) is one of the loudest birds in the world. Its vocalization, sounding something like a human "wolf whistle," reaches 100 decibels (dB) one meter away. In contrast, normal human speech is only about 70 dB or less (100 times less power than 90 dB and 1,000 times less than 100 dB) and even an extremely loud shout by a human hardly reaches 90 dB. Yet humans have more than 1,000 times the overall mass (and presumably the muscle mass) of a warbler. So much power from such small organisms presumably requires a special mechanism. It is an attractive hypothesis that songbirds achieve such intensities of sound by avoiding the wasteful passive filtering of human speech and instead by coupling the vibrations of their syringeal structures with the resonances of their air spaces. I would have no problem accepting the Hooded Warblers I study as virtuosos. Definitive evidence, however, has yet to be mustered.

Another way to increase the distance at which a sound can be heard is to concentrate all of this power in a single frequency at any instant. Provided the appropriate listeners have ears that can discriminate different frequencies, a signal with all of its power in a single frequency at a time has advantages. Because background energy is usually spread over a relatively wide spectrum of frequencies, a signal with all of its energy at one frequency achieves a higher ratio of signal to noise for a frequency-analyzing ear. This advantage accrues at any distance from the source, so it extends the active space of the signal, the area within which it can elicit a response from an appropriate listener. It seems probable that many songbirds have evolved the nearly tonal songs that are so pleasant to our ears just to achieve greater "throw" for distant communication.

In addition to their ability to produce intense sounds, birds also have capabilities for rapidly modulating the frequency of sound in order to produce songs with great complexity. A warbler uses a single frequency at a time to make itself understood to listening rivals over 100 meters away through a forest. The complexity in modulation of this frequency serves in part to distinguish its songs from the acoustic signals of dozens of other organisms—insects, frogs, and other birds—all trying to communicate at the same time with similar frequencies in the same forest.

3

RECEIVING ACOUSTIC SIGNALS IN NOISE

TRANSDUCING SOUND

RECOVERING THE INFORMATION in sounds—whether a frog's croak, a falling rock, or the nightly news—requires an analysis of the patterns in sound. As emphasized in Chapter 1, signals of any sort contain information only insofar as their features have predictable patterns. Computers and other electronics allow us to visualize and to measure the patterns of sounds, but our ears and brains must also analyze these patterns, as must the ears and brains of other animals. Whether the apparatus is an ear, brain, or computer, the analysis of patterns has similar challenges and similar limitations. Furthermore, similar principles apply to the analysis of patterns in other sensory modalities as well. Sound, however, is again a good place to start.

This chapter focuses on a receiver's ability to analyze the frequencies in sounds. Acoustic signals are patterns of frequencies in time and space. Only by analyzing the frequencies in sounds (or at least some of the frequencies) can a receiver extract the information in acoustic signals, and, just as important, only then can it take advantage of signals that counteract noise by concentrating sound in one or a few frequencies at a time.

The actual changes of pressure in air or water as the result of a passing sound wave are so small (less than a billionth of the standard atmospheric pressure) that it takes special structures to register them at all, and even more remarkable ones to respond to them accurately. Until 1876, when Alexander Graham Bell patented the first telephone, humans relied entirely on their ears to detect sound. Microphones made it possible to record sounds by means of electromechanical devices for the first time. A thin membrane, often across a tube enclosed at one end, vibrates weakly as the pressure on the outside of the membrane rises and falls with a passing sound wave. Different kinds of microphones use this vibration to produce a fluctuating electrical resistance or capacitance and ultimately a fluctuating voltage. This voltage is then amplified enough to record nearly the exact pattern of fluctuations on some permanent medium— such as a vinyl record, magnetic tape, or a compact disc—or to drive a loudspeaker to reproduce the sound with more or less fidelity to the original.

Animals must also convert, or transduce, the minute fluctuations of pressure in a sound wave into a more useful form—in this case nerve impulses, small changes of voltage (action potentials) across the cell membranes of their neurons. Just as microphones do, animals all require some structure of extremely low inertia that vibrates with the passing sound wave. In some insects this structure is a fine hair in contact with a nerve cell. In most terrestrial animals, which detect sound in air, the primary structure for transducing sound is like a microphone, a thin membrane covering a closed space. Because the pressure in the space behind the membrane remains constant, the small fluctuations of pressure in a sound wave on the outside are sufficient to induce vibrations in the membrane. In humans the tympanum, or eardrum, which covers the opening of the middle ear, is an example. The middle ear behind the tympanum is effectively a closed space. The connection from the middle ear to the pharynx (the Eustachian tube) collapses except to allow occasional overall adjustments of pressure in the middle ear. As a consequence, the two tympanums on opposite sides of the head vibrate independently in response to a passing sound wave. The two tympanums usually do not vibrate identically, for two reasons. First, a sound wave travels past one ear and then, after a brief interval, past the other (except in the special case of sound arriving from directly ahead or directly behind). Second, the head can cast a sound shadow so the intensity of sound at one ear is greater than at the other. Most large animals, including all mammals, have ears that include tympanums of this sort. Many insects also have vibrating membranes covering enclosed spaces in the wall of the thorax, abdomen, or a leg. In these cases also each tympanum usually has its own closed space, and tympanums on opposite sides of the body vibrate independently as a sound wave passes.

An even more sensitive mechanism has two membranes on opposite sides of a single closed space. In some small birds and frogs, for instance, there is a permanently open passage from the tympanum on one side of he head, through the pharynx, to the tympanum on the other side of the head. It is as if our Eustachian tubes were wide open all of the time. As a result, the two tympanums vibrate jointly in response to a difference in external pressure at the membranes on the two sides of the head. For instance, imagine a sound wave reaching a small frog from its right side. When a peak in pressure reaches the right-hand tympanum, it pushes that tympanum inward. If the acoustic properties of the passage through the head from one tympanum to the other are just right, the air in the passage tends to push the left-hand tympanum outward. The movements of the two tympanums thus vibrate in concert rather than independently. Which direction they move at any instant depends on the *difference* between the pressures at the two membranes. This mechanism of coupled membranes produces an extremely sensitive response to sound and is particularly important for small organisms. The two tympanums on opposite sides of a small bird or frog might after all be separated by less than one centimeter, only 1/10 the wavelength of a 3,400 hertz (Hz) sound in air.

In all vertebrates the vibration of a tympanum is transferred by means of minute bones to a fluid-filled compartment enclosed in the skull, the inner ear. This bone (or in mammals, three bones) induces vibration in the fluid by vibrating a thin membrane in a tiny opening on one side of the inner ear. Because liquids are nearly incompressible, another membrane moves reciprocally and passively in a tiny, membrane-covered opening elsewhere. As the active membrane is pushed inward by the bones of the middle ear, the fluid of the inner ear shifts and the passive membrane pushes outward. By transferring the movement of the tympanum in air to movements in a fluid, the fluctuations of pressure in air have been transformed into oscillating movements of a fluid. In this way the hearing of all terrestrial vertebrates ultimately occurs underwater. Surges of the fluid of the inner ear push tiny hairs in contact with nerve cells on membranes that stretch the length of the inner ear. The nervous systems of all animals use the movements of fine hairs to transduce sound into nerve impulses.

This apparatus does not amplify the energy of the sound. It instead transforms fluctuations of pressure in the air to movements of fluids in the inner ear. Yet this change greatly improves the mechanical efficiency of bending the tiny hairs that trigger action potentials in the sensory cells. It is another elaborate system of impedance matching.

SOUND WAVES IN TIME AND IN FREQUENCY

There remains the challenge of recovering all the patterns in the fluctuations of pressure impinging on the organism. These patterns can be described in two equivalent ways. One is the waveform, the fluctuating pressure at any one point in space as a sound wave passes—in other words, a plot of the instantaneous amplitude of pressure against time. A waveform would have no pattern if the amplitudes of the wave at successive points in time were unpredictable (in other words, random). If these amplitudes were plotted in time, they would form a speckled band with no pattern (aside, perhaps, from some limitation on the maximum). Sounds of more interest to us have pressure rising and falling in ever-changing arrangements. A purely tonal sound, such as that from a tuning fork, would consist of a regular sinusoidal fluctuation in pressure, rising and falling 440 times each second (if the fork were tuned to middle A on a piano). A louder tone would have wider fluctuations in pressure, still in a sinusoidal pattern.

Another way to describe these fluctuations is the spectrum of frequencies that compose the waveform, a plot of amplitudes against frequency, with the amplitudes now indicating the strength of each frequency present during a particular interval of time. For instance, the tone just described consists of a single frequency (440 Hz). Its spectrum is a single point at a frequency of 440 Hz and a strength that indicates its amplitude. In contrast, the speckled waveform of random amplitudes described above is white noise, because all frequencies (within limits) have about equal strength at the same time. It is "white" in the same sense that light is white when it contains all frequencies of visible light at once. Its spectrum would include all frequencies (between a maximum and minimum) present with equal strengths. There are thus two ways to describe a complicated wave, either in the *time domain* as a waveform or in the *frequency domain* as a spectrum (see Figure 3.1).

The amplitude of each frequency in a spectrum is often converted to the *power* of that frequency in order to produce a *power spectrum*. Power is the amount of energy in a unit of time, which influences how much effect a sound has on sensory transducers, such as tympanums, and hence how loud it appears. The power of a wave is proportional to the square of its average amplitude. Likewise, in a power spectrum, the power of each frequency is the square of the amplitude of each frequency.

To be a little more precise, a spectrum is a set of frequencies, each with its own amplitude and phase (its relative alignment in time). The phase of each frequency is determined mathematically by adding a sine wave and a cosine wave, both with the same frequency but with different amplitudes. Often we settle for a simplified spectrum, because humans do not in general hear differences in the phases of frequencies in a sound. Consequently, we routinely compute a power spectrum by multiplying the sine and cosine terms for each frequency. Recall that power is related to the square of the effective amplitude of a waveform, so the power of each frequency in a spectrum is related to the product of amplitudes of the sine and cosine terms for that frequency.

The waveform is the usual way we think of waves. Ocean waves, for instance, rise and fall in time at any one point. It was not until the nineteenth century, however, that people recognized that sound also consisted of waves—pressure in air that rises and falls in time at any one point. The changes in pressure are admittedly rapid but nevertheless produce a pattern of amplitudes in time, much as ocean waves do. The waveform, however, is not the way we naturally think about sound. Instead, a musical score—combinations of frequencies in successive intervals of time—

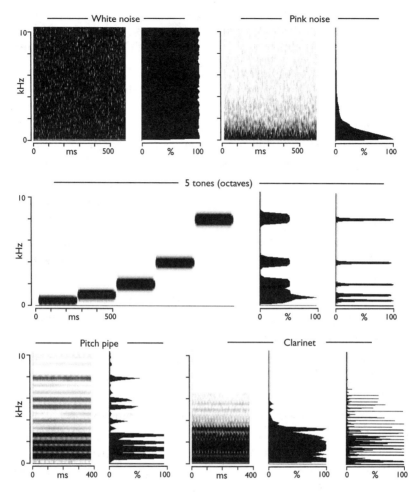

FIGURE 3.1. Spectrograms and spectra of "white" noise, "pink" noise, tones, and harmonics. A *spectrum* shows the frequencies that compose a waveform in a specified interval of time. Here each spectrum indicates the power (as a percentage of the maximum) of each frequency. In some cases, there are two spectra of the same sound but with different frequency resolutions. A *spectrogram* is a plot of frequency against time, with the power of each frequency in each interval of time represented by a shade of gray. Here each interval is 3 ms (thousandths of a second). The horizontal scale shows intervals of 100 ms (0.1 second, or about 33 vertical lines). The vertical axis is frequency from 0 to 10 kHz (thousands of cycles per second). The spectrogram of "white" noise is almost entirely black (as is the spectrum of any one interval), because all frequencies are continuously present in a waveform with random amplitudes. "Pink" noise also results from random amplitudes but with lower frequencies (longer wavelengths like red in a light wave) predominating. The spectrogram of a pure tone is a horizontal line, a single frequency lasting many intervals of time. The spectrogram and spectra of five tones span four octaves of frequencies, each octave twice the frequency of the preceding. Harmonics are equally spaced frequencies, each one an integer multiple of the fundamental frequency, as in the spectrograms and spectra of a pitch pipe and clarinet.

seems more intuitive, although even this way of thinking about sound only developed in the fifteenth and sixteenth centuries. A musical score is a simplified spectrum, really just a rough schematic in comparison to a mathematically precise spectrum of even a single musical instrument's sound. Its convenience for musicians reflects the way humans usually think of sound, as changing frequencies (pitches with varying intensities) rather than changing amplitudes of pressure.

Consequently, even mathematicians and engineers usually analyze sound in terms of frequencies in successive intervals of time. A spectrograph is a device that computes, within specified intervals or times, the frequencies present and their powers. We can choose brief intervals—if rapid changes in sounds are of interest—as in a human voice, a bird's song, or an orchestral performance. In these cases, spectra of brief, rapidly succeeding intervals of time seem more revealing than does a waveform. Mathematically, however, it makes no difference whether the spectrum of a long waveform is computed as a unit or as a series of shorter segments. Either way the spectra (complete with the phases) are equivalent to the waveform and contain the same number of parameters. One could, for instance, compute the complete spectrum of an entire 30-minute symphony. It would be a dauntingly complex spectrum, but it could be converted back to the 30-minute waveform, just as could the spectra of a series of 18,000 successive intervals each 1/10 of a second long. Of course the spectrum of a 30-minute symphony would be nearly impossible for us to understand (even if studied for 30 minutes) or printed over 30 pages of a book (the spectrum would presumably take about as long to print as the musical score or the waveform). This preference for short intervals has nothing to do with the mathematical equivalence of the long waveform and the complex spectrum. Instead, it has more to do with the way our ears work and thus the way we understand sounds.

RECIPROCITY BETWEEN A WAVEFORM AND A SPECTRUM

There is an important consequence of a shift from longer to shorter intervals of time. The shorter the interval of time analyzed, the wider the bands of frequencies included in the computed spectrum. This relationship shows up in the spectrum of any waveform. A constant frequency appears in a spectrogram as a band of frequencies, with maximal intensity at the actual frequency and decreasing shoulders of intensity on either side of this frequency. The rule is, the shorter the time interval of analysis, the wider the band for every frequency in the spectrum (see Figure 3.2).

This relationship is a fundamental principle in any spectral analysis of a waveform: the shorter the interval of analysis the greater the separation of the bands of frequencies in the spectrum. In other words, the greater the temporal resolution, the less the frequency resolution of the analysis. There is no getting around this inverse relationship regardless of whether the mechanism of analysis is electronic, digital, or neural. It is a result of the mathematical equivalence of a waveform (a series of amplitudes in time) and a spectrum (a series of frequencies in a finite interval of time).

This reciprocal relationship between the duration of a waveform and the width of the bands of frequency in a spectrum is sometimes called an uncertainty principle: high temporal resolution and high frequency resolution cannot be obtained simultaneously. This use of the term *uncertainty* is somewhat misleading. It is not, for instance, related to uncertainty in estimating the mean of a small sample of measurements. And it is not affected by the precision with which the amplitudes of a waveform are measured. The frequency of a tonal waveform corresponds precisely to the center of the corresponding band of frequencies in the spectrum.

Instead, the inverse relationship between temporal and frequency resolution is a consequence of the mathematical equivalence of a waveform and a spectrum. In the 1820s the remarkable mathematician Joseph Fourier suggested this equivalence. Any realistic function (those with no ordinates reaching infinity) can be synthesized by adding a series of periodic functions (usually a series of sine and cosine functions). If the function is itself nonperiodic, its synthesis requires an infinite number of sine and cosine functions, each with a different amplitude and periodicity (or frequency, the inverse of period). If the function is restricted to a finite interval (for instance, in time), then its synthesis is dominated by those sines and cosines with periods that fit evenly in the

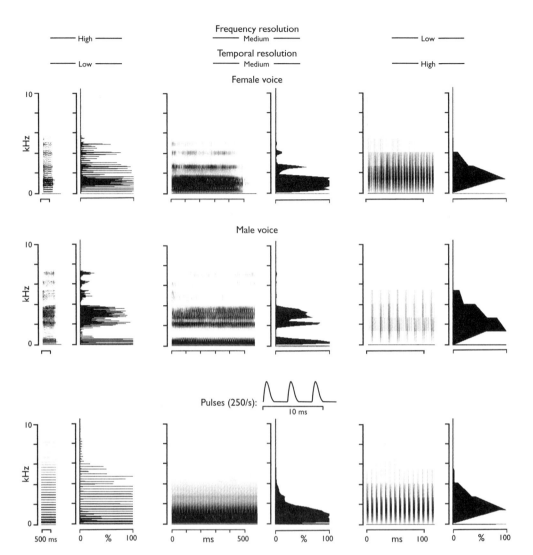

FIGURE 3.2. Reciprocal relationship of time and frequency in the human voice. Each row includes a spectrogram and spectrum (left and right) for three analyses of one sound. The *upper row* presents an adult female's voice pronouncing the English long vowel "ee" ([i] in phonetics), the *middle row* an adult male's voice pronouncing the same vowel, and the *bottom row* a train of pulses. In each row the left-hand pair of plots shows an analysis with a temporal resolution of 23 ms (frequency resolution, 43 Hz). The middle pair has a resolution of 2.90 ms (344 Hz), and the right-hand pair a resolution of 0.36 ms (2,756 Hz). The longer the analysis interval, the lower the temporal resolution and the higher the frequency resolution. In the left-hand pairs, the spectrograms reveal horizontal bands of frequencies; in the right-hand pairs, they reveal vertical pulses of sound in time. In the middle pairs, the analyses produce intermediate results. In each case, the spacing of the bands of frequency contains the same information as the spacing of the pulses (the fundamental frequency in the left-hand analyses is the reciprocal of the rate of the pulses in the right-hand ones). These examples show that the male voice has a slower rate of pulses and thus a lower fundamental frequency in comparison with the

interval. The longest such period is equal to the length of the interval. Its frequency (once per period) is the fundamental frequency of the series. All integer multiples of this frequency (higher harmonics or modes) are separated by an interval equal to the fundamental frequency. If synthesis of the function in this interval requires a frequency near one of the modes, then that mode is present in the spectrum (its amplitude > 0). And that mode has shoulders that extend into the interval between adjacent modes. The frequencies in the shoulders adjust that mode to the limited interval of the original function in time or space. Thus the spectrum includes not only any necessary modal frequencies but also their shoulders.

The width of the shoulders varies with the spacing of modal frequencies. This spacing, we have just seen, equals the fundamental frequency, which in turn is the inverse of the interval of time or space. Consequently, the spacing of modes is wider (and their shoulders broader) when the interval of time or space of the original function is shorter. In other words, the width of the bands of frequency in a spectrum precisely equals the reciprocal of the length of the interval being analyzed. This inverse relationship, for any realistic function, between the length of an interval in time (or space) and the separation of the constituent frequencies, is better called a reciprocity principle rather than an uncertainty principle.

The same can be said of Werner Heisenberg's uncertainty principle in atomic physics. Mathematically this principle is the spatial equivalent of the temporal reciprocity in spectral analysis just described. Each is a different physical instantiation of Fourier's equivalence of a function and its spectrum. Heisenberg's principle arises because the smallest

---

female voice. The formants that distinguish vowels are best seen in the intermediate analyses, where the horizontal bands are the resonant frequencies of the vocal tract, not harmonics (integer multiples) of the fundamental frequency of the larynx. The *bottom row* presents analyses of a train of pulses resembling those emitted by the human larynx. Without the vocal tract, the corresponding spectrograms lack formants. Our hearing has a temporal resolution between 2 and 20 milliseconds, about like that of the middle analyses in each row. As a result, our hearing is suitable for hearing formants in speech instead of pulse trains.

particles comprising matter also behave like waves in space at any particular point in time. Localizing such a particle's position in a tiny interval of space is equivalent to determining the spectrum of a sound in a tiny interval of time. For sound, the modes of the spectrum in this case are widely spaced. For subatomic particles they are also. In the latter case the relevant spectrum consists of spatial, rather than temporal, frequencies. Wide spacing of the modes of spatial frequencies for subatomic particles amounts to less precision in the rates of change between modes (its *velocity*, and hence momentum). Position and velocity of a subatomic particle wave have the same reciprocity that time and frequency do for sound. Both are a consequence of a fundamental mathematical relationship between waves and frequencies.

There is another, equally important point to emphasize about the reciprocity of waveforms and their equivalent spectra. The number of parameters needed to describe a complete spectrum (including both sines and cosines and thus the phases of all frequencies) is exactly the same as the number or parameters required to describe an equivalent waveform. All of the information in a waveform is thus present in the complete spectrum of that waveform, and vice versa. Fourier showed that the conversion applies in both directions. Any function such as a waveform can be analyzed into its constituent spectrum of sines and cosines. In addition, this waveform can be exactly reconstructed from its complete spectrum. This mathematical equivalence of waveform and spectrum becomes important in thinking about hearing.

DIGITAL ANALYSIS OF WAVES

Digital analysis is of course the norm today. To retrieve the spectral content of a waveform, the first step is to convert the waveform to a series of measurements of the amplitude of the wave at closely spaced intervals. Convenient intervals are often fractions of 1/10,000 of a second (for instance, 44,100 times per second for recordings on compact discs). In the 1920s and 1940s Harry Nyquist and Claude Shannon showed that a waveform is completely described by equally spaced measurements of amplitudes at a rate at least twice the highest frequency present in the waveform (the Nyquist frequency). Thus the complete range of audio frequencies heard by humans (about 50 to 20,000 Hz) is captured by digitizing sounds at a rate of 40,000 Hz or more. The standard rate for compact discs (44,100 Hz) was chosen to meet this requirement with a comfortable cushion.

Once the waveform has been converted to a series of measurements of amplitudes in time, the next step is to divide this series into brief intervals of time. An interval of 1/10 of a second would include 4,410 such measurements for recordings on compact discs. The spectrum of frequencies equivalent to the waveform in each such interval of time is then calculated by a procedure called a Fast Fourier Transform (a nifty trick discovered in the 1960s for rapidly computing the Discrete Fourier Transform—"discrete" for separate amplitudes within a brief interval of time). The result is a spectrum corresponding to that particular interval of time. The number of frequencies in the spectrum equals half the number of amplitudes in the interval (but each frequency is represented by both a sine and a cosine term). The separation of these frequencies (modes) equals the fundamental frequency for the interval, and each mode has shoulders extending to higher and lower modes.

In a complete spectrum (to recall our discussion above), each mode is represented by both a sine and a cosine term. Together the two determine the phase of the frequency, its alignment with other frequencies in the spectrum, as well as its power. Nevertheless, spectrograms standardly multiply the sine and cosine terms to produce a power term for each frequency but ignore its phase. The rationale is that human hearing does not, except to a limited degree at very low frequencies, distinguish phase differences. However, other organisms might perceive phase relationships more than we do, a point that becomes important below.

To visualize the result of all this computation, a spectrogram—a plot of frequencies versus time—is often produced. A musical score is also a plot of frequencies (pitches) versus time, but in a schematic way for the convenience of musicians. A spectrogram is more exact. It is obtained by plotting many spectra, each corresponding to a brief interval of time, one after another side by side, much as a musical score proceeds one note after another across a page. The spectrum of any one of these intervals, usually just a fraction of a second, is represented by a single vertical line along which the intensity of each frequency in the spectrum is represented by a shade of gray from white to black, just as several notes in a chord are plotted one above another on a musical score (although in longer intervals of time). When successive intervals are plotted in this way, the pattern of frequencies becomes clear. For instance, a tone (a single frequency continuing for many intervals) appears on a spectrogram as a horizontal band. The width of the band depends on the interval of time analyzed. The actual frequency of the tone would lie in the center of the band. A violin playing a succession of notes on the same pitch would likewise play a horizontal series of notes on a musical score. In contrast, a click, a transient pop, includes all frequencies in a single brief interval, so it appears as a vertical dark line in a spectrogram, corresponding to one interval of time—something like one beat on a snare drum. In general, tones appear as lines on a spectrogram, rising or falling as pitch changes and continuing as long as the tone does; clicks, snaps, and beats appear as vertical lines on a spectrogram, brief instants of white noise.

The result is the same regardless of whether we use computers to calculate Fast Fourier Transforms of digitized amplitudes or an old-fashioned sonograph with analog electronic filters. The shorter the intervals analyzed, the wider the bands of the component frequencies in a spectrogram. As emphasized above, this reciprocity derives from mathematical fundamentals and not from any particular procedure or instrument for analysis. It also applies to our ears.

COCHLEA AS SPECTRUM ANALYZERS

All of these issues concerning the analysis of sounds into their spectra apply to the entire panoply of anatomical and neural apparatus that animals of all sorts use for hearing. In particular, the reciprocity of time and frequency resolution applies to organisms as well as computers. The shorter the interval of time during which any mechanism responds to a sound, the wider the bands of frequencies to which it responds. Furthermore, the spectra of sounds allow animals to hear all of the information in the waveform (with the possible exception of phases).

Mammals and birds have a structure in the inner ear, the cochlea, that performs a decomposition of the waveform of sound into component frequencies. In other words, the cochlea converts a waveform to a spectrum. Birds' and mammals' cochlea have evolved independently from rudimentary structures in the inner ears of reptiles and even more rudimentary structures in those of amphibians. Despite their independent origin, birds' and mammals' cochlea both share a basic arrangement—a tubular cavity in bone, filled with a fluid, and divided into three longitudinal compartments by two thin membranes (see Figures 3.3 and 3.4). These two membranes are the walls of a membranous tube (the cochlear duct) inserted inside the bony tube. One side of the cochlear duct, the basilar membrane, supports clusters of sensory cells; the other side, exceedingly delicate, has no sensory cells. Movements of the fluid, induced by movements of the tiny bones of the middle ear, send waves down the basilar membrane. Because the membrane broadens along its length, it resonates at different frequencies at different positions (the elasticity and stiffness of the membrane at each point along its length result in maximal response to the oscillating fluid at a particular frequency). The cochlea in this way converts the sound wave from a pattern of amplitude with time into a pattern of frequency with position along the cochlea.

The sensory hair cells, in ranks along the length of the cochlear membrane, then respond to the amplitude of the resonance at each cell's respective location and thus to a component frequency in a brief interval of time. Because these neurons have thresholds for response, they roughly digitize the response of the cochlear membrane. The mechanical properties of the basilar membrane determine how rapidly it responds to changes in the waveform. In addition, the biochemical properties of the sensory neurons determine how rapidly each responds to oscillations of the basilar membrane. As a result these properties set an effective interval of time for the analysis of the waveform (often called a time-constant for response). The hair cells thus generate measures of the amplitude of each frequency in successive intervals of time. The width of the band of frequencies to which they respond is influenced, as has already been emphasized, by the length of these time intervals. The cochlea of a bird or mammal thus mechanically extracts the spectrum of a sound in successive short intervals of time. Each sensory neuron

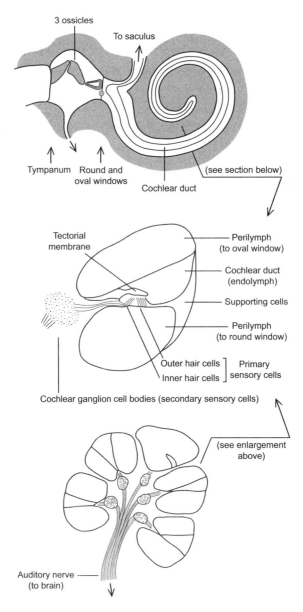

FIGURE 3.3. Human cochlea. The *top* diagram shows the tympanum (eardrum), which responds to pressure waves in the atmosphere, and the three tiny bones that carry its vibrations to the cochlear duct, where the transduction of vibrations to nerve impulses occurs. The cochlea of a mammal spirals much more than this diagram shows. The *bottom* diagram presents a section through the two and a half complete spirals of a human cochlea. In the center are clusters of secondary sensory neurons, which send their axons through the auditory nerve to the brain. In the *middle* is a single cross-section of the cochlear duct within its bony enclosure. The

responds to the waxing and waning intensities of a particular band of frequencies.

Each sensory neuron responds not to a single frequency but to a narrow band of frequencies. Two frequencies falling within this narrow band cannot be distinguished by the neuron. Furthermore, irrelevant noise with frequencies anywhere within this band interferes with detection of any frequency within the band, so every sensory neuron has a *masking bandwidth,* a band of noise that maximally interferes with detection of its characteristic frequency. Experiments with birds and mammals, including humans, reveal that the just-noticeable difference between two frequencies (and the corresponding masking bandwidth) is greater for higher frequencies. The difference in the bandwidths of frequencies corresponds with a difference in the time constants of the sensory neurons, just as the duration of a time interval corresponds with the bandwidth of frequencies in mathematical computation of a spectrum.

The wider bandwidth for discrimination of higher frequencies indicates that the cochlea of a bird or mammal differs from a standard spectral analysis, such as those described above. The time constant for auditory analysis (and the masking bandwidth or just-noticeable difference in frequencies) is not the same for all frequencies. The higher the frequency, the greater the bandwidth of frequencies and the shorter the time constant. The result is somewhat like the recent trend in computational spectral analysis called wavelet analysis, which uses different time intervals for analysis of different parts of the spectrum. Mathematical wavelet analysis allows adjustment of the time intervals for different

---

basilar membrane across the middle supports two clusters of primary sensory neurons. A flap of this membrane (the tectorial membrane) bends the delicate cilia of these neurons and triggers sensory impulses as the basilar membrane vibrates with passing waves. The width of the basilar membrane increases and its stiffness decreases along the length of the cochlear duct, so the membrane resonates with progressively lower frequencies at greater distances from the oval window. The time constant for the response of the basilar membrane at each point along its length (never more than a fraction of a second) determines the interval of time for its analysis of the corresponding frequency. The cochlea thus partitions the waveform of sound into its constituent frequencies within small intervals of time. Adapted and simplified from illustrations in many anatomical textbooks.

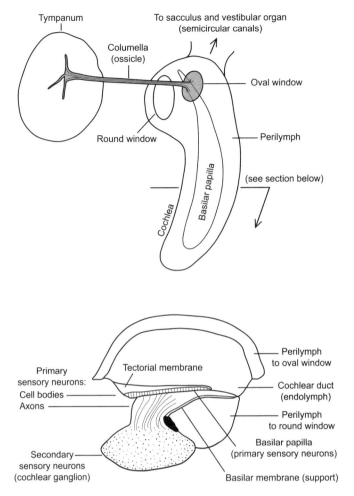

FIGURE 3.4. Avian cochlea. The *top* diagram shows most of the auditory system of a bird, with many components analogous to those of mammals. Vibrations of the tympanum are transmitted to the cochlea by a single ossicle. In birds the cochlea curves but does not form a spiral because it is much shorter than in mammals. The *bottom* sketch shows a cross-section of the cochlea for comparison with that of mammals. A relatively thick membrane across the bony tube includes the basilar membrane and support for the sensory neurons. Just as in mammals, a flap lies over the sensory cells to stimulate them as the basilar membrane vibrates. The primary neurons send their axons to an adjacent ganglion, where secondary sensory cells send axons through the auditory nerve to the brain. Birds and mammals differ in the number of primary sensory cells in any cross-section of the cochlea. Rather than two small clusters of cells, as in mammals, birds have 10 times more cells underneath the tectorial membrane, an arrangement extending along the entire length of the cochlea. As in mammals, the avian cochlea transforms the pressure wave of sound into its constituent frequencies in brief intervals of time. This conversion of a waveform to a series of spectra is subject to the mathematics of Fourier transforms. Adapted from illustrations in Manley (1990).

frequencies in complex ways, while the cochlea of birds and mammals have a simpler pattern of gradually decreasing (to be exact, logarithmically decreasing) frequency resolution with increasing frequency.

Keep in mind that the reciprocity between time constants in the responses of sensory cells to waveforms, on the one hand, and bandwidths in perception of frequencies, on the other, reflects the equivalence of the time domain and the frequency domain. It is the nature of conversion from one domain to the other that high precision in time (short intervals) converts to low precision in frequencies, and vice versa.

This result is fundamental for comparing hearing in different kinds of animals. One particularly instructive comparison is the hearing of birds and mammals. As already emphasized, both have evolved remarkably similar structures for the analysis of sound, including a cochlea. Despite their similarities, the ears of birds and mammals have some definite differences. Birds' cochlea, for instance, are shorter than those of mammals with similar overall size; the cochlea of both birds and mammals increases with the species' overall size. For instance, the cochlea of a house mouse is about 6 millimeters long, that of a starling or budgerigar about 2 millimeters and a domestic pigeon about 3 millimeters. On the other hand, birds have ten times more hair cells for each unit of length of the cochlea. The shorter length of a bird's cochlea should, all else being equal, make the resolution of frequencies less precise, but the greater density of hair cells could compensate by making the resolution more precise. The hearing of birds extends over a narrower range of frequencies than that of mammals. Most birds' hearing decreases markedly above 10 kilohertz (kHz), whereas mammals routinely hear frequencies twice that high and some bats as much as ten times that high or greater (although they do not hear frequencies so low as other mammals). With a cochlea roughly a third as long, but hearing confined to half the range of frequencies and roughly ten times as many sensory neurons at each cross-section of the cochlea, birds would seem to have plenty of processing power to match mammals in frequency analysis despite their shorter cochlea.

The narrower frequency range of birds has often seemed to be a consequence of the shorter cochlea. This conclusion would be correct provided the mechanical properties of the cochlea were similar and the precision of sensory neurons in measuring amplitudes were similar. If the higher density of hair cells does not serve to extend the range of frequencies for each unit of length of the cochlea, does it have some other effect?

This is not the only mystery of avian hearing. Birds produce sounds with amazingly intricate frequency modulation. It is always a source of wonder for a person to listen to a recording of a bird's song played back at 1/4 speed and to realize what intricate patterns of frequency modulation it includes. Our hearing cannot resolve these rapid changes because the temporal resolution of our auditory system is about 1/50 of a second. Playing back a recorded song at 1/4 speed allows us to hear most of the detail, at least in many birds' songs. If the temporal resolution of our ears were four times better, 1/200 of a second, we could hear all of this temporal detail without slowing down a recording. At first thought, it seems that birds must have much greater temporal resolution in their hearing to be able to appreciate the details of their own songs.

Considerable effort has focused on comparing the temporal resolution of hearing in birds and mammals. Measuring an animal's ability to discriminate sounds closely spaced in time is not so easy as it might seem. Recall that any wave has equivalent expressions as a waveform or a spectrum. Analyzing a prolonged sound wave in successive intervals of time, as described above, does both. For relatively infrequent changes (or slow fluctuations) in a sound, the successive intervals of time can register the changes. Slowly fluctuating intensity or frequency is apparent as changes in intensity or frequency from one interval of time to the next. More rapid changes, those fast enough to be incorporated in a single time interval, appear in the spectrum of that interval, not in temporal changes from interval to interval. The duration of the intervals of time thus divide the analysis of the sound wave into the two kinds of representations—rapid fluctuations affect the spectra of individual intervals, slow fluctuations affect the changes in these spectra from interval to interval. For humans, fluctuations in a sound wave slower than 20 Hz (corresponding to a time interval of 1/50 second or $1{,}000/50 = 20$ milliseconds) sound like changes in time. Fluctuations faster than 50 Hz sound like changes in spectra—what musicians call the timbre of a sound.

As a consequence, humans can hear rapid fluctuations in sound waves perfectly well. What we cannot hear as changes in time we hear as changes in timbre. The same must apply to all animals with frequency-analyzing hearing. The time constant of the auditory sensory cells sets the boundary between time and timbre in hearing the details.

This duality in perceiving sound creates a problem when measuring the time constant of hearing. Two clicks, even within the same interval for analysis, would sound different from a single click, but the difference would be in their timbre (spectra), not their timing. Human experimental subjects can be instructed to pay attention to the difference between two clicks in timing or in timbre. Such experiments indicate that humans do not hear changes in timing when the changes occur less than about 20 milliseconds (1/50 second) apart.

To determine this time constant for birds, experimenters are stymied by the impracticality of instructing birds to pay attention to the timing rather than the spectra. The challenge is to devise sounds that differ primarily in the time domain rather than in the frequency domain. One expedient is to present brief gaps in white noise (sound that includes all frequencies equally). Gaps longer than an analysis interval would be heard as brief interruptions in sound. Gaps shorter than an analysis interval would hardly differ from the absence of a gap, because the spectrum of a short gap would consist primarily of white noise and thus would be masked by the continuous white noise. Such experiments indicate that birds can detect gaps in white noise provided they are longer than about 2 or 3 milliseconds. In similar experiments, humans can detect gaps about the same length. Yet there remains the possibility that birds make this discrimination based on the subtle difference in spectra (timbre) rather than the difference in time.

Recently, Robert Dooling and his colleagues have explored another approach to this problem. A special waveform, a particular sweep of frequencies designed to have the same spectrum when played either forward or backward, provides another chance to separate discrimination in the time domain and the frequency domain. Replicates of this waveform can be concatenated so that successive segments either alternate in direction or do not. Consider what happens when these segments differ in length. If birds can discriminate between a concatenation with all segments in the same direction and one with segments alternating, then they presumably hear these segments in the time domain. The intervals for auditory analysis by their ears must be shorter than the segments. If these intervals for analysis were longer than the segments, then they only hear their spectrum, the timbre, which remains constant regardless of whether the segments are reversed. In these experiments birds have shorter time constants for analysis than do humans—about 1 millisecond for birds versus about 20 milliseconds for humans.

This is the result everyone has expected. Birds appear to have higher temporal resolution in their hearing than do mammals, humans in particular. The tenfold additional sensory neurons in the cochlea of birds seem, in some yet unexplored way, to contribute to an increase in temporal resolution. On the other hand, their shorter intervals of analysis are not accompanied by a corresponding increase in the bandwidth of frequencies. In their range of hearing below 10 kHz, birds have an ability to discriminate frequencies that is not much different from humans'. Understanding the relationships between frequency and time in the cochlea of birds and mammals still has a ways to go.

Perhaps an unresolved problem arises from phase differences in spectra. Recall that a spectrum includes as many parameters as its corresponding waveform. Half of those parameters are the amplitudes of sines, and the other half are the amplitudes of cosines. The sines represent the component frequencies; the cosines adjust the phase relationships of the frequencies. Two different waveforms, such as the forward and backward frequency sweeps mentioned above, might have identical amplitudes of the component frequencies (the sines), but they cannot also have identical phases (the cosines). Mammals, including humans, are not so sensitive to phases of frequencies as they are to the amplitudes. On the other hand, in birds the responses of single sensory cells are phase-locked to their characteristic frequencies, for frequencies at least as high as 2 kHz, higher than phase-locked responses in mammals. Perhaps the pairs of sweeps differ enough in the phases of frequency components to allow birds to make a discrimination in the frequency domain rather than the time domain. If so, we would still have no evidence that avian hearing had greater resolution in the time domain than human hearing—despite the temporal intricacy of birds' songs.

Regardless of whether or not birds can hear the details in their songs in time, they would not have to. They could hear the differences in the spectra instead. Because any wave has equivalent representations in the time and the frequency domains, as explained above, it is possible to hear all of the detail in either form, especially if the phases of different frequencies could be compared. It would be possible to discriminate differences in a sound wave either in time or in frequencies. It would also be possible to reproduce a sound by matching sounds in either time or frequencies. Indeed my experience over years of examining spectrograms of birds' songs indicate that we humans can hear all of the detail in the intricate songs of birds—if not in time, then in timbre. I can distinguish birds' songs that differ in temporal details even though I cannot hear the differences in time. If I can see a difference in a spectrogram, I can hear the difference in timbre.

Even though it is amazing to recognize the temporal details in a bird's song slowed down to 1/4 speed, we can nevertheless hear the differences at full speed in the form of differences in the timbre of sound. Even the shape of the notes in a bird's song affects the timbre we hear. Because we are not accustomed to thinking about sound in this way and have a depauperate vocabulary for describing the timbre of sound, it is difficult to learn these distinctions in the frequency domain. Yet with practice humans can distinguish different species' songs by differences in the inexpressible timbre of their notes. Experienced birders and musicians learn this talent routinely, even when they must borrow words from other realms of experience to express what they have learned.

Even if humans can perceive the differences in birds' songs, they certainly cannot produce them vocally or by whistling. This defect, however, is entirely on the production side. We lack the apparatus for producing such sounds and the neural control for modulating them rapidly. In an analogous way, chimpanzees are unable to match the marvelous ability of humans to produce vowels because they lack the anatomy to alter the resonances of their pharyngeal cavities. They might also lack the neural mechanisms to control this anatomy (tongue, glottis, larynx) even if it were properly deployed. Humans, on the other hand, are comparably deficient in our ability to produce the sounds of birds. If we were not deficient in production, our ability to hear the differences in time and timbre would allow us to match our output to birds'. It is possible that the constraints imposed on production of sounds by the syrinx of birds helps to channel the possible sounds they can make. Birds imitating other birds with similar apparatus for producing sound could then match the intricacies in the waveform by matching the timbre in the spectrum. Thus, although we cannot produce birds' songs by listening to their tunes and timbre, the birds themselves, equipped with the necessary anatomy and neural control, should be able to produce songs like other birds' by matching their perceptions of the model in either the time or the frequency domain.

ADVANTAGES OF FREQUENCY ANALYSIS

Why should animals have evolved this astonishingly complex mechanism for frequency analysis? The answer undoubtedly lies in the advantage of encoding information in frequencies. Information could be encoded either as pulses at varying intervals of time (as humans do in Morse code) or in frequencies varying in time. We shall see that frequencies of sound are, as a rule, not altered by propagation of a wave. Their amplitudes are often attenuated to various degrees, but the frequencies themselves are not altered. The timing of pulses, in contrast, can easily be obscured during transmission. If animals' ears responded only to the overall intensity of sound, without discriminating frequencies, the temporal patterns in sound would become obscured during transmission.

In contrast, if a receiver's ears can discriminate frequencies, then a signal encoding information in particular frequencies would leap out of the spectrum of more or less random noise. The energy of noise is usually spread over a wide band of frequencies, while the energy of a signal can be concentrated in one (or a few) frequencies at any instant. Thus signals that encode information in particular frequencies complement receivers that analyze signals in the frequency domain. Ears that convert signals from the time domain to the frequency domain greatly improve the chances of detecting signals that use particular frequencies in the presence of unpredictable noise. Cochleas, in their initial evolution, might well have been more important for overcoming noise than for discriminating complex signals.

Birds have evolved the animal kingdom's most elaborate mechanism for rapidly modulating tonal sounds. Many birds produce one frequency of sound at any instant, but patterns of rapid fluctuations of frequency in time. Because most birds' syringes have a dual structure (each half in the anterior end of each bronchus) they could conceivably sing duets with themselves by producing two harmonically unrelated frequencies at the same time, one with each half of the syrinx. This feat is performed by some birds, but not many, and then usually for only brief parts of any one song. Instead, many birds concentrate all of the energy in their songs in a single frequency at each instant. This procedure creates the most easily detectable signal provided the intended receiver can convert the waveform to a sequence of spectra.

Humans, on the other hand, have evolved the animal kingdom's most complex mechanism for modulating the spectrum of a sound. Human vowels require independent modulation of at least three frequency components at once. We have seen that to accomplish this feat requires a partial uncoupling of the complex resonant spaces of the oral cavity from the vibrations of the vocal folds. The impedance mismatch between the larynx and the pharynx, which accomplishes this uncoupling, has the cost of a significant loss of power. Nevertheless, humans achieve the highest density of spectral information in any animal's vocal communication, presumably a corequisite for the evolution of language.

Some birds also produce complex spectra in their songs. The Zebra Finch, a common subject of laboratory studies of vocalization and hearing, comes to mind. Like humans, Zebra Finches use these vocalizations mostly for close-range communication. Birds and humans both use complicated patterns of frequency to communicate, but most birds use rapidly changing patterns of a single frequency at a time, while humans (and Zebra Finches) use more complex spectra changing less rapidly.

Birdsong is a specialization for long-range communication of sparse information: mostly their species, their individual identity, and certain individual attributes (physiological condition, location, reproductive state, and perhaps other attributes still under investigation). Human language is a specialization for short-range communication of dense information (everything we discuss with language). To take advantage of either kind of signal, receivers require intricate spectral analysis of acoustic signals. Long-range communication faces some major obstacles. Chapters 4 through 6 suggest that the intricacies of birds' songs might assist in overcoming some of these obstacles.

4

TRANSMISSION OF ACOUSTIC SIGNALS

A PIONEERING STUDY

ANIMALS' ADAPTATIONS FOR COMMUNICATION, as Chapters 2 and 3 have shown, include some remarkable mechanisms for producing and receiving signals. They also resolve a trade-off between communication of sparse information at long range, as do many birds' songs, or communication of dense information at short range, as does human language. The basic difference between long-range and short-range communication is the higher level of noise, with all else equal, when a signaler and a receiver are far apart. The increased noise with increased distance comes from one of the three sources of noise identified in Chapter 1—degradation of signals during transmission. The other two sources of noise, irrelevant energy impinging on the receiver and properties of the nervous systems of signaler and receiver, are less affected by distance.

Noise, as explained in Chapter 1, is any source of error in a receiver's responses, so the objective in the present chapter is to consider how transmission of signals over large distances can increase a receiver's chances of error. Changes in a signal as it passes from the signaler to a receiver reduce its intensity and diminish its pattern. The lower intensity and obscured patterns of signals compromise the receiver's ability to distinguish each signal from others and from the irrelevant energy in the environment. The focus thus is on how intensity and pattern in signals decrease as distance increases.

A pioneering report in 1975 by Eugene Morton first directed attention to the degradation of animal's sounds during propagation through natural environments. His doctoral dissertation had focused on adaptations of birds' songs to different natural environments. In this and in a subsequent paper Morton identified most of the important issues in the transmission of acoustic signals and included the first measurements of the degradation of animals' sounds during transmission. He emphasized that adaptations for effective transmission of acoustic signals should be most apparent in sounds used for long-range communication. Communication at close range can have other problems, but those associated with transmission of signals are bound to be minimal. Many birds' songs, however, are used for communication over distances great enough to pose real challenges for transmission. Signalers and receivers are often 10 to 100 meters apart and sometimes even farther.

Morton emphasized attenuation of the intensity of sound but also discussed degradation of the patterns of sound. In the subsequent decades these processes have been repeatedly compared in different natural environments. This chapter examines the ways that attenuation of sound occurs and then the ways that degradation occurs. In each case the changes in sound during propagation can be measured by sophisticated analysis of recordings at increasing distances from a loudspeaker. On the other hand, all that is necessary to make us aware of these changes is some attentive listening outdoors to the sounds of birds, frogs, or insects. Mammals such as primates in tropical forests also provide opportunities, although less convenient for many people.

ATTENUATION

Sound attenuates during propagation in three ways. First, it attenuates because the energy spreads over a progressively greater surface as sound radiates outward from its source. The density of acoustic power (intensity) must thus decrease progressively with distance from the source.

Sound often spreads spherically, at least approximately, which is to say that the acoustic power stretches like an inflating balloon—the larger the radius, the thinner the balloon. The density of power thus decreases as the area on the surface of a sphere increases. This surface is proportional to the square of its radius, so the density of acoustic power drops in inverse proportion to the square of the distance from the source. Increasing the distance from the source by a factor of two decreases intensity by a factor of four—an inverse-square relationship. This change can be expressed in decibels. By definition a decibel (dB) equals −10 times the logarithm of a ratio of two intensities: $-10 \log (I_a/I_b)$. The logarithm of 1/4 is approximately −0.6, so the drop in intensity for a doubling of distance from the source is about 6 dB. As a sound spreads outward from its source, its intensity falls by 6 dB for every doubling of distance from the source, solely as a result of spherical spreading.

In view of the discussion in Chapter 3, it is important to note that if we begin with measurements of sound pressure, rather than sound intensity, we must convert pressures to intensities. Because intensities are proportional to the squares of pressures, we can just square our measurements of pressure before calculating the ratio in decibels. By applying the rules of logarithms we get the same result by computing $-20 \log (p_a/p_b)$, when p_a and p_b are two sound pressures we wish to compare, and $-10 \log (I_a/I_b)$, when I_a and I_b are the corresponding intensities. Even though sound pressures are easily measured with calibrated microphones, it is the corresponding intensities (densities of acoustic power) that explain the amount of work a sound wave can perform per second on, for instance, our ears. Intensity is thus the relevant measurement for understanding hearing.

Second, as Morton emphasized, sound attenuates in natural environments in two additional ways—absorption and scattering. Like spherical spreading, absorption of sound has consequences for attenuation that are relatively easily measured. Sound is absorbed when energy is transferred from the minute movements of particles in the medium as a sound wave passes to movements of other objects. For instance, vibrations induced in vegetation might absorb energy from a passing sound wave, but evidence indicates that this effect is very small. On the other hand, vibrations of the atoms in molecules of oxygen resonate at frequencies of audible sound. They thus absorb a substantial amount of the energy of a sound wave in air. The atoms of molecules of nitrogen, in contrast, do not resonate effectively at audible frequencies of sound. This process of atmospheric (actually oxygen) absorption of sound depends on the frequency of sound. The higher the frequency within the audible range, the greater the absorption. Unlike attenuation by spherical spreading, attenuation by atmospheric absorption is thus frequency-dependent. Throughout the range of frequencies used by animals for communication, including even ultrasonic frequencies used by rodents and bats, atmospheric absorption increases with frequency.

In contrast, absorption of sound by the atmosphere is only slightly influenced by temperature and humidity, at least for frequencies that humans, birds, frogs, and many mammals use for communication (for ultrasonic frequencies used by bats for echolocation, humidity and temperature have more influence on absorption). In addition, unlike the attenuation of sound by spherical spreading, the absorption of sound energy by the atmosphere does not follow an inverse-square relationship. Instead, the same proportion of sound is absorbed by a homogeneous atmosphere with every meter of propagation.

This difference between the attenuation of sound by spherical spreading and by atmospheric absorption has a major consequence for communication. At close range, spherical spreading is the predominant cause of attenuation, but at long range atmospheric absorption becomes predominant. By spherical spreading, intensity drops by a factor of four between 2 and 4 meters from the source and by the same factor between 16 and 32 meters. By atmospheric absorption, in contrast, sound intensity drops at a constant rate with distance and thus eight times as much between 16 and 32 meters as between 2 and 4 meters from the source.

A third source of attenuation comes from the scattering of sound. Scattering is the accumulated reflection and diffraction of waves by solid objects in the path of transmission. In a complex environment like a forest, with many trunks, branches, and leaves of trees, sound reflects from these surfaces innumerable times in all directions. If a source produces a beam of sound—in other words, a sound with most of its energy spreading outward in a spherical angle (a cone in space)—then scattering tends to deflect energy out of the cone so that sound intensity in the cone decreases progressively. If a source produces sound that spreads equally in all directions, in a sphere or hemisphere, then the effects of scattering are much less because energy scattered out of any one spherical angle is compensated by energy scattered into that angle from nearby. Some sound is scattered backward and thus lost to any sound spreading forward, but backward scattering accounts for only a minority of the total energy lost (except when large objects create echoes). Scattering, like atmospheric absorption, produces a constant loss of energy with distance from the source, proportionately the same between 2 and 3 meters as between 102 and 103 meters.

Scattering, like atmospheric absorption, is strongly frequency-dependent. Any wave interacts with an object in its path in a way that depends on the wavelength and the size of the object. An object here is anything that differs markedly in density from the medium (bubbles of air in water scatter sound just as solid objects scatter sound in air). If the wavelength of the sound (which, remember, is inversely related to its frequency) is much smaller than the diameter of the object, most of the energy of the wave is reflected backward. For instance, when we shout beside a large building we hear an echo, and a person on the other side hears little if anything. On the other hand, when the wavelength of sound is much longer than the diameter of an object, the sound wave passes with virtually no disruption. When we shout beside a flag pole, we hear no echo, and a person on the other side hears our voice unaffected by reflection. When the wavelength approximates the diameter of an object (when it is between 0.1 and 10 times its diameter), the distortion of the sound wave by the object becomes complex, with energy spreading in all directions, often in a complex pattern of lobes. For sound in air, where the speed of sound is approximately 340 meters per second, sound with a frequency of 1,000 hertz (Hz) has a wavelength approximating 1/3 meter; 2,000 Hz has a wavelength approximating 1/6 meter; 4,000 Hz, 1/12 meter; and 8,000 Hz, 1/24 meter. The trunks of trees are often about 1/3 or 1/6 meter across, broad leaves about 1/12, and conifer needles or bundles of needles about 1/12 or 1/24 meter across. The wavelengths of frequencies in most birds' songs thus correspond to the dimensions of many objects in forests. For this range of frequencies, sound in forests is progressively scattered by objects of decreasing size. Because leaves predominate over trunks, more scattering results from foliage. Trunks

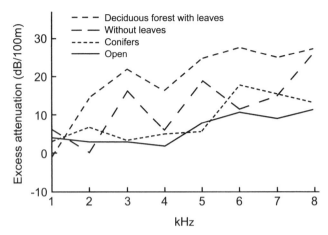

FIGURE 4.1. Measurements of excess attenuation in natural habitats. Excess attenuation of sound (attenuation after subtracting spherical spreading) results from atmospheric absorption (energy absorbed by molecular vibrations of atmospheric gases, especially oxygen) and from scattering (reflection of sound away from the path of transmission). Both effects accumulate linearly with distance of transmission but depend on the frequency of sound. Some typical results for natural temperate environments come from the study by Marten and Marler (1977). Attenuation (the inverse of the intensity of sound) is measured in decibels (so a difference on the vertical axis of 10 means a tenfold decrease in intensity) and is plotted against the frequency of sound in kilohertz (1,000 cycles/second). Most songs of small birds use frequencies between 2 and 6 kHz. In all four environments, excess attenuation increases with frequency, although the increase is not regular in all cases. This irregularity (especially evident in this study for deciduous forests after leaves have fallen) is not consistent from place to place or time to time in any one place. Scattering in particular is likely to vary depending on the exact atmospheric conditions and distribution of vegetation. As a result, exact consistency in the details of excess attenuation is difficult to obtain. Nevertheless, the broad pattern revealed here is consistent: attenuation is greatest in forests in full foliage, except at frequencies at or below 1 kHz. Open environments (grasslands and fields) have the least attenuation at all frequencies, as expected from the near absence of scattering. On a calm day, well above the ground, excess attenuation primarily results from molecular absorption by the atmosphere. Nevertheless, the frequency dependence of atmospheric absorption is enough to make low frequencies best for long-range communication in all environments. Notice also that coniferous forests have low excess attenuation below 6 kHz, a different pattern than in deciduous forests. The smaller leaves (or even clusters of leaves) in evergreen forests scatter only sound with short wavelengths (6 kHz corresponds to a wavelength of about 6 cm). The larger leaves of deciduous forests scatter sound with much longer wavelengths as well as all shorter wavelengths (2 kHz corresponds to a wavelength of about 17 cm). Adapted from Marten and Marler (1977).

and branches of trees scatter energy at all frequencies above about 1 kHz. In temperate forests during winter, this is the only scattering. In broad-leaved forests with dense foliage scattering increases markedly for frequencies above 2 to 4 kHz, and in needle-leaved forests above 4 to 8 kHz (see Figure 4.1).

We probably all recognize that sound attenuates with distance, but might not have noticed the universal influence of frequency on attenuation. As a result of absorption and scattering, attenuation is greater for higher frequencies of sound than for lower ones, and high frequencies attenuate even more in a forest than in the open. A little attentive listening can reveal that distant sounds, especially in forests, seem deadened from a lack of the crispness that results from loss of high frequencies. Morton, and many investigators after him, have reported that birds in forests, on average, use lower frequencies in their songs than do those in open habitats.

SOUND SHADOWS

Morton realized that sound in open environments, more than in forests, is modified by wind. One consequence of wind for the propagation of sound is refraction, the bending of a wave as the properties of a medium change. The important property of a medium is the speed of sound. When a wave of sound encounters different speeds across its direction of propagation, the wave front bends toward the side with slower speed. Refraction changes the direction in which a sound wave propagates.

Just this situation occurs frequently in the layer of air above the ground or vegetation. Because the heat capacity of earth or vegetation is less than that of air, it warms up faster than air in sunlight and cools down faster than air at night. Warm ground or vegetation then warms the overlying air by conduction and convection. The temperature of the air is highest just above the ground or vegetation and then decreases more or less exponentially, faster at first and then progressively more slowly. Above a height of ten meters or so, temperature changes little with further distance from the ground (at least for heights of no more than several

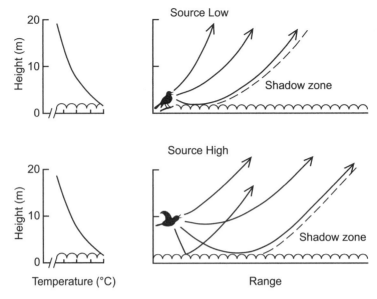

FIGURE 4.2. Temperature gradients above the ground produce sound shadows. On a sunny day, sunlight heats the ground faster than the air. The warm ground then creates a gradient of temperature in the overlying air. Air temperature decreases with height, as shown in the plots of degrees centigrade against height on the left. Because sound travels faster in warmer air, a sound front traveling within 10–20 meters of the ground is diverted upward, as shown in the diagrams on the right. If the gradient of temperature is sufficiently strong, a shadow zone develops—a zone at a distance from the source of sound into which little sound penetrates. As the diagrams on the right indicate, the higher the source above ground (imagine a lark singing 20 or 30 meters up), the farther sound travels before a shadow zone develops. Adapted from Wiley and Richards (1978).

hundred meters). The steep gradient of temperature near the ground results in a correspondingly steep gradient in the speed of sound near the ground. Sound traveling parallel to the surface of the ground or vegetation, as a result, bends upward, as the sound near the surface surges ahead of that higher up. If the gradient of temperature is strong enough, all sound can be redirected upward so that almost none continues forward parallel to the surface. The surface beyond a certain distance from the source is thus in a sound shadow (see Figure 4.2). A receiver there would not hear the source, even if, without the gradient of temperature, the intensity of sound would permit easy hearing.

A gradient of wind speed can have a similar effect. As a result of friction at the surface of the ground or vegetation, the layer of moving air at the surface travels more slowly than layers farther from the surface. Like temperature, the speed of moving air increases approximately exponentially with height. The gradient of speed is strongest near the surface and gradually decreases with height until above 20 meters little further change occurs. Consider what happens when sound is traveling upwind near the surface of the ground or vegetation. A headwind tends to slow the propagation of sound with respect to the ground, but as a result of the gradient of wind speed, sound travels faster upwind near the surface than at a height. Just as with a temperature gradient, sound is refracted upward and a sound shadow can develop upwind. In contrast, sound traveling downwind is refracted downward (sound propagating downwind travels faster at a height than near the surface). As a result, sound intensities near the surface are augmented downwind. Downwind sound spreads from a source with a geometry that is more nearly planar (two-dimensional) than spherical. The effects of gradients of wind speed and temperature on the formation of sound shadows depend on both the steepness of the gradients and also on the height of the source above the surface. A sound shadow develops closer to a source near the ground than to a source at a height. Sound traveling obliquely downward from a high source can eventually become refracted upward, leaving a sound shadow beyond, but total refraction occurs farther from the source than when the source is near the surface.

Anybody can quickly be convinced of the reality of these effects with a little attentive listening on a sunny or windy day in an open field. Minimizing the effects of sound shadows on the propagation of sound no doubt explains why so many birds of open country fly 10 to 30 meters above ground to sing.

TURBULENCE

In reality, sound shadows are not as distinct as we might expect from thinking about sound waves spreading smoothly outward from a source. Instead turbulence in the air refracts and scatters some energy into what would otherwise be an area of shadow. Sound shadows are thus areas of attenuated energy, not areas of silence.

Turbulence in air has other consequences for transmission of sound. It results from air passing over uneven surfaces or around many natural objects that lack streamlining. As a consequence, air develops eddies and swirls that differ slightly in density, temperature, and speed of sound from adjoining air. Sound is deflected by these cells of inhomogeneous air in constantly shifting ways. The effect of a particular cell of turbulence on the passage of sound depends on how strongly its properties differ from surrounding air and on its dimensions in relation to the wavelength of sound. Long wavelengths are deflected or scattered only by large cells of turbulence. Shorter wavelengths are influenced by large cells but also by correspondingly smaller cells of turbulence. As a consequence it is possible to use sound to study turbulence in the atmosphere. Broadcasting a sound that sweeps through a range of frequencies reveals the prevalence of different sizes of eddies in the atmosphere. Such measurements usually indicate that turbulence consists mostly of relatively large cells, meters in diameter. The frequency of turbulent cells decreases exponentially with size. Turbulence thus influences long wavelengths—low frequencies—of sound as well as short wavelengths.

Turbulence makes the overall intensity of sound wax and wane in an irregular way. This irregular modulation of the amplitude (and hence intensity) of sound occurs mostly at rates that we hear in the time domain, as discussed above. It does not take much wind to introduce strong fluctuations in the intensity of sound in an open environment (see Figure 4.3). As a consequence, a listener hears obvious dropouts, moments when sound disappears, as well as moments of clarity.

With some attentive listening in an open environment, such as a wide field or a beach with a gentle breeze, it is easy to become aware of these effects of turbulence on the propagation of sound. Any animal attempting to be heard by distant individuals in such conditions must expect that much of the sound it produces will be lost in dropouts from the turbulence of an everyday breeze. The problem is, a signaler cannot predict which parts of its signal will drop out.

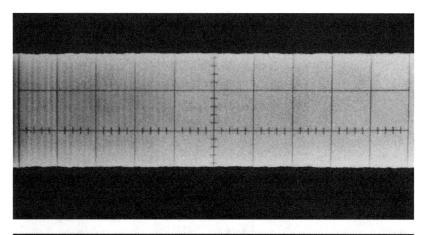

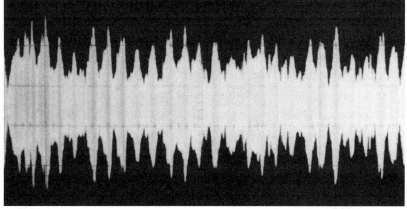

FIGURE 4.3. Fluctuations of the amplitude of sound as a result of turbulence in air. These oscillograms (time on the horizontal axis, amplitude on the vertical) show the amplitudes of sound recorded close to a speaker and at a distance of 60 meters, in a deciduous forest in leaf on a day with hardly perceptible wind. As broadcast, the sound has almost constant amplitude. At a distance, even when atmospheric turbulence is minimal, large fluctuations in amplitude appear, with periods ranging from small fractions of a second to many seconds (the horizontal scale indicates 1.0 second). Turbulent cells in the atmosphere divert sound in ever-changing patterns. With much more turbulence, the fluctuations exceed the dynamic range of standard microphones unless the amplification is turned almost to zero. The effects, which are easy to hear in open environments such as beaches or in large fields, amount to irregular dropouts in a transmitted sound. Adapted from Richards and Wiley (1980).

BOUNDARIES

Attenuation of sound becomes even more complicated near a boundary. This situation occurs, for instance, when the sender and receiver are near the ground or just above a layer of vegetation. In such cases sound traveling from the sender to the receiver is altered by the presence of a nearby boundary between the air and a medium with much different acoustic properties. There are two processes that occur when a wave propagates parallel to a nearby boundary—reflection from the boundary and interactions with surface waves.

Reflection is the easier to understand. If a source and a receiver some distance away are both close to a boundary, such as the surface of the ground, sound emitted by the sender at any instant reaches the receiver by two paths, one straight through the air and the other after reflection from the ground (see Figure 4.4). The reflected path is somewhat longer than the direct path. If the difference in the lengths of the two paths is 1/2 of a wavelength, then waves from the two paths cancel each other at the receiver by destructive interference. A peak in the shorter path would combine with a trough in the longer path. As a result, little energy would remain in the wave. Alternatively, if the two paths were nearly equal in length, such as when the sender and receiver are far apart in relation to their height above ground—say, 100 meters apart but only 1 meter above ground—then the waves from the two paths would hardly be out of sync (shifted in phase). The two waves would then combine to increase the energy at the receiver by constructive interference.

A complication occurs when reflection from the surface involves a 180-degree shift in phase. Then, even if the direct and reflected paths had nearly the same lengths, the reflected wave would again be almost completely out of phase with the direct wave. Once again the two waves would cancel at the receiver and little energy would remain in the wave. Such a dramatic phase shift on reflection is not unusual for porous surfaces such as most vegetation and dry soil. It depends on the acoustic impedance of the surface (its frequency-dependent resistance to the passage of sound). There are few careful measurements of the acoustic impedance of surfaces in natural environments, but the information available indicates that porous surfaces, such as soil and vegetation, often produce

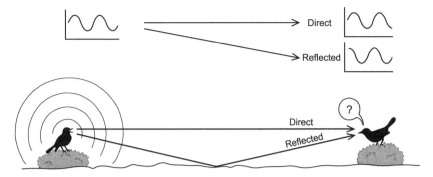

FIGURE 4.4. Interference between direct and reflected sound near a boundary. For sounds propagating near a boundary, such as the ground, a listener receives both direct and reflected waves. If the height above ground is small in relation to the distance from the source, the two waves travel paths that differ only slightly in length. If there is no shift in the phase of the reflected wave, then the two waves are nearly coherent when they arrive (their peaks and troughs nearly coincide), so they interfere constructively (reinforce each other). They thus increase the intensity of the received sound. In contrast, if reflection produces a phase shift in the reflected sound (porous ground such as a forest floor can produce nearly a 180-degree phase shift), then the two waves arrive without coherence (peaks of one coincide with troughs of the other), so they interfere destructively. Intensity is drastically reduced. The geometry of the situation (the heights of signaler and receiver and their separation) as well as the phase shift on reflection determine the outcome. Imagine some possibilities for the geometry and phase shift to realize that space is a complicated pattern of hot and cold spots, places with various degrees of constructive or destructive interference. It has a "standing" wave of intensities in space. Adapted from Wiley and Richards (1978).

nearly 180-degree shifts in the phase of sound on reflection. New snow has a similar effect. In contrast, denser surfaces, such as asphalt and water, produce almost no shift in phase.

Again, a little attentive experience provides routine confirmation of these effects of porous and dense boundaries on propagation of sound. Days with new snow often have an uncanny quiet, no doubt because familiar sounds are so strongly affected by attenuation after reflection from a layer of loose snow. The strong phase shifts of sound waves reflected from snow cancel the direct sound waves. In contrast, on still mornings (hence without turbulence and without temperature or wind gradients) sounds travel surprisingly far across water. In this case, the reflected wave without a shift in phase augments the energy in the direct wave. Furthermore, an inversion of temperature in the air above cool water traps sound by refraction near the surface so that it spreads in two rather than three dimensions from the source and thus attenuates only 3 dB with a doubling of distance (rather than the 6 dB attenuation for a spherically spreading wave). This effect thus also contributes to the surprising carry of sounds over water in these conditions.

For many animals, though, communication with sound at long distances close to the ground (or to the surface of a layer of vegetation) is compromised by destructive interference of direct and reflected waves, as well as by gradients of wind and temperature and by turbulence. It is thus not surprising to find that many birds in open habitats sing from the highest perch they can find, perhaps no more than a shrub projecting above the rest of the vegetation. In forests, birds often sing in treetops, even those species that otherwise forage and nest near the ground.

The effect of reflections from a nearby surface is frequency-dependent. Even when long wavelengths result in cancellation of direct and reflected waves by destruction interference, shorter ones can result in augmentation by constructive interference. As frequency increases further, destructive and constructive interference should alternate as wavelength decreases. For short wavelengths, the expected patterns of constructive and destructive interference rarely appear in measurements of attenuation in the field. The slightest turbulence in the air seems to obliterate the details of these patterns.

Long wavelengths, in contrast, are often conspicuously attenuated during propagation near the ground. Measurements of attenuation over distances of 25 to 200 meters at heights of 0.5 to 2 meters above ground often reveal strong attenuation at the lowest frequencies (presumably a result of destructive interference of direct and reflected waves), somewhat less attenuation at intermediate frequencies (1–2 kHz), and then progressively increasing attenuation as frequency continues to increase (as a result of frequency-dependent absorption and scattering). Morton's measurements of attenuation during propagation of sound in such conditions showed this pattern of reduced attenuation at intermediate frequencies, which he called a "sound window." Because his measurements were all made at a height of about 1.5 meters above ground, he did not realize that this effect becomes negligible when transmission occurs at greater heights.

For a source even closer to the ground—within centimeters, such as calling frogs and a few birds that do not ascend to sing—sound propagating within a wavelength of the surface interacts in complex ways with sound propagating just below the surface. In this situation, usually all but the lowest frequencies are effectively eliminated as a result of cancellation of higher frequencies.

FORESTS AS CONCERT HALLS OR WAVE GUIDES

In an environment with many dense objects, such as a forest, sound emitted by a sender at any instant reaches a receiver some distance away by many different paths. Some sound arrives by the shortest straight path between the sender and receiver. Additional sound energy arrives after scattering or reflection from one side of the direct path or the other. Some of this scattered sound returns from objects farther to the side and thus arrives even later. The later sound, because it has traveled farther, has attenuated more by atmospheric absorption and spherical spreading. Consequently, sound emitted by a signaler at any instant arrives at a receiver with decreasing intensity over an interval of time. In other words, the sound emitted at any instant by the sender arrives at the receiver with a "tail."

In some cases scattering of sound back to a receiver could result in a brief moment during which intensity increases before the tail of decreasing intensity begins. This possibility of a brief increase and then decrease in the intensity of an arriving pulse of sound is especially apparent in enclosed and highly reverberant spaces. Engineers who measure the reverberation of concert halls can detect this effect. It is hard to detect in natural environments where reverberation is so complex and sound is not confined to an enclosed space.

Several measurements in forests indicate that reverberation might produce some amplification of sound as it does in concert halls. Broadcasting tones in tropical forests and then recording them at a series of distances has shown that longer tones are less attenuated than shorter ones. One interpretation of this result is that longer tones allow time for the accumulation of reverberations and thus an amplification of sound such as occurs in concert halls. A comparison of reverberation in concert halls and forests is, however, not obvious.

Forests might behave like concert halls in other ways. In a hall, the many large surfaces that reflect sound produce standing waves. A standing wave results from interference between direct and reflected waves, as a wave reflected from a boundary combines with the wave still arriving from the source. As a result, the reflected and the arriving waves at any spot in the concert hall either mutually augment or diminish in intensity by constructive or destructive interference. For a constant sound, interference between waves establishes a stable spatial pattern of interference, with some places "live" and some places "dead"—in other words, a "standing wave." If the sound continually changes, then the spatial pattern of interference shifts continually rather than standing still, but it is no less prominent at any instant in time. All of these effects occur in natural environments too, although the complexity of reflecting surfaces and of frequency patterns in sounds makes the patterns of interference complicated in both space and time. Nevertheless, at any instant standing waves are potentially prominent. When sound is enclosed within reflecting boundaries, its attenuation through spherical spreading is replaced by this complex pattern of shifting or "standing" patterns of interference.

There is another way that an increase of sound intensity could occur during communication in the interior of a forest. Rather than accumulation of reverberation in a concert hall, an alternative analogy might be augmentation of intensity in a waveguide. Reflection and refraction from the top of the canopy and the ground in a forest, or even from denser layers of foliage above and below a more open stratum of a forest, could produce the effects of a two-dimensional waveguide. A change in acoustic impedance at the top of the canopy, where the small compartments of air within the canopy abut the large ones above, would reflect sound from sources inside the forest back into the forest (see Figure 4.5). A temperature gradient above the canopy would also refract sound downward into the forest. Both of these effects tend to restrict the spherical spread of sound upward and to confine sound within a layer of the forest. The

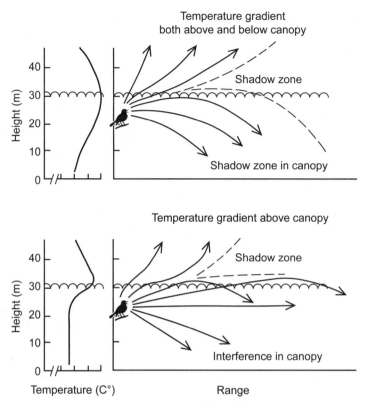

FIGURE 4.5. Complicated sound transmission in and near the canopy of a forest. After sunrise, radiation from the sun heats up the top of the canopy. The heated canopy gradually creates temperature gradients both in the air above the canopy (like the temperature gradient above the ground in open environments) and below the canopy inside the forest. Depending on conditions (compare the results of two temperature gradients in the left-hand diagrams), shadow zones or interference can develop in the canopy. In some circumstances sound could be constrained to propagate within the forest (or between two layers of foliage in a forest). It would thus spread more nearly in two dimensions rather than spherically. A forest would then become a waveguide or sound duct, with many acoustic complexities. Little is known about this possibility. Adapted from Wiley and Richards (1978).

ground in a forest, or even a dense understory, would also reflect sound back into the upper layers of a forest and restrict the spherical spread of sound. As a result, the top of the canopy and the ground, or distinct layers of dense vegetation within a forest, could produce a two-dimensional waveguide—albeit with somewhat "soft" boundaries—but perhaps enough to confine sound to some extent between upper and lower limits.

Waves propagating in waveguides have some peculiar properties, none of which has been confirmed in terrestrial habitats (though the effects are well known in shallow water, where the upper and lower boundaries are more distinct).

Measurements suggest that forests might have enough reflection from the top of the canopy and from the ground to augment the intensity of sound for a receiver. The measurements of attenuation of longer and shorter tones in tropical forests are consistent with this interpretation. In addition, Peter and Mary Sue Waser have shown that loud calls by Grey-cheeked Mangabeys, monkeys of African forests, propagate through the canopy of a rainforest with the least attenuation during a few hours after sunrise. This is the time when a temperature gradient develops above the canopy but turbulence from wind is still relatively slight. Under these conditions, sounds of a male mangabey in the canopy do not escape into the air above but are instead at least partially refracted back into the canopy, where the refracted wave can augment the intensity of the direct wave through the canopy to other mangabeys listening perhaps 100 meters or more away.

REVERBERATION

Quite aside from any reduction or augmentation of the intensity of sound during transmission from a signaler to a receiver is the question of degrading or maintaining its pattern. The pattern of frequencies and timing distinguish a sender's signal from any of the other sources of sounds in the environment. It provides the basis for the receiver to separate relevant and irrelevant sounds and also to determine which of the possible relevant ones it might have heard. The patterns of signals thus are the vehicles of information, the basis for a receiver's discriminations among signals.

Attentive listening, even common experience, shows that the features of sounds from a distance are less easily distinguished. Even when intensity is still adequate, the features appear less distinct. For sound, the main reason for this loss of comprehension is reverberation. The principal effect of reverberation on propagating sound is the smearing of

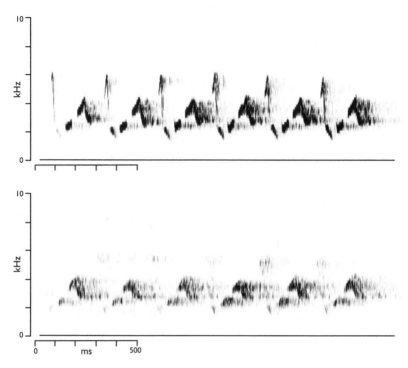

FIGURE 4.6. Reverberation of a complex sound smears the details. Recordings of a Carolina Wren's song, one from nearby and the other from 20 meters away, show the effects of reverberation in a forest. Scattering of sound waves from numerous objects of diverse dimensions produces reverberation. Sound arrives at a listener in a reverberant environment by many paths. Those on a longer path and later to arrive have attenuated more than those on a more direct path. As a result, an impulsive sound arrives with a "tail" of decreasing echoes. Reflection of sound does not change its frequency, so reverberation at any one frequency produces a tail restricted to that frequency. Because scattering is frequency-dependent, higher frequencies have more reverberation than low frequencies. Notice that reverberation at any one frequency does not affect reverberation at others. Closely spaced sounds at low frequencies can remain distinct despite smearing of such sounds at high frequencies. Marks on the vertical axis indicate frequencies (0–10 kHz); marks on the horizontal axis indicate 0.1 second.

sound in time (see Figure 4.6). This effect is easily seen in spectrograms, and it can be heard by attentive listening. Because reverberation depends on multiple reflections of sound from objects in the sphere or cone of transmission, it has more effect on sounds in forests than in open areas or above layers of vegetation.

My former student Douglas G. Richards and I made some comparisons of reverberation in a forest and in open areas in North Carolina. Our measurements indicated that reverberation in forests was most prominent at intermediate frequencies. Because scattering increases with frequency, reverberation increases with frequency, at least for low and intermediate frequencies. High frequencies are attenuated more by atmospheric absorption. So, as frequency increases, sound arriving at a receiver by the longer reflected paths becomes less prominent than sound arriving by the shorter direct path. As a result, reverberation in our forests decreased markedly for high frequencies. The maximum for reverberation occurred around 10–12 kHz. Most songbirds use frequencies below this maximum, so within their range higher frequencies result in more reverberation.

Perception of reverberation depends on frequency-analyzing properties of the receiver's ear. Each sensory receptor is affected by reverberation near its characteristic frequency. Reverberation at other frequencies would have no influence on it. Consequently, for animals with frequency-analyzing ears, only sounds near the same frequency are smeared in perception. To be precise, reverberation affects sound at frequencies in the masking bandwidth, or critical band, of a sensory receptor, as described in Chapter 3.

OTHER MODALITIES

This chapter has now identified the principal ways that sound changes during transmission through atmosphere. The focus has been on frequencies of sound used for long-range communication by most animals. Attenuation of these sounds results from spherical spreading, atmospheric absorption, gradients of temperature and wind, destructive interference near reflecting boundaries, and scattering of directional sound. All of these changes occur because sound is a wave. Similar principles apply to other kinds of waves.

Sound of ultrasonic frequencies such as those used by most rodents and bats are not so relevant here because atmospheric absorption of these sounds is so great that they cannot be used for communication over distances greater than a few meters. The inherently short-range nature of ultrasound, on the other hand, makes it ideal for echolocation (so the sounds emitted by nearby animals do not usually interfere with each other) and avoiding the unwanted attention of predators during communication with nearby social partners or family.

In contrast, sound in water has such low attenuation that communication is possible over much greater distances than in air. As a result of the low attenuation of sound in water, the background noise (irrelevant sound) can become overwhelming in some conditions, such as near (within kilometers) of shipping channels or in waters abounding in snapping shrimp. Sound has over four times greater speed in water than in air, so sound of a particular frequency has a wavelength about five times greater in water. Because reflection, refraction, and scattering of sound depend strongly on wavelength, all of these processes become prominent only when objects are four times larger than in the atmosphere.

The change in acoustic impedance at the boundary between water and air is greater than the change at the boundary between the top of a forest and the air space above. As a result, sound in water is more effectively confined between the surface and the bottom. These boundaries thus create a pronounced two-dimensional waveguide for sound. Waveguides substantially alter the transmission of waves. Already mentioned is the reduced attenuation in a waveguide. No less important is dispersive propagation in a waveguide—*dispersive* because the speed of propagation varies with frequency. The reflections from the boundaries of the waveguide (the surface and the bottom) create vertical patterns of interference that propagate horizontally at speeds that depend on the frequency of the sound. Furthermore, waveguides impose a cutoff frequency below which no sound propagates. For water one meter in depth, the cutoff is approximately 1 kHz, depending somewhat on the characteristics of the bottom. Even in the absence of a waveguide, the frequency-dependent speed of sound is more pronounced in water than in air, and it is more strongly affected by temperature and pressure. The result of all these frequency-dependent alterations in the speed of sound make water a dispersive medium for sound. In air, because all frequencies travel at very nearly the same speed, a pattern of frequencies arrives at a receiver intact with all the frequencies still in the same relative temporal arrangement.

In a dispersive medium, patterns of frequencies do not have this convenient constancy. Those extremely rapid patterns of frequencies in the songs of birds would quickly become a jumble in water. Perhaps to minimize these effects, Humpback Whales "sing" in the ocean in deep water and produce their patterns so much more slowly than do birds that one must listen for many tens of minutes to recognize the pattern.

Ambient sound in water, especially in shallow tropical waters where snapping shrimp abound, can match the intensity of ambient sound in tropical forests. Human activities in the past century have, however, transformed underwater soundscapes at least as much as they have atmosphere ones. The propellers and engines of ships produce high levels of environmental noise for animals communicating by sound in water. Although there are fewer shipping lanes than highways, sound attenuates much less with distance in water than in air.

Light also has the properties of waves. Communication with light has all of the general properties discussed for communication with sound. John Endler has explained the issues for communication with great clarity. For present purposes, it is sufficient to indicate a couple of ways communication with light differs from that with sound. Most important, visual signals rely, with few exceptions, on the reflection or scattering of incident light rather than the generation of light by a signaler. The incident light comes in most cases from the sun, but it is often modified by reflection, scattering, and absorption before it even reaches, for instance, the plumage of a bird that will reflect or scatter the light for a receiver to see. Consequently, all the processes that alter a wave during transmission affect light both before it reaches the surface of a visual signal and afterwards, both on its way to the signaler and during its transmission from the signaler to the receiver.

Furthermore, light waves have much shorter wavelengths than do sound waves even in air—on the order of 10,000 times shorter. Thus the sizes of objects that reflect or scatter light are minuscule in comparison to those that affect sound. Animals use this property of light to "produce" some of their most dramatic colors. Almost all of the blue colors in the feathers of birds result not from pigments but from tiny bubbles of air incorporated in the barbs of their feathers—bubbles less than 1/200 of a millimeter in diameter, just the right size to scatter short wavelengths of visible light (blue) but not long wavelengths (red). On the other hand, the propagation of light in the atmosphere is also affected by tiny particles of dust or water vapor. On humid days, distant objects appear bluish for the same reason the sky looks blue—the short wavelengths of sunlight are scattered toward us and other objects by particles in the sky. North Carolinians are familiar with this phenomenon every humid summer day when they look at the Blue Ridge Mountains on the horizon. In forests, the incident light, mostly reflected from foliage, is predominantly greenish, the frequencies reflected rather than absorbed by the chlorophyll in leaves. Animals using light for communication must work with what they get, often bluish or greenish incident light. The only exceptions are those few organisms that generate their own light for communication, the bioluminescent glowworms, fireflies, and deep-sea fishes.

Other modalities for communication do not share the properties of waves and face other problems of transmission. Chemical signals often serve only for short-range communication. Receivers must get within centimeters (or at least within a few meters) to receive many scent marks. The only way for a signaler to spread such a signal around is to leave many marks, and animals that mark their territories with scents do just that. The alternative is to produce smaller, more volatile molecules that can be borne on the wind to a receiver. The challenge of transmission is the inevitable turbulence in the atmosphere, especially when air moves past objects or over irregular surfaces such as vegetation in a field or forest. Careful study reveals that scents on the wind become irregular filaments in the eddies of the air, not at all the orderly stream we might expect. Receivers face challenges in detecting these filaments and then determining from which direction they have arrived. In fact most receivers of airborne odors respond not to the structure of the odor in the air but simply by moving upwind.

DISRUPTION OF PATTERNS

Regardless of modality, transmission of a signal from a signaler to a receiver always results in disruption of the patterns in the broadcast signal. Attenuation reduces their intensity to a level closer to that of background energy in the environment. In addition, the patterns themselves are degraded by physical processes that tend to disrupt the relationships between the components of the signal. The components become blurred and jittered, their boundaries less definite and their arrangements less predictable. This degradation applies to sound, light, and chemicals. In each case the physical processes are specific to the modality. Even touch depends on deformation of tissues of the receiver's body to reach its sensory cells. In every case, the general result is a decrease in intensity and pattern of signals during transmission. In other words, noise increases.

5

ADAPTATIONS TO NOISE
IN DIFFERENT ENVIRONMENTS

ACTIVE SPACE OF SIGNALS

As a result of attenuation and degradation, there is some distance at which an acoustic signal can no longer be recognized. Recognition requires both detection of a signal amid other sounds in the environment and discrimination between different signals. Both are limited by degradation of the structure of a signal during transmission. The maximal distance at which a signal can be recognized is the radius of an area within which communication can occur—the "active space" of the signal.

The principles apply to a signal in any modality, but acoustic signals provide a good place to start. To measure the active space of an acoustic signal, it is necessary to play an animal's sounds with an intensity appropriate for a signaling individual at different distances from appropriate receivers. The active space is then the maximal distance at which a receiver responds to the broadcast signal. These measurements allow us to ask whether or not the active space of a signal is adapted to the usual distance between communicating individuals.

There have been a number of measurements of the active space of acoustic signals, but the first, by Eliot Brenowitz in 1982, made an important point. The songs of Red-winged Blackbirds in New York State have an active space with a radius close to twice the average diameter of a male's territory. Thus a male singing anywhere in its own territory can be heard by each of its neighbors anywhere within their own territories.

The active space of monkeys' vocalizations in African forests also correspond approximately to the usual spacing between groups, as Peter and Mary Sue Waser have shown. Charles Brown reached similar conclusions for the long-range calls of Blue Monkeys and Grey-cheeked Mangabeys, two rain forest primates in east Africa. The active space of each species' long-range vocalizations has evidently evolved to match the requirements of communication among neighbors. It would be interesting to know if birds' songs in forests also fit this pattern, but so far there are no available measurements.

As one might expect, not all the elements of a complex vocalization, such as a bird's song, have the same active space. At least in one case, these elements have active spaces adapted to different roles in communication. Nightingales in brushy habitats in Europe sing especially complex songs. Each bird has a repertoire of some 200 different phrases, which it sings in sequence, with some variation, over and over. Some of these phrases have much smaller active spaces than others. Those with rapid wide-spectrum trills degrade in structure, particularly as a result of attenuation of high frequencies, at distances less than the average spacing of territorial males. These phrases are especially frequent during close-range interactions of territorial males near their mutual boundaries. In contrast, phrases with the species' characteristic piping notes transmit without much degradation in structure to distances well beyond the average spacing of territorial males. These phrases are frequently sung at night in early spring by males that have not yet attracted a mate. Marc Naguib and his colleagues, who made these measurements, suggest that an especially large active space for these phrases increases the chance of attracting newly arriving females. Nightingales migrate at night, so males singing these piping songs at night can be heard by arriving females even at a distance. In this case there is no limit beyond which a vocal signal's utility declines—the farther it carries the better.

In choruses of frogs, the active space of a male's call can be surprisingly small. This situation makes it clear that background noise sets the horizon for a signal's active space. Any one male Green Treefrog in the southeastern coastal plain of the United States produces its mating call in the midst of a chorus of hundreds or thousands of other males. Anyone who has experienced this phenomenon knows what it is like to be immersed in inescapable noise. A female searching for a suitable mate must first detect individual males in this din of potential mates. By playing back calls to a female in a small circular arena, it is possible to compare her responses to the background sound of a chorus (noise) and a single male's call (a signal) mixed with this noise. In three species of frogs tested so far, the intensity of a single male's call must exceed that of the chorus by at least 3 decibels (dB) for a female to approach it consistently. Presumably she cannot detect a single male's call in the sound of the chorus unless it is at least this intense from her vantage. How close to a male does a female have to be for his call to stand out this much? By measuring the intensity of a chorus from a random point in the middle (but not close to an individual male) and then measuring the intensity of a male's call at a distance of one meter (and remembering that sound from a small source attenuates by spherical spreading at a rate of 6 dB for each doubling of distance), a ranologist can determine that a female must be within one to two meters of a male to detect him. Males in a dense chorus are often about a meter apart, so a female in a chorus might hear between three and ten individual males in a comparable area around her position. The high level of background noise restricts the active space of a male's call to a small circle. To sample more males, females must move around, often risking an encounter with predatory snakes also attracted to the chorus. Males must call as loudly as possible to maintain an area as large as possible in which to attract a female.

Maximizing a signal's active space is not always advantageous. In close-range communication with offspring, parents, mates, or other social partners, a large active space for signals might just attract inappropriate receivers, such as predators or competitors, without compensating advantages for appropriate receivers. It would again be interesting to know the active space of such conversational signals, but no measurements are yet available.

RANGING THE SOURCE OF SOUND

Attenuation and degradation of signals can also serve a useful purpose by allowing a receiver to judge its distance from the signaler. Signals used to regulate spacing between individuals are candidates for this purpose. For instance, the advertising songs of territorial birds often evoke responses from territorial neighbors, particularly when boundaries are under negotiation. In these circumstances, it could pay for a singer to promote ranging of its songs by listeners. That way, a neighbor could determine whether or not the singer had trespassed on its territory (as will be discussed further in Chapter 14). The question is, can a listening bird use the attenuation or degradation of songs to judge the distance to a singer? (See Figure 5.1.)

My student Douglas Richards was the first to test the possibility that territorial birds could judge the distance to a singing neighbor by the degradation of elements in songs. He and I felt that loudness alone was not likely to be a reliable indicator of distance, because the signaler could actively alter the intensity of its songs and because intensity for a listener would change significantly depending on incidental circumstances, such as the orientation of the singer, or its position near reflecting surfaces, such as the ground or the top of the canopy. Instead, more consistent indications of distance might rely either on reverberation of rapidly repeated elements at the same frequency (such as a songbird's trill) or on differences in the attenuation of high and low frequencies in a song.

In his experiments, Richards recorded the songs of a Carolina Wren and then broadcast them from a loudspeaker just inside the boundary of a neighboring wren. If the broadcast song sounded like it came from the location of the speaker, he expected the subject would respond as if its boundary had been transgressed, so the subject would arrive promptly at the speaker ready for a challenge. A song recorded at close range should sound, when broadcast, as if it came from close to the location of the speaker. The same song recorded at long range, with its accumulated reverberation and frequency-dependent attenuation, should sound when broadcast as if it came from farther away than the actual location of the speaker. That is, a clean song should appear to come from within the listener's territorial boundary, but a degraded song should appear to come

Great Carolina Wren.
TROGLODYTES LUDOVICIANUS, Bona. 1 M.& 2. Female.
Dwarf Buck eye. Æsculus Pavia.

FIGURE 5.1. John James Audubon's illustration of a singing Carolina Wren. A male in full song has chosen the highest perch available, while a female, apparently insensitive, forages nearby. Like many woodland birds, this species is drab in appearance but remarkable to hear. Each male sings about 40 rollicking patterns, all with the same characteristic rhythm as the song illustrated in Figure 4.6. Potential listeners include territorial neighbors, often 50 to 100 meters away through a forest, as well as its mate or potential mates. Exaggerated signals like the songs of Carolina Wrens evolve in response either to distant listeners or nearby reluctant (fastidious) ones. From an engraving by Robert Havell for John James Audubon's *Birds of America* in the Mangum and Josephine Weeks Memorial Ornithological Collection, North Carolina Collection, Wilson Special Collections Library, University of North Carolina at Chapel Hill.

from outside—provided the listener could judge the distance to a signer by reverberation and frequency-dependent attenuation. Richards prevented the listener from using overall intensity to distinguish the broadcasts by presenting both kinds of songs at the same intensity. Sure enough, the listener (in repeated trials) arrived predictably in response to the clean (short-range) recordings but not in response to the degraded (long-range) recordings. He thus demonstrated that Carolina Wrens could use reverberation and frequency-dependent attenuation to judge the distance to a singer, irrespective of the overall intensity of a song.

A subsequent student, Marc Naguib, repeated these experiments with many refinements. When he played back only a single reverberated song, he often observed the wren fly beyond the speaker a short while later, just as expected if it had judged the degraded song to come from farther away than it actually had. He found that wrens could use either reverberation or frequency-dependent attenuation separately to judge distance. Overall intensity helps as well. Perhaps most surprising was his comparison of wrens' ability to range songs in late winter, when the deciduous forests in North Carolina have no foliage, and in spring, when foliage is thick. Reverberation in particular is much greater when forests are in foliage in comparison to when they are not. Wrens were approximately equally capable of ranging in these two seasons. Evidently they adjust their criteria for estimating distance, presumably by some trial-and-error experience as the seasons change.

It is important, however, to realize that ranging sounds by their accumulated reverberation, their frequency-dependent attenuation, and their absolute intensity is unlikely ever to be exact. Scattering of sound during transmission is too complex to afford precision. Furthermore, any use of degradation or attenuation to judge distance requires that the listener have an antecedent memory (or perhaps innate awareness) of the properties of clean, nondegraded sounds. Frequency-dependent attenuation can only be judged by comparing the overall spectrum of the signal near the source to the spectrum of the signal received. Reverberation is perhaps more easily judged without comparison, but awareness of the rates of repetition of elements in the original signal would help.

An ability to range vocalizations might serve even for dynamic adjustments of individuals' spacing. This possibility arose during James

Whitehead's study of the calls of Mantled Howling Monkeys in Costa Rica. These monkeys live in small stable groups in the canopies of forests, where they feed on fruit and selected leaves. Groups occupy areas that overlap the areas of neighboring groups. When two groups happen to approach each other in a zone of overlap, there is much calling back and forth. These encounters rarely come to blows, because if they get too close, both of the groups move away from each other. Whitehead used playbacks of the males' barks, one component of their calls in these situations, to show that males in nearby groups respond to each other reciprocally. Playbacks that simulated a male moving away from an encounter (clean barks followed by reverberated barks) evoked reciprocal movement away from the playback by a listening group. The converse situation—playbacks that simulated a male moving toward an encounter (reverberated barks followed by clean ones)—evoked the opposite response, a listening group moving toward the playback. Neighboring groups of howling monkeys, out of sight of each other but interacting by means of their loud vocalizations, thus use changes in the degradation of their calls to judge the direction of the neighbors' movements and respond reciprocally.

These interactions raised the possibility that howling monkeys might modify their vocalizations in order to manipulate their neighbors' movements. Do they ever produce "noisy" barks, ones that seem to be reverberated, in order to appear farther away from a neighbor than they actually are? Might they trick a neighboring group by falsely simulating their own withdrawal from an encounter when actually they stay put? This trick might allow them to continue feeding without competition. Whitehead noticed that his howling monkeys sometimes produced rough barks, ones that sounded (and appeared in spectrograms) like barks that had been reverberated by distance through the forest. He suspected that these rough barks were especially likely when a male's group had found a good place to feed in a zone of overlap. Yet a few trials with playbacks of these calls failed to reveal any change in the behavior of nearby groups.

These preliminary observations raise a possibility that needs more attention. Do monkeys, or other animals, ever adjust their vocalizations to appear farther away from rivals than they actually are? Deception of any sort is difficult to investigate, because, as emphasized later, receivers should cease to respond to signals when responses are disadvantageous on average. Can animals make adjustments to their signals to deceive listeners occasionally? One possibility is manipulation of a signaler's apparent distance, by producing vocalizations that appear to include more or less reverberation or high-frequency attenuation.

THE COCKTAIL PARTY EFFECT

Not all acoustic signals have evolved for long-range transmission. There are also signals for communication with nearby individuals. Parents, offspring, mates, and the members of any compact social group in general often engage in more complex communication than individuals at a distance from each other. In particular, as described in Chapter 2, the human voice as a vehicle for spoken language serves primarily for close-range communication. In these circumstances, attenuation and degradation of signals during transmission is not an important source of noise in communication. In dense congregations of individuals, each involved in communication with just one or a few others, noise levels are high, but the noise comes from other individuals of the same species trying to communicate nearby with similar signals, a *cocktail party effect,* and not from degradation or attenuation of a signaler's own signals.

Noise in aggregations of individuals can increase as a result of positive feedback. When pairs or small groups of individuals are engaged in separate conversations, speakers can make themselves heard more easily by speaking more loudly. This increase the intensity of vocalization in the presence of background noise, called the Lombard effect, has been documented in humans for over a century and is now known to occur in a wide variety of vertebrates. Of course, if every speaker in an aggregation raises its voice, the background noise for everybody increases. The overall rise in noise to some extent nullifies the advantage for each individual speaker.

Another way to counteract the rise in noise in aggregations is to use directional hearing. As explained in Chapter 3, most animals have the ability to determine the direction of arriving sounds. Many can use this ability to focus attention on sounds from a particular direction—roughly a spherical angle or sector. By thus focusing on sounds of interest, listeners can exclude most of the noise from attention. Directional hearing in effect serves to increase the signal-to-noise ratio of signals of interest. There are other ways to differentiate simultaneous sources of sounds as well. Their spectra, the timing of elements, or redundancy in the sequences of elements all are known to allow humans to segregate sounds from different sources. Evidence suggests that many vertebrates, from frogs to mammals, have similar abilities. Any set of associated features of the sounds from a single source can serve to differentiate it from other sources with other associations of features.

Penguins have provided an especially interesting example of cocktail-party effects in birds. The two largest species, King and Emperor Penguins, make no nests, so parents and young have no fixed location at which to meet. Mates must locate each other in order to take turns incubating the single large egg held on their feet. When the egg hatches the young huddles with hundreds of others in a dense crèche. Parents returning from sea with food must find their own young in this aggregation of calling young. Playbacks have shown that mates and parents can accomplish this feat even when the calls they hear are less intense than the background of other calling adults and offspring. The reports from a team led by Thierry Aubin and Pierre Jouventin suggest that these penguins can use both redundancy in the calls of their young, as a result of repeated similar notes, and also an ability to localize their calls in order to recognize their young in this extraordinary level of noise. Smaller penguins, which build permanent nests within small, close-packed territories, have simpler calls. These calls might thus not provide the same possibilities for redundancy and localization. The fixed nest instead serves as a spatial focal point that allows young and parents to get close to each other before recognition is necessary by means of vocalizations.

Determining the direction of sound is just one aspect of a more complex process called auditory scene analysis. This process associates the successive sounds from any one source. Only in this way can multiple concurrent sources of signals be discriminated. It is thus the only way that a receiver can focus attention on one source in the presence of others. The analysis is comparable to object recognition in vision. Scene analysis is needed for more than communication in noise, but it is easy to understand that it helps for communicating in a crowd. What features of the steady stream of stimulation reaching an individual's sensory receptors allow it to recognize discrete entities, whether visual objects or sources of sounds?

Colocalization of sounds can help with this analysis. Other features of signals also contribute. Temporal relationships, such as absence of overlap, characteristic cadences, and idiosyncratic details of timing, can all contribute to the formation of auditory "objects." Common spectral properties of sounds can too. Human voices include cues characteristic of individuals, such as fundamental frequency and idiosyncrasies in the formants of vowels. For communicating in noisy aggregations, any features of acoustic stimulation that differentiate signals from separate sources can allow a receiver to focus attention on one source at a time. Auditory scene analysis cannot eliminate noise in communication in aggregations but it can effectively reduce it.

Even without nearby interfering individuals, close-range communication is likely to be noisy because participants, as Chapter 10 explains, are not expected to adopt tactics that completely avoid noise. It turns out that all communication is expected to occur in noisy conditions, whether it involves complex information at low intensities over short distances or simple information at high intensities over long distances.

ACOUSTIC ADAPTATION HYPOTHESIS

To the extent that natural environments differ in transmission of sound, the organisms that occupy these environments might differ in features of their signals for long-range communication. Chapter 4 alluded to adaptations that might improve transmission, but it is now appropriate to consider the evidence more thoroughly.

Chapter 4 explained how the transmission of acoustic signals differs among natural environments. The density and dimensions of foliage strongly influence scattering of sound (and the resulting reverberation and frequency-dependent attenuation). Forests should thus differ in predictable ways depending on the presence and density of foliage. Furthermore, needle- and broad-leaved forests should differ as a result of the different dimensions of foliage. Even greater differences should characterize forested and open environments. As a result of these differences, each species might be expected to use sounds for long-range communication that are optimal for the environment it inhabits.

Over the past four decades there have been many attempts to confirm the prediction that signals for long-range communication transmit best in a species' usual habitat. Often studies have played sounds from a loudspeaker and recorded them at a series of distances. These studies have employed a diversity of protocols. Some have used natural sounds, such as birds' songs or other animals' calls, and others simpler sounds, such as tones with frequencies used by animals. Some studies have compared songs of different species in one or more habitats, and some have compared songs of the same species in different habitats. They have used different speakers and microphones, different distances, and various habitats. The equipment and habitats have often not been specified in detail. They have tested a variety of hypotheses or variants of hypotheses about optimal sounds for long-range transmission. As recent reviews by Giuseppe Boncoraglio and Nicola Saino and by Elodie Ey and Julia Fischer have emphasized, the results have been mixed.

The various predictions for optimal sounds have often been grouped as a collective *Acoustic Adaptation Hypothesis*. The suite of predictions has become focused on a standard set: in comparison to open habitats, in forests (habitats with complex vegetation structure) songs propagate better when they have low frequencies, narrow bandwidths, less frequency modulation, long notes, and long intervals between notes. For instance, Boncoraglio and Saino, Ey and Fischer, and Eugene Morton all settle on similar hypotheses for investigation. Attention has particularly focused on two predictions: birds in forests should use lower frequencies than those in open habitats, and they should have longer intervals between "consecutive elements" of songs. In the first overall review of these studies by Boncoraglio and Saino, the frequencies of songs were found to differ consistently between closed and open habitats. Maximal, minimal, and peak frequencies and the range of frequencies in songs were lower in closed habitats. The difference was particularly pronounced for peak frequencies. In contrast, their analysis did not support a prediction that intervals between notes should differ between closed and open habitats.

Many of the studies included in these reviews have not investigated carefully formulated inferences about long-range communication. For instance, greater attenuation of high frequencies in forests than in the open does not imply that the optimal frequencies for long-range communication are higher in open habitats. As explained in Chapter 4, low-frequency sounds attenuate less than high-frequency ones in all natural environments. The only exceptions are low frequencies within one to two meters of the ground. The difference between habitats in attenuation of high frequencies does not affect the lesser attenuation of low frequencies in any habitat. That birds in forests use low frequencies for long-range communication is thus not surprising; the surprise is that birds in open habitats use high frequencies.

It is sometimes argued that high frequencies are affected less than low frequencies by turbulence in open habitats, but this argument is not correct. As Chapter 4 noted, most turbulence consists of cells large enough to produce dropouts of all frequencies normally used by animals. Only infrasonic frequencies (and perhaps very low sonic frequencies) would have wavelengths long enough to propagate without much refraction by turbulence. Consequently, some other advantage of high frequencies must be sought to explain their use in open habitats. Two come to mind. First, it is easiest for small birds to produce high-frequency sounds. For any rate of expending energy, they can produce high frequencies with greater intensities than low frequencies. Doubling the intensity of a sound at the source doubles its active distance and quadruples its active area. It is possible that the greater intensity possible for higher frequencies is enough to compensate for their greater attenuation by atmospheric absorption but not enough to compensate for additional attenuation from scattering in forests.

Second, it must be easier to produce rapid modulations of amplitude or frequency with high-frequency sounds than with low-frequency sounds, just as it is easier to play a rapid scale on a violin than on a bass viol. This limitation is more a result of physiology than physics. Rapid modulations require rapid changes in the rate or amplitude of vibrations of the sound source. At low frequencies of sound the rate of change can approach the time required for one vibration of the source, so the inertia of a more massive source vibrating at a low frequency constrains the rate at which the frequency of vibration can change (for any input of power). Rapid modulations, such as the tinkling songs of many larks, pipits, and longspurs, all groups of birds with long evolutionary associations with open habitats, is only economical with high frequencies.

In open environments, where turbulence produces random dropouts of sounds, it might be advantageous to produce such tinkling songs. The portion of a signal heard between dropouts, perhaps just a snippet of the total, could then be complex enough for a listener to judge the species or identity or versatility of the singer. Either the more complex modulation or the greater intensity possible with high frequencies could provide an explanation for the tendency of small birds in open habitats to use higher frequencies than those in forests, despite the basic advantage of low frequencies in reducing attenuation in both habitats (see Figure 5.2).

Because most studies of adaptations for long-range acoustic communication have focused on forests, one of my former students, Michael Green, set out to examine predictions for communication by birds in open habitats. Realizing that reverberation from reflections predominates in forests and amplitude fluctuations from turbulence in the open, he decided to focus on the contrasting consequences of reverberation and turbulence. These studies led him to the Great Plains in North Dakota, where many birds have long evolutionary histories in grasslands. He surmised that birds communicating over long distances in atmospheric turbulence should sing in a way that is resistant to dropouts. In these expansive grasslands, the tinkling songs of pipits and longspurs, the rapid trills of sparrows, and the elaborate flourishes of meadowlarks might convey enough information about the singer's identify to evoke responses even if substantial portions of a song were missing. We thus wrote a program to impose sinusoidal modulations of amplitude on recordings of these species' songs. Green then investigated the effects of these dropouts for about half of each song. When he played these modified songs to territorial Baird's Sparrows in North Dakota, they responded as strongly as when complete songs were played. Artificial reverberation of their songs at a level appropriate for transmission through 50 meters of forest, on the other hand, almost

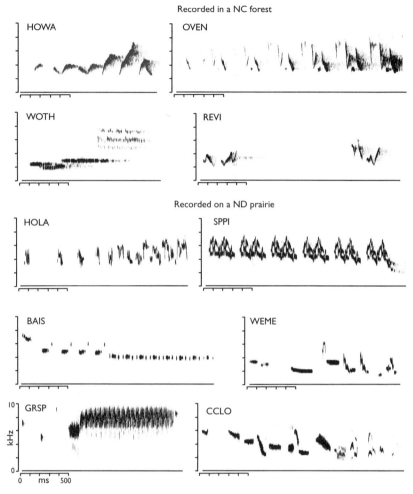

FIGURE 5.2. Comparison of birds' songs in open and in forested habitats. The top four spectrograms show songs of typical birds in deciduous forests of eastern North America (Hooded Warbler, Ovenbird, Wood Thrush, and Red-eyed Vireo). The remaining six songs come from birds typical of grasslands on the Great Plains (Horned Lark, middle section of a longer song; Sprague's Pipit, final half of a song; Baird's Sparrow, Western Meadowlark, Grasshopper Sparrow, and Chestnut-collared Longspur). The rapidly repeated notes of many grassland birds create buzzy or tinkling songs to our ears. Some of these species also have songs that are much longer than woodland birds'. The more slowly modulated and more widely spaced notes of forest birds sound more melodious to us. In forests, rapidly repeated elements at any one frequency are hard to distinguish as a result of reverberation from vegetation. In grasslands, rapidly repeated elements might let enough get through between dropouts from turbulence in the air. Reverberation in forests also makes it difficult to record "clean" examples of songs. The songs from forests shown here were all recorded with highly directional microphones within 8 meters of the singing bird, whereas the songs from prairies were recorded from 20–30 meters away. Spectrograms from recordings by Michael Green.

completely eliminated responses. The next step was a test of whether exactly the opposite would happen with species adapted for life in forests. Although our long experience with the songs of warblers in forests of North Carolina suggested this result, the definitive experiment still remains one for the future.

The difference between needle-leaved and broad-leaved forests also needs more attention. The smaller dimensions of needles (even clusters of needles) in comparison to those of broad leaves, as discussed above, should produce a clear difference in frequency-dependent scattering of sound. Because the attenuation of sound in forested habitats comes from increased scattering of partially beamed sounds, attenuation from scattering in needle-leaved forests should occur only at relatively high frequencies of sound (wavelengths between one and two centimeters correspond to frequencies between 5 and 10 kHz, high for most birds' songs). In broad-leaved forests, with larger dimensions of foliage, attenuation from scattering should become important at lower frequencies and thus affect a broader range of frequencies (a wavelength of ten centimeters corresponds to a frequency of 3 kHz). Studies reviewed above have observed this difference in transmission of sound in needle- and broad-leaved forests, but there is still no definitive comparison of the songs of birds in these two kinds of forest.

Still another prediction that needs more careful attention concerns the intervals between notes. Reverberation in forests tends to blur the intervals between notes, so long-range signals in forests lose much of their temporal complexity. Listeners would have difficulty making subtle distinctions between species, individuals, or virtuosity of singers on the basis of reverberated signals. Attempts to confirm this prediction have seldom distinguished clearly between the intervals between *notes* in a song and the intervals between *repetitions of equivalent frequencies*. When listeners have frequency-discriminating hearing, the relevant intervals are between successive sounds at the same frequency.

When checking this prediction, it is necessary to focus on bands of frequencies corresponding to the listeners' ability to distinguish frequencies. As discussed in Chapter 3, listeners with frequency-analyzing ears have critical bands for discrimination of frequencies, bands that correspond roughly with the just-noticeable differences for frequencies.

Frequencies farther apart than the width of the critical band are perceived as different; those closer than the width of the critical band are perceived as the same. Many birds produce rapid sequences of notes, each of which sweeps from high to low (or low to high) frequency over an octave or so. Often a bird repeats these notes with almost no separation between them (the low-frequency end of one note is followed almost immediately by the high-frequency beginning of the next). Nevertheless, repetitions of equivalent frequencies (for instance, at the beginnings of two successive notes) are separated by larger gaps.

My own analysis of North American birds' songs, which focused on intervals between equivalent frequencies, indicated that rapid repetitions in songs occurred, as predicted, more often in open habitats than in forested habitats. By broadcasting and recording sounds at different distances in a forest, Marc Naguib confirmed that rapid trills (repeated notes at the same frequency) were obscured by reverberation more than were slow trills. For communication in forests, reverberation makes it disadvantageous to use rapidly repeating frequencies (but not necessarily rapidly repeating notes). Studies that have focused on notes, rather than frequencies, have produced less consistent results.

This review of the Acoustic Adaptation Hypothesis shows that several of its aspects need further investigation. Its various components require more careful formulation. Furthermore, as it is currently formulated, the Acoustic Adaptation Hypothesis does not include all of the predictions possible. For instance, there has been almost no investigation of the consequences of atmospheric turbulence for long-range communication in open habitats. Nor has much attention focused on comparisons of signals for long- as opposed to short-range communication. Finally, as already discussed, there is the possibility that including some degradable features in long-range signals could have advantages— for instance, for allowing receivers to judge the distance to the signaler. If so, we cannot expect consistently strong confirmations of the predictions of this hypothesis as it is currently investigated.

In view of the many exceptions to predictions of the Acoustic Adaptation Hypothesis, it is perhaps not surprising that recourse has been made to meta-analysis, a statistical summary of the results from all of the studies combined. The individual studies provide a noisy signal, and averaging, the basic technique of meta-analysis, is a way to improve detection of a signal. It, however, sidelines the effort to understand the exceptions. Clarification of the inferences, as outlined above, might reduce the exceptions. In general, it is a mistake to focus on confirming predictions rather than the assumptions and inferences. Nevertheless, the patterns confirmed so far—despite the exceptions, the confused inferences, the uncertain techniques, and the reliance on meta-analysis— indicate that animals' long-range signals do often differ among habitats in ways that adapt them for optimal transmission and ultimately for detection and discrimination by listeners. Birds and other animals thus reduce the noise added to their long-range signals as a result of degradation during transmission. But they do not eliminate it.

ADAPTATIONS TO URBAN NOISE

Noise is among the manifold impacts of humans on wildlife. The noise of traffic is easily heard kilometers from the nearest highway. Neighborhoods adjoining major highways construct elaborate barriers to attenuate some of this noise for the humans who live nearby. Such barriers are not however constructed to relieve birds and other animals that communicate near highways. There is also noise from machinery, such as mining operations, even in otherwise remote places. Furthermore, in the world's seas, noise from ships and underwater seismic studies can reach high levels, in part because of the much lower rates of attenuation of sound in water.

In the past decade the effects of these sorts of anthropogenic noise have attracted a lot of attention. Noise from traffic is "pink," as mentioned earlier, so called because low frequencies are most intense and higher frequencies progressively less so. Numerous reports have shown that birds sing songs with higher frequencies in the presence of traffic noise. They also often sing louder songs, and some studies have found differences in length and complexity of songs.

These reports raise some interesting questions about animals' adaptions to background noise. One question is whether the observed changes result from each individual learning to adjust it songs in the short term to the level and nature of noise. Short-term learning could result from trial and error in communication with neighbors, mates, or offspring. If the increase in the intensity of birds' songs results from such short-term learning it is another example of the Lombard effect, discussed above in the context of the cocktail party effect. Evidence suggests that such individual learning, a form of physiological adaptation, can explain at least some of these shifts.

Another possibility is that birds develop their songs at an early age by learning from adults that they can hear clearly. Experiments with young birds have shown that they preferentially learn clean songs rather than degraded ones, so perhaps less degraded songs are better models for learning than more degraded ones masked in part by environmental noise. A study of White-crowned Sparrows in San Francisco by David Luther indicates that the songs of these birds have gradually shifted to higher frequencies as traffic has increased in the city over three decades. Learning of songs by young birds from generation to generation might explain this shift, but experiments to support this possibility are still needed.

A final possibility is that urban populations have evolved songs with greater intensity and higher frequencies as a result of natural selection. This sort of change is evolutionary—rather than individual physiological—adaptation. There is evidence that populations of the same species of bird can differ genetically and also vocally. Indeed, the Acoustic Adaptation Hypothesis discussed in the preceding section assumes that the adaptations of birds' songs to different habitats result from evolutionary adaptation. Although a reasonable hypothesis, the necessary genetic experiments have yet to show that the changes in song are associated with genes. Presumably, adaptations to noise might occur in any or in a combination of these three ways—by short-term learning, by intergenerational learning, or by genetic changes. Either individual (physiological) or evolutionary (genetic) adaptation might occur in any case.

This issue is related to a second one, the degree to which different features of song can be adjusted separately. There are suggestions that changes in frequencies in songs in some species occur as a side effect of changes in intensity. Actually all features of songs are possibly interrelated through the biomechanics or the neurobiology of sound production. If so, there could be limits to how much individuals can learn to adjust each feature separately. There could likewise be limits to how fast evolution by natural selection could change each feature separately. These limitations all amount to genetic constraints on individual plasticity in a particular environment. Further investigation of adaptations to urban noise should lead to a better understanding of these constraints on communication in noise.

6

REDUCING NOISE, ENHANCING PERFORMANCE

CONTRAST

Chapters 4 and 5 have explored some of the ways that the features of signals affect their suitability for communication in different environments. One focus has been features that increase the overall intensity of a signal for a receiver, either by increasing the power of a signal at the source or by reducing its attenuation during transmission. Another focus has been features that minimize degradation of the structure of a signal during transmission. The issue in this case is not increasing the overall power of the signal but preserving the relationships among the parts of a signal. The temporal or spatial relationships among the components of a signal produce a pattern that differentiates any one kind of signal from other signals and from irrelevant energy in the environment. A bird's song is, for instance, a pattern of frequencies and amplitudes in time. Obscuring these relationships with reverberation or dropouts makes it more difficult to differentiate these patterns and thus to distinguish a species' songs from irrelevant sounds.

The power and the pattern of a signal thus both serve to distinguish it from irrelevant stimulation of a receiver's sensory receptors. The term *contrast* can refer to all features that differentiate a signal from irrelevant energy in the environment. Environmental noise is, as emphasized earlier, not the only source of noise in communication, but it is an important component of the noise. *Contrast* is a general term for features of a signal that reduce the consequences of environmental noise.

There are two other attributes of signals that reduce errors— redundancy and predictability—both of which refer to the relationships between two or more signals or between signals and other events. Redundancy is a relationship between different signals or their parts and predictability a relationship between a signal and any other event. These three attributes of signals—contrast, redundancy, and predictability—are the only ways that signalers can reduce receivers' errors. Receivers have their own ways to reduce errors—tuning and averaging. It becomes clear in this chapter that the signaler's and receiver's tactics for reducing error require close coordination.

Contrast between the signal and the background or between two different signals has a direct effect on the receiver's ability to detect a signal or to discriminate between signals. Contrast in this context is the ratio between some feature of a signal and the same feature of background energy in the environment. It is the signal-to-noise ratio not only in overall intensity, but in the intensity of any feature of the signal or in the strength of any pattern in the parts of a signal. The features of the background and their patterns thus provide the denominator, and thus the reference, for measuring a signal's contrast.

A signal's contrast differs from its detectability. Detectability—as used in Signal Detection Theory—is a measure of the separation of signal and noise in the output of a receiver; contrast is the separation in the input. Detectability, as explained below, depends on the difference in the probability (or average level) of response in the presence and absence of a signal. These differences in the output of a receiver are affected by the differences in the input but also by the properties of the receiver itself. The tuning, sensitivity, and variation in a receiver's sensory receptors, its decisions to respond, and its generation of responses all affect the detectability of a signal.

Greater contrast increases the detectability of a signal, provided the receiver responds to the increased difference between signal and noise. To study contrast, it is necessary to measure the properties of signal and noise, perhaps with sophisticated equipment. To study detectability, it is necessary to measure the behavior of receivers.

The influence of contrast on detectability of signals is a common experience. As a rule, louder sounds are easier to hear in a noisy background than are quieter ones. Some of the first experiments on signal detection by humans accurately measured this difference. A person's ability to detect a brief tone in the presence of "white" noise increases as the intensity of the tone increases. The relationship, we might not notice, is approximately logarithmic, so that doubling the intensity of a tone increases its detectability in noise about the same amount no matter what the initial intensity. This logarithmic relationship between the intensity of stimulation and the strength of response is often called Weber's Law, after Ernst Weber, its discoverer over a century ago. The term *law* suggests greater invariance in this relationship than measurements actually suggest. In fact, human detection of the intensity of sound departs significantly from a logarithmic relationship particularly at high frequencies, although it is approximately logarithmic over most of the range of human hearing. This logarithmic perception is the justification for using decibels to measure intensity.

Animals increase the contrast of their communicatory signals by choosing times and places with low levels of background energy. Birds often adjust the timing of their songs to avoid overlap with neighboring individuals of the same species or nearby individuals of other species, although overlapping a rival's song can also be a threatening signal. Birds also often sing more at dawn than at other times of day. There are a number of possible explanations for this choice of time, but one possibility is that environmental noise from wind and rain is often lower at dawn. The time just after sunrise is also optimal for the transmission of monkeys' calls in the canopy of African rainforests. Peter Waser's measurements, described in Chapter 5, suggest that refraction by temperature gradients both above and below the canopy tend to focus sound in the canopy just at that time. Of course, many individuals and many species might make the same choice to take advantage of the potential diminution of noise. Indeed, the dawn chorus of birds in some cases must raise the level of noise for any individual's signals enough to nullify most, or at least some, of the advantages of diminished noise from wind.

Animals that must communicate in especially noisy conditions often have signals that avoid masking by noise as much as possible. Near rushing or falling water, for instance, signals are often intense and high-pitched. Intensity increases contrast with any background sound but, as described in Chapter 5, high frequencies attenuate faster than low frequencies. Nevertheless, because falling water produces sounds primarily at low frequencies, high-pitched signals contrast more even though they attenuate more rapidly with distance than would signals with lower frequencies.

In tropical forests, insects and frogs together often produce a nearly continuous band of sound between 4 and 8 kilohertz (kHz). It is thus not surprising that most birds of tropical forests use frequencies below 4 kHz for their long-range songs. On the other hand, small primates in American tropical forests often have vocalizations with dominant frequencies above 8 kHz. Insects dominate the sound in these environments by day and night, so birds and primates must adjust their signals to minimize overlap with insects. Because of their relatively small bodies insects cannot effectively produce intense sounds below 6 or 8 kHz. In addition, they communicate at shorter ranges and often have denser populations than do birds. Birds in tropical forests must communicate with each other from scattered positions in a throng of insects. As a result, if both used the same band of frequencies, insects' signals would mask those of birds more than birds' signals would mask insects'. Insects win in the scramble for bandwidth for acoustic communication.

Wind also produces intense sound, with intensity decreasing exponentially with frequency and sources distributed evenly in space. Birds of open country often use high frequencies, as described in Chapter 5. In forests, wind generates intense wide-spectrum sound as a result of the rustling of moving foliage. Rain on foliage produces even more intense sound. Both wind and rain generate sound over a wide range of frequencies and from numerous sources finely distributed in space. The wide spectrum and nearly uniform distribution of sources even surpass these properties of sound from insects. In my experience, birds in forests simply stop singing in anything more than a slight breeze or a fine shower.

CONTRAST BETWEEN SPECIES' SIGNALS

Contrast also applies to the distinctiveness of signals by different species and even individuals within a species. The distinctiveness of a signal depends on the differences between the features of the signal (or their relationships) and those of signals from other species or individuals that are irrelevant for the receiver. Species of birds that communicate over long distances in any one area almost always have contrasting songs with distinctive features. It seems plausible that the songs of different species might even differ as much as possible in their acoustic features. Imagine different species' songs plotted in "acoustic space." This space would have multiple axes corresponding to independent combinations of the parameters of their songs (their principal components in a statistical analysis). To maximize contrast among their songs, coexisting species might not occupy this space randomly but instead spread out to maximize their differences.

One of my students, David Luther, investigated this possibility for 80 species of birds that had loud songs in rainforests at a locality in southern Amazonia. When all species' songs were plotted in the same acoustic space, there was no indication that they were spread out more than randomly. If the species singing in each of four strata of the forest were plotted separately, however, the species in each stratum had songs that were statistically overdispersed in the acoustic space (more evenly distributed than randomly). Within any one level of the forest, the different species' songs contrasted with each other.

The songs of birds thus provide some evidence that signals have evolved to increase the contrast between different species that share the same location for communication. Do differences in contrast of signals evolve in conjunction with listeners' ability to distinguish between them? Even if an influence of contrast on the detection of signals is easily demonstrated for humans and other animals, there remains the question whether or not evolution results in an association between greater contrast in signals and easier discrimination by receivers. An opportunity to investigate this question arises in the case of individual recognition.

CONTRAST BETWEEN INDIVIDUALS' SIGNALS

An ability to recognize different individuals by their songs requires finer discriminations than an ability to recognize species by their songs. Individual recognition presents some challenges for communication. Each individual's songs must have properties that diverge from those of other individuals. Yet the songs of all individuals of the same species must share enough properties to allow a listener to focus on the songs of their own species. Characterization of signals by species should thus take precedence over characterization by individual.

There is then a trade-off between the advantages of standardized signals for all individuals of the same species and any advantages of individualized signals to promote recognition of individual signalers. Suppose, for instance, that frequency of a song was used for recognition of both species and individuals. To promote quick identification of species by listeners, signalers of any one species should converge on the same frequency. Yet for quick identification of individuals, they should diverge. No doubt this trade-off is part of the explanation for the complexity of many birds' songs, as different components of a song can carry different kinds of information. To maximize the chances of correct responses, however, contrast would have advantages for both species-specific and individual-specific components. Contrast among species would favor stronger standardization of each species' songs. Contrast among individuals of any one species would favor differentiation of each individual's songs within a species.

These patterns have never been examined quantitatively, but some predictable patterns seem likely. Among North American warblers that occupy large territories in the understory of forests, such as Ovenbirds and Prothonotary, Swainson's, and Kentucky Warblers, individuals tend to produce a single song pattern with the species' distinctive features. Stereotypy of songs within a species is pronounced, although it does not preclude some fine individual distinctions. Among those species that occupy smaller territories in denser populations, such as Hooded Warblers, American Redstarts, and many species in secondary forests, individuals have multiple song types. Shorter distances of communication, and consequently less distortion during transmission, might allow less stereotypy within species and more diversification among individuals.

These warblers all learn many components of their songs, as do nearly all oscine birds (order Passeri, the true songbirds). In contrast, insectivorous birds of forests that do not learn their songs, such as those in the families of tyrant flycatchers and related tropical suboscines (suborder Tyranni), have songs with less individual distinctiveness (see Figure 6.1). In these flycatchers individuals develop normal songs without learning from other members of their species. In fact, individuals acquire normal songs despite experimental isolation that precludes ever hearing another individual of the same species. Learning has two consequences for development. On the one hand, individuals can match each other's songs more or less exactly. On the other hand, individuals can improvise more widely. When individuals do not learn songs, they differ only slightly from others of the same species but normally do not exactly match them either. The songs of suboscine species fit this pattern. They have differences in their songs, but the differences among individuals are slight in comparison to the songs of most warblers and other oscines that learn their songs. It seems plausible that the development of unlearned behavior might be about as variable as the development of anatomy. For most birds other than oscine songbirds, individuals vary about as much in their acoustic signals as they do in their anatomical features.

Can these suboscine birds recognize these slight differences in other individuals' songs? Does the difference between suboscines and oscines in the contrast between individuals' songs result in a difference in their abilities to recognize individuals by their songs? I had an opportunity to compare recognition of individuals' songs by a suboscine bird, the Acadian Flycatcher, and an oscine, the Hooded Warbler. A graduate student, Renee Godard, had shown that the warbler had spectacular capabilities for recognizing individuals of their own species by their songs. Their songs had all the usual indications of learning. Each individual sang a repertoire of patterns, some closely matching neighbors' and others clearly distinctive. They could recognize each of their neighbors' songs. They could even remember their previous year's neighbors' songs for the eight months each autumn and winter during which they did not sing while

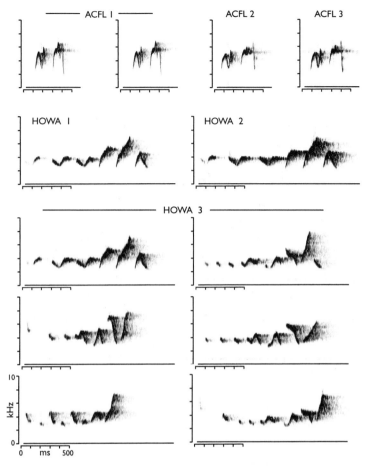

FIGURE 6.1. Songs of oscine and suboscine songbirds show the difference in repertoires, matching, and individuality that typify birds that do and do not learn their songs. An Acadian Flycatcher sings a single pattern most of the time (it uses a variant in the dim light before sunrise). Each individual's song closely resembles the songs of all other individuals of this species, but not quite exactly. The first two songs in the top row were sung several hours apart by the same individual; the next two songs come from two additional individuals, all near Chapel Hill, North Carolina. In contrast, a Hooded Warbler, like many true songbirds, sings a repertoire of patterns, some of which match those of its neighbors almost exactly and some of which are clearly individualized. The spectrograms in the second row show almost exactly the same pattern sung by two different individuals. The first song in the next row is also the same pattern sung by a third individual. All individuals sing one of several widely shared patterns, such as this one, repeated monotonously during the days before they acquire a mate in the spring and less consistently thereafter. The remaining songs, all from the third individual, represent about half of a normal repertoire for an individual of this species. Each individual often sings a varied mixture of these patterns once it acquires a mate. The flycatcher (suborder Tyranni) does not learn its song; the warbler (suborder Passeri) does.

they migrated to Central America and back. Their memory and specificity for recognition of other members of their species foreshadows recognizably human capabilities.

The flycatchers are about the same size, have territories similar in size, and occupy the same strata of the same forests in which the Hooded Warblers had been studied. The Acadian Flycatcher, however, showed no signs of learning its songs. Their songs, all very similar, had slight but consistent differences between individuals. Neighbors had no greater similarity in their songs than did distant individuals. This seemed to beg the question, can they also recognize their neighbors? Closely matched experiments with playbacks showed that these two species differed markedly in their ability to recognize their territorial neighbors. The flycatchers only distinguished individuals' songs after prolonged playbacks, but the warblers made distinctions among individuals often within seconds. In comparison to the flycatchers, the warblers incorporated more contrast among individuals' songs without sacrificing contrast among species, apparently because they learned the complex features of their songs. Furthermore, a facility in individual recognition was associated with this greater individuality in songs. Chapter 14 describes how these warblers use this capability to negotiate their territorial relationships with neighbors. The flycatchers, on the other hand, must have social relationships that do not rely on quick recognition of neighbors. Contrast and recognition of signals seems to evolve hand in hand.

CONTRAST IN VISUAL SIGNALS

Contrast is expected for signals in other modalities as well. Vision provides clear examples. Among some birds of tropical forests, males clear courts for use when displaying to females. Clearing leaves and debris from the court enhances the contrast between a male's plumage and the background, as shown by John Endler and Marc Théry's measurements of reflected light. Birds of the canopy of tropical forests are brighter and more diverse in coloration than those that inhabit the forest understory. William Dilger's classic experiments on the five spotted thrushes in North American forests confirmed that reduced contrast in the appearances of species was associated with a lack of discrimination based on appearance. His experiments combined playbacks of the distinctive songs of the five species with taxidermic mounts of the species in all possible arrangements, and he presented these combinations of songs and mounts inside 10 to 20 territories of each species. The territorial birds quickly recognized their species' songs but attacked the accompanying taxidermic mounts indiscriminately. Contrasting songs of the five species provided signals for recognition of species. Their uniformly dull coloration presumably has the advantage of reducing conspicuousness to potential predators, but this lack of contrast in coloration precluded recognition of species by their appearance.

REDUNDANCY

In addition to increased contrast, redundancy is another way to improve the chance of a receiver's response in noisy conditions. In its simplest form redundancy consists of repetition of a signal. If a receiver misses one signal, repetition provides a second chance. Redundancy thus counteracts the effects of both environmental and internal noise by reducing the chances that the receiver misses a signal. The sources of noise—distortion of a signal during transmission, confusion with background energy, variation in activity of the receiver's nervous system—can combine to mask a signal enough that the receiver fails to detect its presence reliably. Repetition of a signal even moments later might coincide with a fluctuation in distortion, background, or neural activity that could leave the occurrence of the signal unmasked and allow a receiver to detect it. Claude Shannon's original mathematical analysis of communication showed that redundancy serves to reduce noise—in other words, receivers' errors.

Many animals repeat nearly identical signals, sometimes almost tediously, at least from the viewpoint of an observer with an optimal vantage. Birds, for instance, sometimes sing the same pattern five or more times per minute for scores of minutes on end. Insects and frogs can drone for hours on end with almost no change. Visual displays are often repeated equally monotonously. Indeed, inasmuch as the visual appearance of an organism constitutes a signal, some visual signals are effectively continuous.

Redundancy can take more complex forms than simple repetition. Two features of a signal with a predictable association also produce redundancy. In a birds' song, a series of distinctive notes sung in a regular pattern are just as redundant as a single repeated note. In either case, a receiver detecting part of a complex signal can predict the remainder, so detecting any part of such a signal is as good as detecting the entire signal. Birds' long-range vocalizations are commonly called songs for the very reason that they share the characteristic of human songs, a repeatable pattern of notes. Even the longest songs produced by a nonhuman animal, those of Humpback Whales, are repeated with remarkable precision and thus are exercises in prolonged redundancy.

Furthermore, redundancy can apply to spatial patterns as well as sequential patterns. A visual pattern with components in a stable arrangement in space shares the essential features of redundancy in time. The arrangement of the components of a signal in space or time also need not be absolutely constant. Any arrangement with some predictability provides a measure of redundancy. The more such components there are and the tighter their correlation, the greater the redundancy. The temporal or spatial association of signals, or parts of signals, differs from contrast, the magnitude of the difference between signals or their parts and any background energy or matter in the environment.

This broad view of redundancy leads to questions about the difference between multiple parts of signals and multiple signals. In the past few decades there has been some debate about the evolution of multiple signals. Superficially it seems unnecessary for signalers to duplicate signals to evoke similar responses. Nevertheless, many animals perform elaborate displays, with multiple visual, acoustic, and sometimes olfactory components. There have thus been proposals that additional signals serve as amplifiers or modifiers of a basic signal. They might augment or alter the effect of the basic signal on a receiver, for instance, if they indicate additional information about the signaler's condition. It is, of course, plausible that multiple signals can convey more information, but it is also true that they can have advantages in communication even if they do not. In the latter case the advantage would come from the advantages of redundancy in reducing receivers' errors—in other words, noise in communication.

There is a clear way to distinguish between multiple signals that can convey additional information and those that can reduce receiver's errors. The former should occur in varied combinations, the latter in stereotyped patterns. Chapter 1 introduced the way that multiple signals convey information by the unpredictability of their associations. Redundancy, on the other hand, counteracts the different forms of noise by the predictable associations of components of signals. Two signals with highly predictable association are usually considered two components of a single signal. Notice that the components of a signal have two consequences for communication. On the one hand, as just discussed, it is the predictable association of the components of a signal that provides redundancy to counteract noise. On the other hand, the predictable components of a signal form a pattern that distinguishes the signal from others. Thus, when many different signals are used to convey precise information, such as the identity of many individuals or the words of human language, the pattern of each signal requires commensurate complexity. Redundancy is thus predictability in the components of signals that reduces a receivers' errors but conveys no more information. It is predictability that changes the behavior of receiver in the presence of noise but has no influence on a receiver's behavior in the absence of noise.

The association between components of signals creates a trade-off in communication. Completely correlated, and thus redundant, components can increase the chances of a response from an appropriate receiver, but they collectively constitute only one signal. The correlation across time or space reduces the possibilities for transmitting separate signals in the same time or area. Redundancy thus reduces the number of separate signals transmitted in space or time. Even the duration of a constant sound, for any length of time greater than the temporal resolution for hearing by intended listeners, is a form of redundancy, which limits the rate of sending different signals. The duration of a visual display beyond the time for a receiver to take one look likewise limits the rate of sending signals. The decrease in the rate of signaling as a result of redundancy is balanced by any increase in the chance of a response by an appropriate receiver. To maximize the number of different responses in any period of time thus requires an optimal balance between the disadvantages of a lower rate of sending separate signals and the advantages of a higher probability of response to repeated (or correlated) signals.

The same principles apply to spatial configurations (or other simultaneously presented components of signals). The extent of an area of constant color, for instance, contributes to redundancy, as does the predictable association of areas with different coloration. To maximize the specificity of a response to a visual signal thus requires an optimal balance between the advantages of a complex signal with many components of restricted extent (necessary for a receiver to make more precise discriminations among signals) and the advantages of a simpler signal with more extensive components (necessary to reduce a receiver's errors in noise).

Evidence that repetition evolves in response to noisy communication comes from Jordan Price's 2013 study of the displays of New World blackbirds, orioles, and oropendolas. These birds compose a monophyletic family, Icteridae, which comprises an extraordinary diversity of mating systems. At one extreme, some have prolonged social monogamy, in which one male and one female remain together for at least an entire breeding season and work together to rear a brood of nestlings. At the other extreme are the oropendolas and some grackles, which practice the most extreme polygyny among songbirds. In these species, one male can obtain most of the matings with most of the females in a large colony. Genetic analyses of parentage confirm that these males can fertilize most of the eggs of as many as 40 or 50 females. Such males have less time to interact with individual females than do males in species that engage in prolonged monogamy. As described in Chapter 8, females choosing mates often behave as if they have high thresholds for response. In other words, they seem coy, seldom responding or even appearing to show interest in males' signals. In order to elicit responses from coy females more rapidly, polygynous males could use more stereotyped and more repeated displays. As noted above, both stereotypy and repetition are forms of redundancy in signals. In Price's study of the Icteridae, an analysis that controlled for the phylogenetic relationships of the species showed that both stereotyped and repeated displays by males have evolved repeatedly in polygynous species.

The stereotypy of displays has seldom been measured. It is clear, however, that both the variation in any one component of a signal and the correlations among the components varies considerably. Noisy communication should assure that stereotyped repetitive signals recur in situations requiring prompt responses by receivers with high thresholds. Furthermore, repetition and stereotypy of signals should evolve in conjunction with high contrast of signals. The spectacular displays by males of many polygynous species fit this expectation. Another situation that is worth investigation is monogamous migratory species. In these species, a male and female must establish or reestablish a strong relationship as quickly as possible after arrival from wintering quarters. Studies of migratory songbirds often show that starting to nest as early as possible increases reproductive success. With constraints on the time available for communication, redundant (including stereotyped) signals and contrasting signals are expected to evolve.

PREDICTABILITY

Predictability is the third distinctive way signals might reduce receivers' errors in communication. Just as predictability in the relationships of components of signals increases the chances of responses by receivers, predictability in the occurrence of signals has the same effect. Experiments with humans have repeatedly shown that prior knowledge about any property of signals makes it more likely that they are detected. Knowledge of when, where, or what is likely to happen makes it more likely that an observer will not miss it. For instance, consider humans detecting faint sounds in noise. When a brief presentation of a light is used to indicate when a faint tone might (or might not) occur, humans make fewer errors—of omission or commission—in detecting the tone.

I have called this second signal an alerting signal—an easily detected signal that alerts a receiver to the possible occurrence of a less detectable one. The second signal could be less detectable because it is more complex and thus more informative. In order to allow recognition of the correct species or a particular individual, a bird's song must include enough complexity to differentiate it from other species' or individuals' songs. The precise structure of a sequence of notes could require the receiver's attention. An alerting note, a signal easily detected in part because it does not include complex information, could serve to attract a receiver's attention to the occurrence of the less detectable, but more informative, signal. An alerting signal thus by its simplicity conveys less information (allows less precise identification by a receiver) but is more likely to elicit a generalized response of attention. It must have a predictable association with the informative component, but it probably need not precede it. Because of the way that memory works, with short-term storage preceding selective consolidation in long-term storage, an alerting signal might immediately follow an informative signal. The alerting component would then function as a signal to preserve the preceding segment of short-term memory.

Many birds' songs begin with simpler, louder notes and then progress to more complex ones (although in a few cases, such as the Hooded Warbler's songs, the order seems reversed). Eastern Towhees show this pattern of one or several simple notes at the start, then a complex trill which consists of a rapid series of syllables, each a complex combination of two or three brief notes (see Figure 6.2). We humans hear the trill as a pleasant jingling. We do not resolve each of the quick frequency sweeps and instead hear the trill primarily in the frequency domain, as explained in Chapter 3. Experiments by my student Douglas Richards explored the possibility that the simple introductory notes served as an alerting component for the more complex informative trill. His hypothesis was that the trill had the information for recognition of the species, so that a listening towhee would only respond if the trill had been correctly recognized.

Richards's experiment demonstrated how to test for an alerting signal. He played back introductory notes and trills separately and in normal combinations. In some cases the notes and trills were degraded by artificial reverberation and in other cases they were "clean." He found that introductory notes alone, whether degraded or clean, never evoked a strong response from territorial towhees. Clean trills did, degraded ones did not. Thus, trills contained the information for recognition of species, but towhees only picked up on it when trills were not degraded. Since all of the playbacks were presented well inside the subjects' territories, presumably they quickly noticed the clean trills but did not notice the

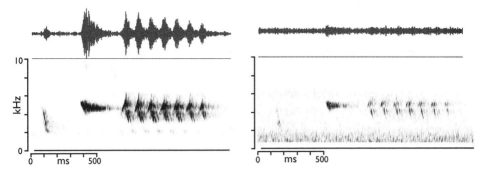

FIGURE 6.2. Alerting component in the song of an Eastern Towhee. The mnemonic for this species' song is "Drink your teeeeeea." The specified beverage is a rapid series of complex syllables. Each syllable consists of several notes, and each note lasts only a few thousands of a second. The introductory imperative, in contrast, consists of two simple notes (or sometimes just one). Douglas Richards showed that playing a clean recording of the full song evoked strong responses from nearby territorial male towhees, typical for territorial songbirds. Playing just the "teeeeeea" did also, but playing the "drink your" evoked no response. When Richards artificially reverberated songs before playing them to birds in the field, then only the full song evoked a strong response. Neither the introduction nor the complex ending got any response. He concluded that the ending alone contained the information that allowed a towhee to recognize the song as a rival male's. The ending is all it took to get a response from a nearby male with a clean recording. Yet with a reverberated recording, such as a song might sound from a distant towhee, this same ending evoked no response. Nevertheless, all Richards had to do to get a full response was to include the introduction in the reverberated recording. Towhees must have trouble recognizing a reverberated ending when it arrives without warning, but an introductory "Hey, listen up" focuses a towhee's attention enough to allow it to retrieve the message. The introductory notes (or note) is thus a general alert, while the complex ending carries the specific information.

degraded ones. When Richards combined an introductory note with a trill, as in a natural song, the degraded trill then evoked a quick response regardless of whether the introductory note had been artificially degraded. Thus, the introductory notes were relatively unaffected by degradation (they were after all simple notes) but they potentiated responses to degraded trills. The introductory notes of towhees' songs therefore did not include information for recognition of the species, but they did attract a potential listener's attention to the immediately following trill. The trill did convey the information, but only when undegraded or when introduced by an attention-grabbing alert.

Alerting signals might even consist of conspicuous signals produced by another species. This possibility was first suggested for several small monkeys of African rainforests. Monkeys inhabiting the canopies of rainforests often use vocalizations to regulate the spacing between nearby social groups of the same species. Some monkeys—especially the larger ones, such as Grey-cheeked Mangabeys and Blue Monkeys—produce loud calls easily heard for long distances. Often the calls have two parts. The first is simple and especially loud, the second less intense but more complex. This arrangement is just like that of the towhee's songs, in which a more complex, information-rich component follows a simpler but easily detectable component. Monkeys with such calls would make good subjects for experiments like Richards's on towhees. Even more striking though is the tendency of smaller species to produce their calls immediately after a larger species that shares the same forest canopies. The smaller species' calls are less intense and often complex. At least four species are reported to "echo" other species' conspicuous calls in this way. Presumably the loud calls of the larger species could serve as alerting signals for the less intense calls of the smaller species.

Fairy-wrens in Australia also use the conspicuous calls of a larger species as alerting signals. In this case experiments confirmed that the fairy-wrens could evoke a response from others more reliably after such an alert. Furthermore, evidence indicated that the fairywrens' responses did not convey any information about or to the larger species.

The timing of a signal is not the only prior knowledge that can improve a receiver's performance. When a signal must be classified in a number of ways as well as simply detected, it helps to know how many and even exactly which kinds of signals require classification. Humans perform better in experiments with foreknowledge of this sort about what signals to expect. Familiar signals are more likely to evoke responses than unfamiliar or otherwise unexpected ones.

All of these examples of an improvement in receivers' performance result from greater predictability of potential signals. Prior knowledge of the timing or location, as well as restrictions on the number, of expected signals all have the same effect. It seems that anything that makes a signal more predictable to receivers improves the chances of a response.

Predictability influences a receiver's performance presumably because it simplifies the problem of classifying input from its sensory receptors. Classification of input by receivers is essential to all forms of communication, a point to be emphasized in Chapter 14. Input requires classification at the most basic level as signal or noise. In addition, signals might require classification by species and individual as well as by the nature of an appropriate response. The more complex the classification necessary for an appropriate response because of more numerous or less predictable signals, the greater the chances of error by a receiver. When fewer signals are predictably associated with fewer situations, they become more likely to evoke appropriate responses. On the other hand, just as contrast and redundancy in signals take energy or time and thus limit the amount of information transmitted at a time, predictability of signals also limits the amount of information by reducing the variability of signals. In every respect, communication must find a balance between information and efficacy in the presence of noise.

RECEIVER PSYCHOLOGY

The properties of stimulation that make responses more likely have received a lot of attention from psychologists. Although these studies do not usually frame the problem in terms of communication in noise, they come to some related conclusions. This psychological perspective was prominent in an influential 1991 paper by Tim Guilford and Marian Dawkins, who marshaled evidence for "receiver psychology" in communication. They distinguished three psychological aspects of responses to stimulation—detectability, discriminability, and memorability. Features that enhance these aspects of responses to signals should be favored by natural selection, because they increase the probability of responses by receivers and thus the benefits for signalers. Guilford and Dawkins also emphasized that receivers must benefit on average from their responses.

Preceding chapters in this book have already discussed examples of detectability and discriminability. Memorability is the additional psychological component needed for learning. Detected or discriminated signals must be remembered for an individual to acquire a learned response to these signals. These three aspects of the psychology of responses thus have a hierarchical relationship. Memorability, as Guilford and Dawkins recognize, depends on the first two aspects, detectability and discriminability, and discriminability is itself secondary to detectability. As appropriate for receiver psychology, all of these three psychological aspects of responding to stimulation depend on properties of the receiver. Detectability, as was explained above, is a measure of a receiver's probability of responding to a signal. Contrast, on the other hand, is a property of signals accessible to an independent observer who is properly equipped.

Nevertheless, most of Guildford and Dawkins's examples of receiver psychology are properties of signals rather than receivers. They only contribute to receiver psychology if we assume there is a property of the receiver's sensors corresponding to each of the properties of signals. For instance, stripes are more easily learned than other patterns only if an organism has sensors tuned to detect, discriminate, or memorize stripes. The prevalent salience of eye spots for many organism's responses, probably even humans, requires receivers with specialized mechanisms for detecting these patterns. Because many animals are susceptible to visual illusions familiar to humans, they presumably share with humans some general aspects of sensory filtering. The lengths of lines depend on their context, and the brightness and hue of colors is affected by the sharpness of their boundaries and the properties of adjoining patches of color. The strength and duration of learning is influenced by predictability of reward or punishment associated with particular signals and with the intensity, familiarity, prior associations, and context of the signals. Novelty enhances aversive learning, perhaps because novel signals do not have prior associations for the receiver that must require extinction before a new association forms.

Two particularly remarkable features of receiver psychology are latent learning and peak shifts. Latent learning is subconscious learning. For instance, humans exposed to two voices simultaneously usually attend to only one of them. Afterward they often profess not to remember what the other voice said. Yet when they are asked to guess answers to questions about the ignored voice, they are more often correct than expected by chance. Furthermore, they often immediately shift their attention when their own names occur in the "ignored" voice. One's name is of course a familiar signal for most people, and familiarity of a signal improves its detection, as discussed above. In addition, however, people might well develop a learned feature detector for their name for processing signals at higher levels of their nervous systems.

Peak shift often follows from learning a discrimination between two signals, particularly if one of them is consistently punished. If the subject is then tested with a wider range of signals, it often reveals a preference centered not on the rewarded signal itself but on signals with properties even farther removed from the punished signal. This widespread feature of learning is explained by Signal Detection Theory. If the consequences for responding to the punished signal (a false alarm) is lower than the payoff for not responding to the rewarded signal (a missed detection), then it can pay for a receiver to minimize false alarms and to maximize missed detections—in other words, to adopt a high threshold for response. It pays to shift responses to signals as far removed as possible from the punished signals. In this case, the noise that complicates discrimination of the two signals is mostly the receiver's own errors in remembering the correct response. It is, in other words, a source of error in the associative mechanism of a receiver.

Guildford and Dawkins do not attempt to classify their examples of receiver psychology. It is clear that many of the examples require specialized sensory filters. The properties of the sense organs themselves or the associative neural mechanisms that categorize stimulation must make matched filters for signals of importance for the receiver. Some of these filters are tuned to general properties of signals that contrast with environmental noise, such as stripes or intense stimulation with particular wavelengths (pitch of sound or hue of color). Some are tuned to special properties of signals that are likely to have significance for many organisms, such as eye spots or signals directed toward the individual (names or individual-specific features) or the individual's species (species-specific features). Other examples cited by Guilford and Dawkins are properties of signals that optimize a receiver's performance in general (contrast, alerting components, or familiarity) or that are a result of optimizing performance (peak shift).

RITUALIZATION OF SIGNALS

The three principal ways a signaler can improve a receiver's detection of signals—by increasing contrast, redundancy, or predictability—all come together in a concept advanced by early ethologists—ritualization. They proposed that the movements, coloration, sounds, and other components of animals' displays had evolved to improve the efficacy of communication. Further development of this idea stalled in the 1970s as views of the evolution of communication changed radically. The problem was that early ethologists often seemed to assume that communication evolved as a mutualistic exchange of information between signalers and receivers. In their classic rebuttal of this argument in 1978, Richard Dawkins and John Krebs pointed out that we cannot assume that communication always or even mostly evolves for mutual benefits. Instead, behavior should evolve to promote each individual's best interest—even at the expense of its partner.

This perspective led many investigators to argue that animals' signals evolved not to convey information but to manipulate receivers. The present book, now many decades later, takes the position that this opposition of information and manipulation is a false dichotomy. At the time, however, it seemed that information was out and manipulation was in. Ritualization went out with the rejection of information. As an adaptation for efficient transfer of information, it no longer seemed to be an important feature of animal communication.

As introduced in Part II, any behavior that elicits a response from another individual, without providing all of the power for that response, is a signal that conveys information from the actor to the recipient. As for the transfer of information, it is immaterial whether this interaction is a case of manipulation (advantageous to one party but not to the other) or cooperation (advantageous to both parties). In both cases, noise— errors by receivers—makes a difference. The incomplete appreciation of noise in animal communication at the time was a second reason for the loss of interest in ritualization. Discussions of efficiency in communication stumbled for the simple reason that efficiency makes little sense except in the presence of noise. Whether manipulative or cooperative, communication should evolve to increase efficiency by minimizing the effects of noise. This proposal is the subject of most of the rest of this book. Part II starts again from scratch in order to assure that the basic concepts of communication in noise are clear and then proceeds to analyze how efficiency in communication is expected to evolve, from both the signaler's and the receiver's points of view.

For now it is enough to indicate that ritualization is indeed a useful concept that summarizes a signaler's options for improving the probability of detection, and hence of a response, by an appropriate receiver. Ritualization in the heyday of ethology was characterized by exaggeration, stereotypy or standardization, and rhythmic repetition of a display. Displays were recognized as signals that evoke responses from receivers, such as mates, rivals, offspring, parents, predators, or prey.

A way to summarize this chapter is to reformulate the features of ritualization. Exaggeration and rhythmic repetition are just another way of expressing contrast and redundancy. Stereotypy or standardization is redundancy as well, in the form of predictable relationships in time and space among the components of a signal. The early discussion of ritualization did not mention alerting components of displays, but they should be added. The conclusion of this chapter thus could then be that communication in noise, whether manipulative or cooperative, should promote the evolution of ritualization in signals.

Ritualization brings the same trade-off already noted above between the rate of transmission and the performance of the receiver. The more a display (signal) combines elements in standardized spatial and temporal relationships, the less it can use these elements to encode information. Encoding information requires variable relationships among signals or components of signals. Whenever a response advantageous for a signaler requires communication of more information to a receiver, there arises the trade-off between the rate of transmission and the efficacy of transmission in noise. At a noisy cocktail party or in any other noisy gathering, this trade-off no doubt constrains us humans to vapid conversation. Over megaphones in crowds we are constrained to platitudes. Likewise, in general, noise favors ritualization—efficiency—over depth of information.

PART TWO

EVOLUTION OF SIGNALERS
AND RECEIVERS

PART I PRESENTED a case for pervasive effects of noise in communication, especially the relatively well-studied case of acoustic communication in animals. The sources of noise are manifold. Their consequences can be measured but also in many cases easily noticed. One lesson from Part I is that noise is anything that results in errors by receivers. To conclude that noise is pervasive in communication is to conclude that errors by receivers are also.

A second lesson is that noise—errors—in communication has received relatively little attention. Most research has concentrated on the degradation of signals during propagation from signalers to receivers. A secondary focus has been interference from irrelevant signals, especially those of other species in the same signal space. Recently the focus has shifted to effects of anthropogenic noise on animals. Only a small fraction of research on noise has attempted to measure errors of receivers and then only the decrement in receivers' abilities to detect signals correctly. It is not that communication has been neglected. The problem instead is that animals' responses are normally investigated in the absence of (or with minimal) noise. The usual aim is to study responses to "clean" signals, as nearly as possible like the ones issuing from the signaler. Investigators make their recordings of acoustic signals after approaching a signaler as closely as possible and then recording its signals with directional microphones or parabolas. Many have attempted to measure the advantages a receiver obtains from responding to an appropriate signal or a signaler obtains from eliciting a response from an appropriate receiver. It has also been widely appreciated that eavesdroppers can impose costs on a signaler, as when predators or parasites intercept signals to locate hosts or prey. This agenda has left a void in our knowledge of communication.

This incomplete agenda has persisted primarily because a coherent conceptual framework for understanding communication in noise has been ignored by most biologists. Signal Detection Theory provides this framework. It was developed originally in the 1950s as a procedure for analyzing psychophysical experiments. To understand, for instance, the threshold at which humans can detect a particular sound involves presenting weaker and stronger signals to a series of subjects. For each subject the experimenter wishes to determine the weakest signal that the subject can detect. The subject is given instructions to make a response whenever the signal is heard. The problems bedeviling this research before the 1950s were the inconsistent responses of any one subject and systematic differences in the responses of different subjects. It became apparent that subjects did not differ in the sensitivity of their sensory receptors so much as in their criteria for detecting a signal. Some only responded when they were quite sure, and others to any slight possibility of a signal. Signal Detection Theory (in combination with Decision Theory) eliminated both of these problems in psychophysical experiments. Despite some controversies about details of the procedures for optimal results, this theory has now permeated the study of human responses to stimulation. It has, however, not been widely applied to understanding the evolution of communication.

Applying Signal Detection Theory to the evolution of communication reveals that the void in our understanding of communication is in fact much greater than naively expected. In particular, the number of parameters required to characterize communication is greater than now realized. None of these parameters is inaccessible to measurement, but the lack of a conceptual framework for analyzing communication in noise has resulted in neglect for many of them. A new framework for understanding the evolution of communication based on Signal Detection Theory provides a more comprehensive way to understand communication. Its implications extend to many facets of behavior, even including human behavior and thought, as addressed in Part IV of this book.

The goal of Part II is to develop this framework. The treatment is largely theoretical. For me this exercise requires a new direction. A field biologist like me finds nothing more extraordinary than discovering, with my recorder and loudspeaker deep in a forest at daybreak, that a bird— a tiny concentration of vibrant energy—will allow me to interject a phrase or two into the stream of its communication with others of its species. A long absence from the field in the suffused aura of computer monitors seems contrary to my impulse to experiment, to measure, and, fundamentally, to observe. Nevertheless, the birds have challenged me, episodically over several decades, to return to the problems of noise in communication. I have tried to plumb ever more deeply the theory of signal detection. The following result is a framework for investigating communication in noise. It might not be the only possible framework for this enterprise, but at least it is now comprehensive. I have endeavored to keep it as simple as possible. It includes no purely abstract, unmeasurable parameters. To my surprise, it brings to light some startling, almost unsettling, conclusions. It thus bears examination.

7

SIGNALS, RECEIVERS, AND EVOLUTION

WHAT IS A SIGNAL?

To PROCEED with a discussion of communication in noise, it is necessary to return to the three components of communication introduced in Chapter 1. Signals, receivers, and signalers all need more explication than they have previously had. Of these three components, signals are the key to understanding everything else. Most of Signal Detection Theory, as explained in the following chapters, follows from the basic properties of all signals. Any signal is a pattern of energy or matter that evokes a response from a receiver more often than randomly and without providing all of the power for the response. All of the components of this specification are important.

First of all, a signal is a pattern, an arrangement of particular features of energy or matter. It might be a bird's song, a pattern of frequencies of sound (pitches) in a period of time. It might be the coloration of a coral reef fish in a particular physiological state, such as a dominant male, a pattern of surfaces each reflecting certain frequencies (colors) of light. It could be the shape, location, and intensity of a touch. It could be a mixture of moderately complex molecules in the pheromone of a female moth.

Next, to qualify as a signal in communication, the pattern of energy or matter must evoke a response in a receiver. In many cases it is obvious that a signal does not always evoke a response from any particular receiver, nor from every appropriate receiver. There are two reasons why signals might not always evoke obvious responses. First, a response need not be immediately overt. It can instead be a change in the state of a receiver that can influence its behavior in the future. It can be a change in hormonal or neural state. It can also be a memory that alters future behavior. Second, a response need not occur inexorably, only predictably. It must, in other words, evoke a response from some receiver more often than randomly. The probability of a response (in some appropriate period of time) must be greater following the occurrence of a signal than in the absence of a signal.

To grasp this probabilistic nature of signals, it helps to distinguish *instances* of communication from *systems* of communication. A system of communication consists of all signals over a period of time (appropriate for the particular study). An instance of communication consists of a single signal. Some systems of communication might consist of a single instance of communication, but often they include repetitions of signals. There is no requirement that every instance of a signal must evoke a response. Yet in any system of communication signals must evoke responses more often than by chance. If a receiver happens to perform some behavior at intervals without relation to another individual's actions, then these actions are not signals for communication.

There are situations in which it is useful to think of "reverse signals," actions by one individual that *reduce* the probability of responses by a receiver. In some cases these signals serve to make the signaler less likely to be detected. Camouflaged coloration is an example. Mimicry is another, to be discussed further in Chapter 11. The same specification for a signal holds in these cases. The reverse signal alters the behavior of the receiving individual. In this case the receiver's response to the signaler becomes less likely than in the absence of the signal. A change in the behavior of a receiver is the requirement for a signal.

Finally, a signal is a pattern of energy or matter that does not provide all of the power for the response. For example, consider an object falling from overhead, perhaps a rock dislodged from a steep slope. If somebody pushes me out of the way, that action is not a signal, as it provides the power to produce my response of moving aside. If my friend instead shouts, "Heads up!" and I leap aside, then those words are a signal to me. Furthermore, if I hear the sound of the clattering rock and I leap aside, that sound is also a signal. Note that in either case a signal must provide enough energy to affect my sense organs, the cochlea in my ears in the case of sounds, although not enough to effect my movement to the side.

Biologists have wrestled with a definition of a signal. Most reject the sound of the falling rock, because this sound had not evolved to produce my response. Indeed, the inanimate rock, with no means of reproduction, cannot evolve at all. Instead, such patterns of energy or matter are separated as "cues," which include any environmental stimulus that might evoke a response from an organism. This distinction is not pursued here because, from a receiver's point of view, it is irrelevant. For a receiver it makes no difference whether or not a signal comes from an evolving signaler or not. All of the features of signal detection apply. Nevertheless, the source of a signal has a special importance. If the source of a signal is a living organism, there is the possibility that signaling and receiving might evolve in concert. Understanding the coevolution of signaling and receiving is the objective of the remaining sections of this book. For a start, this chapter focuses on the dilemma faced by receivers when signals, as just defined, occur with noise.

WHAT IS A RECEIVER?

From the fundamental properties of a signal, it is clear that any receiver requires three components. First, it must include appropriate *sensors* to detect a signal. As described in Chapter 3, the physiology of animals' sensory cells often reveals that they are exquisitely sensitive to minuscule, sometimes almost unimaginable, stimulation. Because a signal does not provide the power to produce a response beyond its delicate effect on sensors, a receiver also requires a *decision-making mechanism*. A receiver must decide whether or not the activity of its sensors justifies a particular response. It is sometimes assumed that only humans can make decisions, or only organisms, but the nature of signals makes it clear that all receivers, animate or inanimate, include mechanisms that determine what response occurs and how much activity in a sensor is needed for this response.

In electronic receivers, these decision-making mechanisms are electronic switches, sometimes called gates, a graphic term for the permissive selection of a response. A simple form of decision maker is a threshold. Many electronic switches are thresholds. Responses of organisms, including humans, also occur when the intensity of a stimulus exceeds some threshold. In such cases, when the activity of a sensor reaches a set level, the mechanisms for a response are triggered. If the activity does not reach this level, these mechanisms are not triggered, so no response occurs. It turns out that the form or complexity of a switch does not change a receiver's fundamental predicament, a point discussed further below.

Finally, once the mechanism for a response is activated, it must generate the power to make the response. When the response involves movements of limbs and locomotion, a substantial amount of power can be required. Receivers thus require a third mechanism, an *amplifier*. The slight power of a signal is amplified to create enough power for a response. It is important to realize that a response need not be overt or immediate. It might involve storing a memory or otherwise altering the internal state of the receiver in a way that would alter the probability of future actions by the receiver but would generate no immediate overt movement. Even if the power to maintain a memory or other change in physiological state is less than the power to produce an overt action, some amplification of the power in the signal still must occur.

WHAT IS NOISE?

The fundamental properties of any signal have so far led to the fundamental properties of any receiver. To incorporate noise in a system of communication, it is only necessary to suppose, as Claude Shannon did, that the receiver's sensors have input from two sources (see Figure 1.1): one is the source of the signals, the other the source of noise. In the ab-

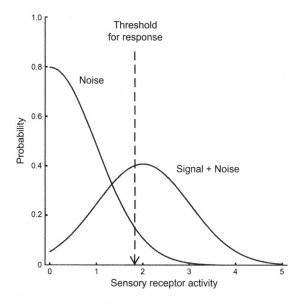

FIGURE 7.1. Both noise and signals stimulate a receiver's sensory receptor. Stimulation of a sensory receptor produces a Probability Density Function (PDF) of nerve impulses. In this diagram such functions are represented by curves (truncated normal or error functions). The peak of each curve indicates the most probable level of activity; the width of the curve indicates the variability of activity. When a signal arrives, its stimulation is added to stimulation by the noise. As a result, the most probable level of activity in the receptor increases. The receiver must then decide when it is appropriate to respond. It might, for instance, set a threshold for response based on the activity in its sensory receptor. If activity exceeds the threshold, then respond; if not, then don't. All figures in this chapter are adapted from Wiley (2013), A receiver-signaler equilibrium.

sence of one or the other source, the receiver's sensors are subject only to the remaining one. This scenario is realistic for noise that consists of extraneous input of no interest to the receiver, either energy generated by inanimate features of the environment (wind, rain, falling water) or energy in the signals of other irrelevant individuals of the same or different species. In addition, any spontaneous activity of the sensor itself adds to any activity produced by noise from external sources.

Figure 7.1 illustrates the situation for a receiver. In the absence of a signal, its relevant sensory receptors are activated by noise in the form of energy from irrelevant sources. This activation of the sensor no doubt varies in an unpredictable (random) way around some modal level. The figure plots the probability that the activation of the sensor in any small period of time reaches a particular level. It assumes that the activation of the sensor by noise is usually low, although there are times when noise could briefly raise the sensor's activation to high levels. The unpredictability of noise would result in progressively lower probabilities of activation away from the mode.

A signal would also activate an appropriate sensory receptor. If the signal were subject to variable amounts of degradation during transmission, then its activation of the receptor would also include variation. The spontaneous activity in signaler's sensor might also contribute some variation to the signal. So, just as with noise alone, it is reasonable to suppose that a signal alone would produce some modal level of activation in a receiver's sensor and some variation around this mode.

When such a signal occurs in combination with noise, the stimulation of the receiver's sensor is the sum of the stimulation from the two sources, noise and signal. It would thus have an average level higher than the level for either signal or noise alone, and it could have wider variation. If the signal had acquired its own variation from the signaler or during transmission, this variation would combine with variation in extraneous noise. The activation of a sensory receptor is usually not a linear response to its stimulation (a logarithmic function is usual). Nevertheless, the overall distribution of activity for signal plus noise would resemble that for noise alone, but displaced to a higher level of activity on average.

The receiver's dilemma arises from overlap in these two distributions of activity in its sensor. If the distributions for noise alone and for signal plus noise overlap at all, the receiver will not be able to separate the two completely. To make a decision when to respond, as explained above, a receiver could adopt a simple threshold of activity in its sensor. It would respond only when activity in its sensor reached this threshold. Presumably it could place this threshold anywhere, from 0 to a high level. With a threshold of 0, it would respond every time it checked its sensory receptor, regardless of the occurrence of a signal. With a high threshold, it would hardly ever respond, even when a signal occurred, but it would minimize the chances of responding to noise alone.

A THRESHOLD AS A CRITERION FOR RESPONSE

Examination of a receiver's situation shows how unavoidable is the dilemma it faces. Whenever the distribution of a sensor's activity in the presence of a signal with noise overlaps this distribution in the presence of noise alone, the receiver faces a quandary in setting its criterion for response. A higher threshold eliminates more false alarms, responses when no signal is present, but it concomitantly eliminates more correct detections, when a signal is in fact present. The dilemma applies to complex criteria as well to thresholds, as discussed more below. First, consider the situation for a simple threshold.

The receiver's dilemma is apparent in Figure 7.1, a schematic plot of the probability that a receiver's sensory receptor has a particular level of activity in any brief interval of time. In the absence of any stimulation, a receptor often has some irregular spontaneous activity that varies from moment to moment. As explained earlier, this internal activity of a receptor is one source of noise. Irrelevant external stimulation is a second source of noise. Both increase the activity of the receptor. The combination of spontaneous activity and irrelevant stimulation is summarized by a *probability density function (PDF)* of a receptor's activity *in the presence of noise alone,* the probability that the activity in the sensor reaches a particular level in any small interval of time when no signal occurs. The shape of the PDF presumably varies with the nature of the noise, but it must often resemble a normal error function—a bell-shaped distribution of random values. The modal (most likely) level of activity in the absence of a signal would often be near 0. Nevertheless, a receptor's activity in the presence of noise alone could have a long tail to the right (to include occasional higher levels of activity). To the left, on the other hand, activity is always truncated at 0 (a receptor's activity cannot have negative values). Even if truncated in this way, the sum of probabilities across all levels of activity equals 1.0 (the receptor always has one and only one level of activity in each interval of time)—this sum is the area under the entire PDF (its integral from 0 to infinity).

When an appropriate signal arrives, the sensory receptor receives additional stimulation, which is combined with the stimulation from inappropriate sources and spontaneous activity to produce a new probability density function of activity *in the presence of a signal plus noise* (again see Figure 7.1). This PDF should have a higher mode of activity than the mode for noise alone. Like the PDF for activity in the presence of noise alone, the area under the PDF in the presence of a signal plus noise equals 1.0. The signal arrives with its own variation, as a result of variation in its production by the sender and variation introduced by the degradation of the signal during transmission to the receiver. The variation in the signal combines with the variation in stimulation from inappropriate sources and with the variation in spontaneous activity in the sensory receptor, so the PDF for signal plus noise could have wider variation than the PDF for noise alone.

To decide when to respond, a receiver might adopt a threshold for response. With this simplest possible criterion. the rule for a decision to respond is simple—if the level of activity in its sensory receptor is above the threshold, then respond; otherwise, do not respond. When a receiver checks its sensory receptor, all it "knows" is the level of activity in the receptor, either above its threshold or not. An independent observer, someone other than the receiver or the signaler, might recognize four situations—activity above threshold and signal present, activity above threshold but no signal present, activity below threshold and no signal present, activity below threshold but signal present. A receiver, however, only knows two situations—activity above its threshold and activity below its threshold. An independent observer is a receiver also, but one with a special vantage or special apparatus. This observer faces its own dilemma of setting its thresholds for recognizing different situations, but depending on its vantage or apparatus it might distinguish the presence or absence of a signal more accurately, even if not perfectly. Although two receivers might differ markedly in the properties of their receptors, each confronts its own problem of overlapping PDFs for noise alone and signal plus noise.

A receiver's dilemma arises whenever the PDFs for noise alone and signal plus noise overlap at all. A diagram makes this dilemma clear (see Figure 7.2). Wherever a threshold is located, it results in four possible outcomes each time a receiver checks its sensory receptor (in other words, pays attention). If a signal is present and activity in the sensor is above the threshold, the receiver responds (a *correct detection* of the signal). If the signal is present, but by chance the activity of the receptor is below

		Receiver's decision	
		Response	No response
Signal	Present	CORRECT DETECTION	MISSED DETECTION
	Absent	FALSE ALARM	CORRECT REJECTION

FIGURE 7.2. Four possible outcomes when a receiver checks a sensory receptor. When a receiver checks (pays attention to) the level of activity in its receptor, there are only four possible outcomes. These outcomes are an exhaustive and mutually exclusive classification of the possibilities. They result from two possible states of the world (signal present or not) and two possible states of the receiver's criterion for decision (in this case, activity above threshold or not).

the threshold, the receiver does not respond (a *missed detection* of the signal). If no signal is present and activity of the sensor is below threshold, there is no response (a *correct rejection* of the possibility of a signal). If no signal is present, but by chance the activity of the receptor is above threshold, the receiver responds (a *false alarm*). Notice that although the receiver only knows two states of its receptor, sensory activity above or below threshold, it experiences four possible outcomes.

For any fixed threshold and for any PDFs for noise alone and signal plus noise, the probabilities of these four outcomes are fixed. For instance, the area above the threshold and under the PDF for signal plus noise is the total probability that a receiver correctly detects a signal by responding (see Figure 7.3). In contrast, the area below the threshold and under this PDF is the total probability that a receiver misses a detection of the signal. The areas above and below the threshold and under the PDF for noise alone are the total probabilities that a receiver correctly rejects the possibility of a signal or responds when there is no signal (a false alarm).

The receiver presumably has the option to locate its threshold anywhere it chooses. It might, for instance, choose a higher threshold. It would then reduce the total probability of false alarms. In Figure 7.3, the area under the PDF for activity in the presence of noise alone makes this point clear. A higher threshold would also decrease the chance of a correct detection of a signal (the area above threshold under the PDF for a signal plus noise). It would also increase the chance of a missed detection

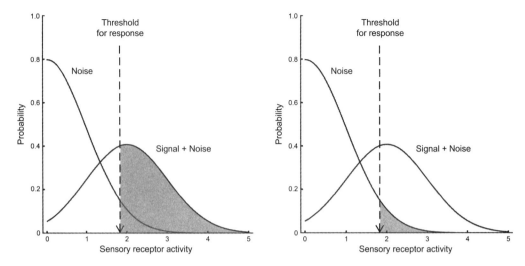

FIGURE 7.3. A threshold sets the probabilities for the four outcomes. A correct detection happens whenever a signal occurs and the activity of the sensory receptor exceeds the threshold. Summing the probabilities (integrating the Probability Density Function, PDF) from the threshold to infinity (the hatched area) yields the total probability of a correct detection, p_D (*left*). Whenever the PDF for a signal plus noise and the PDF for noise alone overlap, the probability for a correct detection is less than 1. Integrating the PDF for noise alone over the same interval (from the threshold to infinity) yields the probability of a false alarm, p_F (*right*). Integrating the two curves in the interval below the threshold (from 0 to the threshold) would give the probabilities of a missed detection (PDF for noise plus signal) and a correct rejection (PDF for noise alone).

of a signal. Finally, it would increase the probability of correctly rejecting the possibility of a signal when none was present. In other words, a change in the location of the threshold changes the probabilities of all four outcomes for the receiver.

Furthermore, the receiver cannot attain ideal performance regardless of the location of its threshold. Raising its threshold decreases the chance of one kind of error (false alarms) but increases the chance of the other kind of error (missed detections). It also increases the chance of one kind of correct reaction (correct rejections) but decreases the chance of the other kind of correct reaction (correct detection of a signal). Lowering its threshold has the opposite effects on a receiver's performance.

Notice that a receiver cannot adopt separate thresholds for occasions when a signal occurs and when it does not, because it has no way to know when a signal occurs except by checking its sensory receptor. It might have another receptor that is better at separating signals and noise, but then that receptor would face the same dilemma even if it could separate signal and noise somewhat better. The original receptor would then become superfluous. Any receiver might thus be expected to use receptors that optimized its performance.

Consequently, for any such receptor, it might be expected to optimize the location of its threshold. Yet provided its sensory receptors cannot absolutely separate signals and noise, a receiver faces the fundamental dilemma of signal detection—the impossibility of simultaneously minimizing both kinds of errors. It cannot simultaneously minimize false alarms and maximize correct detections. As the analysis of signal detection unfolds in the following chapters, it becomes apparent that it does not normally pay for communicators to evolve ideal performance. If so, communication is never expected to escape the dilemma of signal detection—it becomes a condition of life.

The next step in analyzing signal detection in communication is exploring how a receiver might optimize its performance. Before continuing, it is necessary to emphasize the generality of these conceptions of a signal and a receiver.

SIGNALS AND RECEIVERS IN MANY CONTEXTS

So far the discussion has used a threshold level of activity in a single sensory receptor to introduce the issues of signal detection. Not all organisms respond to signals in such a simple way. The activity of many sensory receptors is integrated in higher neural centers, which can fabricate complex criteria for responses. Even the interactions of receptors can contribute to criteria for response. Reciprocal inhibition between sensory cells serves to sharpen the discrimination and detection of particular kinds of stimulation. Furthermore, most sensory receptors themselves do not have simple thresholds for response and instead respond with increasing activity to increasing stimulation (high-pass filters) or with maximal activity to optimal stimulation (band-pass or tuned filters). These processes can occur at successively more complicated stages of neural organization—all the way to the sophisticated criteria for responses as a result of human cognition. All of these criteria reduce to two basic kinds: one-sided (high- or low-pass) or tuned (band-pass) filters. A threshold is a simple case of a one-sided filter. A preference for chocolate is perhaps a complex case of a tuned filter. Note that chocolate comes in a great variety of compositions and concentrations. And in recent years synthetic and artificial flavors have appeared. No matter how refined my taste (complex my criterion), I probably cannot discriminate the presence of chocolate with complete certainty. There will be missed detections and false alarms, as well as correct detections and correct rejections!

All one-sided filters face a trade-off much like the one already discussed for simple thresholds. All tuned filters face an analogous trade-off, in this case by setting the width of the pass band. These filters can take more or less selective forms (wider or narrower bands for accepting sensory activity as an indication that a signal has occurred). The choice of a wider or narrower pass band has the same consequences that the choice of a higher or lower threshold has. Whenever the PDFs for levels of activity for noise alone and signal plus noise overlap, a narrower pass band decreases the chance of a false alarm but also decreases the chance of a correct detection and increases the chance of a missed detection. A wider pass band has the opposite consequences. This trade-off is the same as the widely recognized trade-off in electrical engineering between the sensitivity and the selectivity of filters. Highly selective filters (few false alarms) have low sensitivity (many missed detections).

What has just been said about nervous systems has great generality. It in fact encapsulates all communication. It applies to all communication between organisms, including all human communication. It also applies to all machines fabricated by humans for the purpose of enhancing their communication, from drums and horns to electronic devices such as telegraph, telephone, television, radio, computer, and Internet. All the mechanisms in electronic devices for increasing sensitivity to signals, separating signals from noise, decoding signals, and amplifying them are analogous to communication by organisms.

The analogy between telecommunication and human communication is so alluring that it is often taken as the model for humans. Decoding signals in electronic applications often takes the form of attempting to reconstruct signals at the receiver to match as closely as possible those at the signaler. Actually, humans do not necessarily "decode" signals into replicas of the signal or even the signaler. Reconstructing a signal is just a special case of associating signals with responses.

Responses can take many forms. In addition to immediate overt responses, there are covert responses and those that do not become overt until much later. Covert responses consist of changes in the state of the receiver, which potentially alter its responses in the future. If these changes are impossible for us to perceive without special tools or vantages, they are covert. Our only way to study them might require tests with subsequent signals to determine whether or not the first signal had changed the receiver's internal state. In the brains of humans and other animals, covert responses are often memory or other physiological changes. A test is often needed to confirm a receiver's memory of a preceding signal. In computers, as well, signals can change memory in ways that that require testing with subsequent signals for confirmation. We teachers routinely use examinations to probe the covert responses of our students to instruction.

Note that these principles also apply to communication within an organism, between the cells of its body or between components of those cells. Molecular biologists speak routinely about signaling within and between cells. If this field has demonstrated anything in the past 40 years, it is the complexity and pervasiveness of molecular signaling in all organisms. All of these cases fit the same pattern discussed here. A signal is a molecule released in one place, which attaches to another molecule, a receptor, in another place, perhaps on the surface of another cell or at a different location in the same cell. This attachment initiates changes in the conformation of the receptor and one or several other molecules that it contacts. Eventually, if all goes well, some appropriate change in a distant molecule occurs. Oxidation of a molecule of adenosine triphosphate (ATP), or another high-energy molecule, often provides the power for the responses to a molecular signal. Just like human language or signals between animals or electronic devices, molecular signals satisfy the same requirements for a change in the behavior of a receiver without providing all of the power for that change. Molecular receivers face the same trade-offs among four possible outcomes every time a sensor (molecular receptor) is deployed. A molecular switch can be open-ended or tuned. It can include great complexity. The possibility of noise is always present, either in the form of extraneous energy or molecules that resemble a signal or in the form of spontaneous activity in the receptor molecules themselves.

CAN COMMUNICATION BE OPTIMIZED?

The line of reasoning so far has led from recognition of the essential feature of signals, insufficient power to produce a response, to recognition of the three essential features of a receiver, sensor, switch, and amplifier. All of these concepts generalize easily to the most diverse forms of communication. The following chapters continue a focus on the most basic features of communication—the detection of a signal by means of a threshold for response. Even in this basic case, communication presents a challenge provided the signalers and receivers are living organisms. How should the components of communication evolve? Are there optimal signals? Or optimal criteria for responses? Does the evolution of signaling alter the optimal criterion for responding to those signals? Does the evolution of responding to signals alter the optimal form of those signals?

Optimization requires a goal. Human engineers can optimize electronic filters in order to maximize some measure of performance in communication in relation to the cost of development and manufacture. Any optimization of the behavior of organisms (and its neural and other physiological mechanisms) is the result of both learning (by alterations in the states of individuals) and evolution (by natural selection in populations). Consequently, before analyzing the optimization of communication between organisms, even humans, it is important to be clear about some crucial features of evolution. What is the goal? What is the measure of success? What is an error? Furthermore, because evolution applies to genes, what do genes have to do with communication? The remainder of this chapter addresses these fundamental issues.

A good definition of evolution is a change in the frequencies of alleles or their arrangements in a population from generation to generation.

Alleles are different states of a gene, a section of DNA that produces a particular protein or that regulates the production of a protein. Notice that *gene* is an ambiguous word. We often refer to *gene differences* when we actually mean *allele differences*. For instance, to say that humans differ in their genes is not strictly correct. All humans (with very few exceptions) have the same genes, but each human is genetically unique (except identical twins) because of differences in the alleles of some of those genes. To rephrase our previous statement, evolution occurs as alleles (states of genes) in a population change from generation to generation.

Modern genetics has made it clear that inheritance can occur in complex ways. Offspring inherit not only a set of alleles from each parent, but also particular arrangements of those alleles, chemical modifications of the proteins accompanying DNA, constituents of the protoplasm of the egg, and various bits of parasitic DNA. Organisms, and humans in particular, also "inherit" behavior learned from their parents. Such "traditional" behavior is "culture." All of these epigenetic mechanisms can modify the conditions for the transmission or even mutation of alleles. Conversely, the expression of each of these epigenetic mechanisms is influenced by an individual's alleles. Because these epigenetic mechanisms are more labile than alleles, I prefer to take changes in frequencies of alleles and their arrangements as the defining feature of evolution. The interactions between alleles and the various epigenetic mechanisms of an individual's development are the subject of the next section.

To argue that communication might evolve in a particular way is to argue that certain alleles (genes) affecting behavior relevant to communication might accumulate in a population. The focus is on alleles that affect this particular form of behavior. But do genes have anything to do with communication? Have any such genes been found? What about nature as well as nurture? What about learning?

INTERACTION IN DEVELOPMENT: GENES INFLUENCE, NOT DETERMINE

Behavior routinely raises questions about nature or nurture. These questions are especially frequent about communicatory behavior. Do genes determine behavior, especially complex learned behavior, or does culture? Actually, this issue arises for any feature of any organism. All features of organisms develop as a result of processes that involve both particular genes and a particular environment. The resulting features of an organism are called its phenotype (as distinct from its genotype, the alleles it is born with). In the past century some groups of people have concluded that genes are the most important influence on the development of organisms, including humans, and have promulgated political systems based on this assumption, some of them epochally vicious. Almost concurrently, other groups of people have concluded that the environment is the most important influence on the development of organisms, including humans, and have promulgated political systems based on this contrasting assumption, some of them equally vicious. It has become routine nowadays to assert that, in contrast with these extreme views, development actually results from an interaction between particular genes and particular environments.

Gene-environment interaction proves to be easy to say but difficult to comprehend. Simple causation seems to be a much easier concept for the human mind. It is not clear why we should have this tendency. It seems unlikely that simple causation, rather than interaction, would promote a better understanding of complexities in the natural world or in social relationships. Understanding nature and society would surely have advantages for humans, both currently when environmental complexities overwhelm us and in early evolutionary stages when hunting, gathering, and agriculture were the occupations of most humans. Perhaps a tendency to think in terms of simple causation evolved because, no matter how important it might be to understand nature, it has been even more important to allocate blame in our social relationships. Blame requires causation rather than interaction. Nevertheless, interaction is essential for understanding the natural world in all its forms.

Gene-environment interaction means that the same gene (allele) in different environments produces different phenotypes. Also, different genes in the same environment produce different phenotypes.

Even today those who study behavior sometimes stop short of a full awareness of this interaction of genes and environment. It is easy to feel that biology is all about genes and the social sciences are all about culture. Biologists in recent decades have indeed focused a lot of effort on the influence of genes on the working of cells. They often study "model" organisms, ones whose genetics are easily investigated because they are easily bred in laboratories, in controlled environments. If the object is to study the effects of genes, the procedure of choice is to compare individuals with different alleles in the same environment. Yet this procedure does not provide information about the interaction of genes and environment. If instead the object is to study the effects of the environment on the development of phenotypes, the procedure of choice is to compare individuals with the same alleles in different environments. Identical twins assigned randomly to different environments provide a design for such a study. In the social sciences, laboratory animals bred for genetic homogeneity are often compared in different environments. Like the ideal procedure for studying the effects of genes, the ideal procedure for studying the effects of the environment on development does not provide information about the interaction of genes and environment.

A test for genetic influences on phenotypes is a breeding experiment. For such an experiment it is necessary first of all to measure differences in the performance of individuals on some task. Many measures of behavior in convenient animals have provided the basis for breeding experiments. Various expressions of aggression, sexual behavior, and learning are particularly easy to measure in laboratory mice or rats, for instance. With these measures it is then possible to choose high- and low-performing individuals to produce the next generation. Highs are mated with other highs, and lows with other lows. To isolate the influence of alleles in such an experiment we must measure the performance of all individuals on the same task and allow these individuals to develop in the same environment. If the parents' alleles influence their phenotype, the progeny of highs eventually, over a span of generations, score higher on average than the progeny of lows. As this process is repeated from generation to generation the average scores of the highs and the lows diverge progressively. If we have carefully controlled the individuals' environments, then the rate at which they diverge is a measure of the influence of alleles on performance, in this particular task and in this particular environment. This experiment does not identify which alleles influence the behavior under study, but it does lead to the conclusion that some alleles of some genes do.

With this sort of experiment, laboratory rats can, for instance, easily be bred for quick learning. If the subjects are tested, for instance, for speed in running a particular kind of maze, the faster and slower learners can be identified. It is then routine to discover, as the result of a breeding experiment, that this form of learning is influenced by genes (actually the alleles of genes). It often happens that the alleles influencing performance on one kind of learned task differ from those influencing performance on other learned tasks (different kinds of mazes, discriminating different kinds of stimulation, finding hidden food or places to rest). In other words, animals bred for high (or low) performance on one task often are back to normal performance on the second task. If a task is changed or the environment is changed, it is not possible to predict whether the same alleles have the same influence on behavior in the two circumstances. Learning is influenced by genes, but not in a simple way. Nevertheless, all kinds of behavior, sex, aggression, parental and social behavior, as well as all kinds of learning, when subjected to measurement and investigated by breeding experiments, reveal an influence of genes as well as an influence of the environment. There is no reason to expect that the features of communicatory behavior are any exception.

If we completely understand the interaction of genes and environment in the development of organisms, a conclusion that genes never determine any feature of a phenotype is indeed correct. Yet that is not the only firm conclusion possible. An individual's environment never determines, any more than its genes ever determine, any feature of its phenotype. An interaction of genes and environment in development means that *four statements are concurrently true:* (1) no feature is *determined* by genes; (2) no feature is *determined* by the environment; (3) all features of phenotypes are *influenced* by genes; and (4) all features are *influenced* by the environment.

If all aspects of individuals' phenotypes are influenced by genes, all can evolve by natural selection. In the present case, the question is whether or not the genes that *influence* communicatory behavior, whether learned or not, evolve to optimize noisy communication.

EVOLUTION BY NATURAL SELECTION

Plant and animal breeders have for millennia demonstrated that even the most extraordinary attributes of individuals are influenced by genes and susceptible to selection. Nevertheless, selection cannot produce a population with just any phenotypes from any starting population. Some phenotypes have, to my knowledge, never arisen in plants or animals. No organism uses a wheel for locomotion, an appendage with an axle for a joint. Even at a less extreme level, some features of organisms are less easily altered by selection than others. A plausible reason is the complexity of the interactions of different genes in the development of a normal individual. Changing the influence of a gene on any one feature often results in changing its influence on other features. These ancillary effects of a gene (called pleiotropy by geneticists) can constrain the results of natural selection. A change in the influence of a gene on any one feature might require compensating changes in additional genes. To produce a new phenotype, natural selection often must both change the effects of one part of the genome and stabilize the concomitant effects on the rest. Genes with fewer pleiotropic effects are easier to change because they require fewer compensating changes. Because it is always more difficult to coordinate multiple changes at once, rather than one change at a time, the pleiotropic effects of genes can retard or inhibit the effects of natural selection. In other words, complexities in the current genetic influences on development of phenotypes create genetic constraints on evolution. Studying these constraints is an important objective of evolutionary biologists. Nevertheless, because the same adaptations often appear in different phylogenetic lineages, it seems that—sooner or later—natural selection often produces phenotypes that approach (even if they might never reach) optimal performance.

In exploring the evolution of communication, this book assumes that natural selection eventually produces new genotypes that maximize the performance of individuals. Ultimately "performance" of individuals is their ability to leave offspring to future generations. Those alleles of genes that are associated with phenotypes with higher performance propagate more rapidly and thus accumulate in a population, and those associated with lower performance propagate more slowly and dwindle or even disappear. Alleles that spread have higher "fitness" than those that do not. Strictly speaking, an individual organism does not have *fitness* (although many people, including biologists, routinely refer to the fitness of individuals). Instead, alleles have fitnesses. Those associated with individuals (phenotypes) that perform better have higher fitness. Alleles promoting the development of individuals with higher performance have higher fitness.

This perspective leads directly to clear thinking about natural selection, a difference in the spread of alleles in a population as a result of differences in individuals' performance (reproduction and survival) associated with those alleles. All components of this definition of natural selection are crucial. To say that natural selection occurs when there are differences in the spread of alleles in a population is to say that natural selection is a mechanism of evolution. Evolution, as mentioned earlier, is a change in allele frequencies in a population. Other than natural selection, there are three other mechanisms of evolution: mutation, genetic drift, and gene migration. Natural selection differs from these three because changes in allele frequencies result from differences in individuals' survival and reproduction.

Unlike mutation and drift, which affect the frequencies of all alleles in a population, natural selection influences only those alleles associated with differences in individuals' survival and reproduction. Also unlike mutation and genetic drift, natural selection has predictable influences on the frequencies of alleles. It provides a mechanism for optimizing the phenotypes in a population. By studying features of organisms (phenotypes) that affect the performance of individuals, we can identify targets for natural selection. Natural selection should move populations toward those targets, sooner or later.

OPTIMIZING BY NATURAL SELECTION

Optimizing the phenotypes of individuals in a population thus results from optimizing the survival and reproduction of individuals in the population. For two reasons, this objective is not so simple as it might seem. First, neither survival nor reproduction alone is enough for an individual to pass on its genes to the next generation. Any organism must both survive and reproduce to pass its alleles. Second, survival and reproduction cannot usually change independently. The time, energy, and risks of reproduction can reduce an individual's chances of survival. And the time and energy for survival can reduce the success of reproduction. Each individual must make trade-offs in the course of life between expenditures on itself and expenditures on its offspring. These trade-offs influence the spread of its alleles in the next generation. Furthermore, if two individuals with different features (two phenotypes) make different trade-offs between reproduction and survival, they might leave different numbers of their alleles (including any that influence their respective trade-offs) to the next generation. If we assume, as argued above, that any feature of an individual develops as a result of an interaction between its genes and its environment, then natural selection on alleles depends on the allele-spreading performance of phenotypes associated with those alleles.

A measure of allele-spreading performance is a phenotype's survival multiplied by its reproduction—survival × reproduction. By multiplying the two, the product incorporates the interaction of survival and reproduction in spreading alleles. As a rule, therefore, survival × reproduction provides a measure of natural selection on a phenotype. Alleles associated with a phenotype that have higher values of this measure spread more rapidly in a population than those with lower measures. If two phenotypes differ in survival × reproduction, then alleles associated with these phenotypes differ in their spread to the next generation.

Actually, survival × reproduction is only an approximate measure of natural selection on phenotypes. It is important to consider why. Because all features of every individual are expected to develop as a result of an interaction of its alleles and its environment, both survival and reproduction of any phenotype depend on environment as well as alleles. As a result, natural selection on a phenotype in one environment might differ from that in another environment. To complicate matters, the social interactions of individuals might be part of the environment that influences development of individuals. If so, survival or reproduction of behavioral phenotypes, and thus natural selection on alleles associated with them, might change from generation to generation. The evolution of optimal signaling and receiving during communication, explored in subsequent chapters, exemplifies this sort of change. The optimal phenotype for receivers (the optimal threshold for response) depends on the optimal phenotype for signalers (the optimal exaggeration of signals). So the survival × reproduction of each party changes as the behavior of the other party changes.

Second, multiplying the probability of survival and expected reproduction does not incorporate the influence age. Evolution results, at the most basic level, from differences in the rate of spread of alleles in a population. Differences in generation time associated with different alleles affect their rates of spread, even if reproduction, once it starts, and survival are otherwise unchanged. Alleles associated with early reproduction have shorter generation times in a population, and thus higher rates of spread—if all else is equal. Therefore, to calculate an exact rate for the spread of alleles associated with a phenotype it is necessary to know survival and reproduction at every age for that phenotype. Nevertheless, when comparing individuals of the same age or when comparing individuals with survival and reproduction that do not change with age, survival × reproduction can alone provide an approximate measure of natural selection.

This measure of natural selection can identify the optimal phenotypes in a population, optimal in the sense that no other phenotype has higher survival × reproduction. In recent decades this optimum has been called an Evolutionarily Stable Strategy, a behavioral phenotype that cannot be invaded by any other phenotype (at least in the set of possibilities under consideration). A phenotype might consist of a mixture of two responses, with definite probabilities. This possibility is particularly useful in understanding the evolution of social behavior when the consequences of an individual's acting in one way depends on the way its partner responds. For instance, basic calculations of natural selection on the spread of aggressive behavior and nonaggressive behavior show that, under the right conditions, a mixture of the two is evolutionarily stable. A population entirely composed of meek individuals ("Doves") is easily invaded by aggressive phenotypes ("Hawks"). A population entirely composed of aggressive individuals can be invaded by meek individuals provided aggressive ones suffer high enough costs from the high probability of encounters with each other. A particular mix of aggressive and meek behavior cannot be invaded by phenotypes that produce any other mix.

Similar analyses, based on Game Theory, have been applied to the evolution of cooperation, as opposed to cheating, and to the evolution of honesty and deception in communication. In the discussing the evolution of communication in noise, the objective is again a combination of phenotypes that cannot be invaded by any others, but the situation is different from these earlier game-theoretic analyses. In communication, as already emphasized, signalers and receivers are completely dependent on each other. Yet each optimizes its own behavior in a particular way. The problem is to identify, if possible, the combination of signalers' features and receivers' features that produce a joint optimum, one where neither party can do better provided the other does the best it can. At this stage in the discussion it is not even clear that such a joint optimum might exist. Nevertheless, it is important to acknowledge the goal of the discussion—to explore the possibility of a joint optimum for signalers' and receivers' phenotypes, a joint optimum that is a target toward which natural selection tends to move a population of individuals communicating in noise.

Knowing the target of natural selection does not always help much in predicting the rate of natural selection, in other words, its dynamics. This focus is just one of two ways to use mathematics to make predictions about the course of evolution by natural selection. One is to compute the optimal phenotypes for a population under specified conditions, as suggested above. The other is to compute the changes of genotypes in a population from generation to generation. One examines the target, the other the process. Both require mathematics that is complicated enough to require simplifying assumptions. In the end, although both make assumptions, both contribute to a mathematical description of evolution, a point explored further in Chapter 11.

The procedure in this book follows the former path, an investigation of the target of natural selection by deriving the optimal phenotypes expected for communication in noise. Like other analyses of optimal phenotypes, its strength is investigation of the results of behavioral interactions between phenotypes, for instance, when the performance of an individual depends not only on its own phenotype but also on the phenotype of the individual it encounters. Furthermore, in the case of communication in noise, the behavioral interaction of signaler and receiver changes as evolution proceeds, so natural selection on both parties is never constant during evolution. An analysis of the phenotypic target of evolution in this situation is complex enough. An analysis of the genotypic dynamics will have to wait for someone else.

The following analysis of noisy communication takes survival $\times$ reproduction as a measure of natural selection on the phenotypes of signalers and receivers. It assumes that natural selection might produce the changes in genotypes that could result in phenotypes (individuals) that maximize performance in communication. In other words, it assumes that evolution eventually follows the course of optimizing this measure of the spread of alleles.

8

OPTIMAL RECEIVERS AND SIGNALERS

PERFORMANCE OF RECEIVERS

IN CHAPTER 7 an introduction to Signal Detection Theory led naturally to a procedure for evaluating a receiver's performance at any level of its threshold for response and in any level of noise. A review of natural selection then suggested a way to evaluate its effect on any receiver's performance. These two steps together indicate a way to predict evolution of the performance of receivers by natural selection for any specified level of noise.

Recall that, in the presence of noise, any receiver faces a trade-off in setting its threshold. Increasing its threshold for response to a higher level of activity in its sensory receptors reduces false alarms but also reduces correct detections. There are also concomitant consequences for missed detections and correct rejections. In fact, either of these two trade-offs is sufficient to summarize the receiver's predicament because each of the trade-offs follows from the other. The trade-off between false alarms and correct detections is particularly salient in many cases; these outcomes are thus used in a standard diagram of a receiver's performance, a Receiver Operating Characteristic (ROC).

An ROC is a two-dimensional plot, the probability of a false alarm on the horizontal axis, and the probability of a correct detection on the vertical axis. Each axis thus has a range from 0 (the probability of an event that never occurs) to 1.0 (the probability of an event that always occurs). A diagonal line from the origin to the upper right corner connects all points at which the probability of a false alarm (p_F) equals the probability of a correct detection (p_D).

Estimating these probabilities requires recording a receiver's responses to many presentations of noise, with and without a signal. By now, six decades into the application of this theory, many people have volunteered as subjects for such tests. For instance, to determine the minimal intensity of sound for hearing, a person is asked to respond many times, either positively if the person thinks a signal could be heard, no matter how faintly, or negatively if the person thinks no signal occurred. Sometimes the person judges that something was heard even though no signal had been presented. The number of these false alarms divided by the total number of times no signal occurred is an estimate of the probability that the person's response was a false alarm. In a similar way, the number of correct detections divided by the total number of presentations of a signal provides an estimate of the probability of a correct detection. The probabilities of false alarm and correct detection depend on the conditions of the test and in particular on the level of the signals in relation to the level of the noise in the background.

For any one person, with a particular threshold, one series of these tests provides one estimate of the probability of false alarms and one of the probability of correct detections—one point to plot on an ROC. Because the presence of a signal normally increases the activity in a receptor, p_D is usually greater than p_F. Therefore any one experiment usually produces a point above the diagonal of the ROC.

Different people subjected to such a study might well have different thresholds for positive responses. Some might opt for positive responses on the least suspicion of a signal; others might hold out for the most assurance of a signal. Alternatively, any one person might have different thresholds for positive responses in different circumstances. Either way, a new point on the corresponding ROC is generated for each person (or situation) with a different threshold. At very low thresholds $p_D = p_F$, and both equal 1.0—the receiver always responds when it pays attention, regardless of the presence of a signal or not. At very high thresholds

$p_D = p_F$ also, but both equal 0—the receiver never responds. For intermediate positions of its threshold, the receiver's p_D and p_F lie between 1.0 and 0, with p_D normally greater than p_F, as indicated above. If the Probability Density Functions of the receptor's activity are smooth normal (bell-shaped) curves or truncated normal curves, such as those in Figure 8.1, a series of experiments on a receiver with different thresholds produces a curved plot of p_D versus p_F in a smooth convex line (its ROC) from one end of the diagonal to the other.

The better the receiver's performance, the greater the convexity of its ROC. If a signal with noise produces activity in the receiver's sensor that is well separated from activity with noise alone, then the difference between p_D and p_F is greater. Consequently, the receiver's ROC, as it varies its threshold, lies farther from the diagonal where $p_D = p_F$. This square plot allows us to summarize the possibilities for a receiver's performance, for any level of noise, any level of the signal, and any level of its threshold.

The most interesting aspect of this plot, however, is its use in predicting how a receiver should behave when correct detections and false alarms have different values for the receiver. In studies of human sensory perception, the values of different outcomes can be influenced by offering rewards or imposing disappointments. If, for instance, the penalty for false alarms is absolutely greater than the reward for correct detections (perhaps by small monetary debits and credits), then receivers should raise their thresholds to reduce the chance of false alarms, even though they also have fewer correct detections. In general, the rewards and costs attached to each of the four outcomes should alter the optimal threshold for a receiver wishing to maximize the overall reward.

THE UTILITY OF A THRESHOLD

This overall reward for a receiver is expressed by its expected utility (symbolized by U). In a landmark publication in 1953, John van Neumann and Oskar Morgenstern proposed a rational way to make any decision by maximizing its expected utility. Decision Theory, as it is now called, supposes that a person anticipates several possible outcomes of a decision to respond (do something) or not. Furthermore, these outcomes can

ROC as thresholds change from high to low

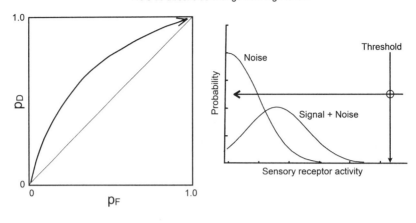

ROC shifts as signal and noise separate

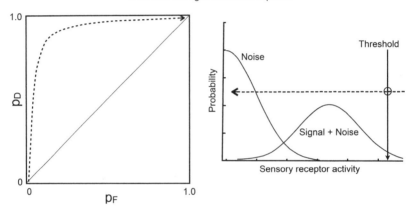

FIGURE 8.1. A Receiver Operating Characteristic (ROC) describes the performance of receivers with different thresholds. It plots the probability of a correct detection by a receiver (p_D, *vertical axis*) against the probability of a false alarm (p_F, *horizontal axis*). To determine these probabilities it is necessary to present a series of trials with noise alone and with a signal plus noise. If a receiver has a high threshold in these trials, we obtain low values for both of these probabilities, with the probability of correct detection higher than for false alarm. These two probabilities plot at one point in the lower corner of the ROC. If another receiver (or the same one, perhaps on another day) has a lower threshold for similar trials, we can again determine these probabilities, which are now both higher. The probability of correct detection is now even higher than that for false alarm. The result is another point for the ROC. If similar trials are presented to many receivers (or to the same receiver in different situations), the many resulting points determine a line curving from probabilities near 0 for both correct detection and false alarm (lower left-hand corner of the square) to probabilities near 1.0 for both (upper right-hand corner). The ROC thus characterizes the possibilities for receivers as their thresholds for response vary from high to low (as indicated on the right). If the separation of signals and noise also

be organized in a mutually exclusive and exhaustive classification. In other words, one and only one of the possible outcomes occurs when a decision is made. If so, then the expected, or average, reward of this person's decision is the sum, across all possible outcomes, of the reward for each outcome times its probability of occurring.

For instance, a shopper deciding whether or not to buy an automobile on sale might consider the consequences. The auto might be economical or it might be a "lemon." This shopper might (1) buy it and find that it pays off in comparison to keeping an older car, (2) buy it and find that it is a "lemon" that costs more in the long run, (3) not buy it and later regret passing up an economical choice, or (4) not buy it and later find out how lucky the decision was. Each possible outcome would have its particular consequences for the shopper's bank account in the long run. A decision might of course depend on the information available at the time. If this information is imperfect, any of the four possible outcomes might occur. The shopper must thus adopt some criterion for making a decision based more or less consciously on the input available. The utility of the shopper's criterion for this decision (or the average payoff from making many equivalent decisions) is the sum of the payoffs for the four possible outcomes, each weighted by its probability of occurrence. In general terms, this utility is expressed thus:

$$U = p_1 Y_1 + p_2 Y_2 + p_3 Y_3 + \cdots$$

or

$$U = \Sigma\, p_i Y_i \text{ for all i outcomes.}$$

In this expression Y_1 is the payoff for the first category of outcome, Y_2 the payoff for the second category of outcome, and so forth; p_1 is the probability that the first kind of outcome occurs, and so forth.

---

changes (compare the upper and lower plots), the ROC shifts away from the diagonal of the square (lower plot). The greater the separation of signal and noise, the closer the ROC approaches the upper left-hand corner (perfect performance, perfect detection of signals and no false alarms). The ROC thus compactly summarizes the receiver's performance in noisy communication. All figures in this chapter are adapted from Wiley (2013), A receiver-signaler equilibrium.

Decision Theory combined with Signal Detection Theory can address the case of a receiver's decisions when responding to signals in noise. The probability of a correct detection is the probability that a signal occurs (whenever the receiver checks its receptor or, in other words, pays attention) multiplied by the probability that the receiver responds in this situation (p_D). If we let the probability that a signal occurs be p_S, then the probability of a correct detection is p_S multiplied by p_D. The probability of a false alarm is the probability that a signal does not occur $(1-p_S)$ multiplied by the probability that the receiver nevertheless responds in this situation (p_F), or $(1-p_S)\,p_F$. So the receiver's expected utility is

$$U_r = p_S p_D D + p_S p_M M + (1-p_S)\,p_F F + (1-p_S)\,p_R R.$$

In this equation U_r is the expected utility for a receiver, p's are probabilities of the four possible outcomes, and D, M, F, and R are the payoffs for the four outcomes (correct Detection, Missed detection, False alarm, and correct Rejection). The mutually exclusive and exhaustive classification of outcomes assures that these products and sums of probabilities describe the situation: whenever a receiver checks its input, a signal either occurs or does not occur; each time, the receiver either responds or not. The probabilities for successive situations multiply, those for alternative situations add. Because $p_M = 1 - p_D$ and $p_R = 1 - p_F$, this expression becomes

$$U_r = p_S(p_D D + (1-p_D)M) + (1-p_S)\,(p_F F + (1-p_F)\,R).$$

With algebra we can expand this equation to obtain

$$U_r = p_S p_D(D-M) + p_S M + (1-p_S)p_F(F-R) + (1-p_S)R$$

and rearrange it to get

$$p_D = [(1-p_S)\,(R-F)\,/\,p_S(D-M)]\,p_F$$
$$+\,[p_S(R-M)-R+U_r]\,/\,p_S(D-M).$$

This expression has the form of a straight line, $y = mx + b$, with p_D on the y axis and p_F on the x axis. When p_D is plotted against pF, the slope is the coefficient of p_F,

$$(1-p_S)\,(R-F)\,/\,p_S(D-M).$$

The right-hand term in the equation above,

$$[p_S(R-M)-R+U_r] / p_S(D-M),$$

is the y-intercept of the line. It includes the payoffs for three of the outcomes and the probability that a signal occurs. These values characterize different situations faced by a receiver and thus are constants for each situation. It also includes the expected utility, U_r, which is not a constant because it depends on where the receiver sets its threshold, which in turn determines its probabilities of correct detection and false alarm.

If, in our imagination, we set U_r to some constant, this line indicates combinations of p_D and p_F that yield this particular expected utility with these particular values for payoffs and the probability of a signal. The line is thus an indifference line, a line along which combinations of p_D and p_F make no difference in the value of the utility. In other words, these values of p_D and p_F that result in the same value of U_r. If we change the value of U_r, the y-intercept of the indifference line changes. The larger the utility, the higher the y-intercept becomes (the line slides upward). A large U_r of course is good for the receiver, so we can imagine increasing the value of U_r until the line touches the receiver's ROC at only one point. Increasing U_r any farther is impossible for the receiver, because the indifference line would no longer intersect the receiver's ROC, so the receiver would not be able to find a location for its threshold that would yield this higher U_r. When this indifference line just touches the ROC tangentially, the receiver obtains the highest utility for any position of its threshold (see Figure 8.2). To maximize its utility, it should thus choose the position for its threshold corresponding to this point on its ROC.

This procedure has thus predicted the optimal location of the threshold for a receiver in a particular situation. A "situation" is defined by the level of noise, the probability of a signal, and the payoffs for the four possible outcomes of a decision to respond or not. The signal-to-noise ratio and the receiver's threshold determine the probabilities of the possible outcomes.

An interesting result falls in our lap at this point. Notice that where the indifference line becomes tangential to the receiver's ROC depends on the slope of the indifference line. For a low slope (more nearly horizontal indifference line) the tangential point lies closer to the upper right corner of the plot, closer to $p_D = p_F = 1.0$, corresponding to a low threshold

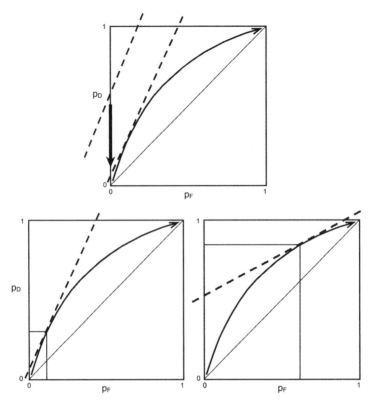

Figure 8.2. Indifference lines for the performance of a receiver in noise. The dashed lines plotted across each Receiver Operating Characteristic (ROC) show indifference lines. An indifference line is the linear relationship between the probabilities of correct detection and false alarm when all else is constant including the utility (the average payoff) of the receiver's threshold. The *y-intercept* of this line depends on the value of this utility, and the *slope* depends on the four payoffs and the probability that a signal occurs. By finding the indifference line that is just tangential to the ROC (as indicated in the top diagram), we also find the highest utility that is possible for a receiver. In Figure 8.1 we saw that different levels of a receiver's threshold correspond to different points along the ROC. Thus the threshold that corresponds with the point of tangency by the indifference line is the threshold with the highest possible utility. As the diagram indicates, a steep slope for the indifference line makes a high threshold optimal (bottom left), and a gentle slope makes a low threshold optimal (bottom right). The slope of the indifference line (therefore the payoffs and probability of signals) determines whether a high or a low threshold is the best strategy for the receiver.

for response. For a high slope (more nearly vertical indifference line), the opposite occurs: the tangential point lies closer to the lower left corner of the plot, closer to $p_D = p_F = 0$, corresponding to a high threshold for response. Thus the slope of the indifference line predicts whether a receiver does better to choose a high threshold or a low threshold in order to maximize its expected utility.

Once more examining the expression for the slope of the indifference line,

$$(1 - p_S) (R - F) / p_S(D - M),$$

we are lead to some remarkable predictions about how receivers should behave in different situations. First, notice that the probability of a signal has a strong influence on where a receiver should locate its threshold (for any one set of payoffs for the four outcomes of the receiver's threshold for response). If all other factors are constant, a lower value for p_S produces a higher slope of the indifference line. As concluded above, a higher slope means a higher threshold yields the highest possible utility. Rare signals thus promote high thresholds for response. In this case, low thresholds result in too many responses when a signal is absent, too many false alarms. In the converse situation, a higher value for p_S produces a lower slope of the indifference line and thus favors a lower threshold for response. In this case, high thresholds result in too many missed detections.

The optimal threshold for a receiver also depends on a ratio of payoffs,

$$(R - F) / (D - M).$$

The numerator is the difference in payoffs between correct and erroneous responses in the absence of a signal, and the denominator is this difference in the presence of a signal. The influence of this ratio on the slope of the indifference line, and thus on the optimal threshold for a receiver, leads to some striking conclusions about communication.

RECEIVERS IN CONTRASTING SITUATIONS

Another thought experiment is called for. Imagine contrasting situations for communication, mate choice and predator warning. In the first case, a female searching for an optimal mate might confront a noisy situation. In most animals, females are relatively cautious in mating, and males often compete for their attention. A female seeks a mate of the same species and preferably one with high qualities, either good genes for her progeny, low immediate risk for herself, or commitment and capability for helping to raise her progeny. Responses to signals from such males count as correct detections, and responses to signals from other species or from suboptimal males of her own species are false alarms. In this case, signals from optimal mates, more or less degraded and attenuated in transmission, have to be distinguished from noise, often similar and also degraded and attenuated in transmission, from suboptimal mates. The consequences of a false alarm for a female in these circumstances is potentially severe. In many species a female only mates a few times or produces only a few offspring, so a mistake has a proportionately large effect on her success in reproduction. There is likely to be a large difference in such cases between the payoff for a female's correct detection and that for a false alarm.

On the other hand, a correct rejection of a suboptimal male's signals means that she must spend some time continuing to search for a mate. Searching always involves some time and energy, and often some risks of injury or death, for instance in any encounter with a predator. In many species these costs in survival, when multiplied by eventual success in finding an optimal mate, might be small in relation to the costs of mismating with a suboptimal male. Furthermore, a missed detection for such a female would also entail additional searching, with costs often similar to those following a correct rejection, and with similar chances of eventual success in finding an optimal mate. So the difference in the payoffs for a correct rejection and a missed detection, in such cases, are likely to be low. Furthermore, both of these payoffs are likely to be lower than that for a correct detection of an optimal mate—and much higher than the payoff for a false alarm in responding to a suboptimal mate.

In this case the slope of the indifference line should equal 1.0 or less. In the expression for its slope, $R - F$ and $D - M$ might be approximately equal if the advantage of an optimal mate and the disadvantage of a suboptimal one were roughly the same in relation to the payoff for either a missed detection or a correct rejection. In some cases the cost of a false alarm might be especially high, so that $R - F$ would be greater than $D - M$, and the slope would thus be greater than 1. Mate choice, as outlined here, would thus favor a moderate or a high threshold for a female's responses to potential mates, in accordance with the moderate or high slope of the indifference line.

Receivers in this situation would appear to be choosy, often "passing up" suitable and even advantageous mates (missed detections). Females of many species have also been called "coy," another allusion to their apparent "reluctance" to choose a mate. To a third party with access to information about optimal and suboptimal mates, such a female would appear finicky, even arbitrary, as well as reluctant. I have proposed that the behavior of such receivers might be called adaptive fastidiousness. They appear arbitrarily selective in their responses. Nonetheless, they have optimized their thresholds for response in situations when missed detections have minor disadvantages but false alarms are calamitous.

Predator warnings provide a contrasting situation. In this case a signaler that spots a nearby or attacking predator gives a warning signal that other individuals might respond to by taking cover or fleeing. A correct detection by a listener is thus likely to have a high payoff for a receiver. The noise in this situation, aside from any degradation of the signal itself or irrelevant background energy, is the possibility of deceptive signalers. It is now clear that such deception occurs in animal communication, at least in a variety of birds and mammals. In the case of several monkeys, reports indicate that subordinate individuals sometimes use fake predator warnings to interrupt matings by dominant rivals. In other instances, a subordinate use fakes predator warnings to interrupt feeding by dominants long enough for the subordinate to steal some food. The latter situation occurs in a number of birds also. Chickadees, for example, are small birds that like to feed on large seeds during the winter. They have a special way to hammer open these seeds without needing a large bill with strong leverage. Most birds using large seeds are thus considerably larger than chickadees. The problem for chickadees, then, is that these large species occupy feeding sites with large seeds and cannot be displaced by hungry chickadees. During the winter, bird-eating hawks frequent areas where smaller birds feed, so seed-eating birds are often on alert and indeed are often on edge for any possibility of a nearby hunting hawk. An alarm call for a flying hawk, a characteristic sound produced in a similar way by many different species, causes pandemonium among feeding birds; they dive for cover into the nearest dense bush. Everybody who has spent long periods carefully watching feeders for birds in the northern hemisphere has noticed occasions when the characteristic warning call is heard, all feeding birds dive for cover, and a chickadee promptly arrives to steal a seed or two.

The cost of such a false alarm for a feeding bird is probably minimal—a few minutes lost for feeding, some energy used for a dash to cover. A missed detection of an actual warning call, on the other hand, could be fatal; a correct detection would be likely to have a substantial payoff. As before, a correct rejection of a fake call would have little consequence for a feeding bird—a bit more time feeding and energy conserved—approximately life as normal in the absence of any communication.

In situations where predator warnings can save lives but some individuals produce fake warnings, $R - F$ would probably be small, but $D - M$ could be large. As a result, the slope of the indifference line for a receiver such as a feeding bird would be low. A low slope means a low threshold for response in order to obtain the highest possible expected utility. Receivers in this situation are thus especially susceptible to false alarms in order to minimize, at all costs, the possibility of missed detections. Such a receiver would appear edgy and vulnerable to false signals. Yet, despite the apparently careless performance of such a receiver, it is well adapted to the noisy situation it faces. I have proposed calling this situation adaptive gullibility—*adaptive* because the receiver has optimized its performance, and *gullibility* because (to a third party with access to more information) the receiver appears prone to respond to deceptive signalers.

PRELIMINARY REFLECTIONS ON COMMUNICATION IN NOISE

A sobering revelation from the preceding analysis is the number of parameters that affect a receiver's adaptations for communication in noise. Furthermore, only a few of them have ever been estimated for any case of communication! It is currently impossible to tell if mate choice by any species actually fits the predictions for adaptive fastidiousness because many of the payoffs and probabilities have never received attention. Likewise our ignorance about adaptive gullibility and predator warning: again, the payoffs and probabilities can only be surmised at the moment.

Chapter 9 discusses the costs and benefits (in other words, the payoffs) of such behavior as mate choice and deception. There are estimates of the benefits of correct detection and costs of false alarm in these cases. In some cases of mimicry, the costs or benefits of all of the four payoffs have received attention. Nevertheless, the parameters of communication in noise introduced in this chapter are mostly unexplored, so the scenarios for adaptive fastidiousness and adaptive gullibility remain hypotheses for verification. This problem extends beyond animal behavior. Discussions of the nature and evolution of human language uniformly ignore the consequences of noise. The problem arises from the direction of research rather than from practical difficulties. Of the crucial parameters introduced above, none seems particularly difficult to estimate.

Before proceeding, pause to absorb how noise in communication creates trade-offs for receivers. Trade-offs are crucial in this context because they prevent an organism from maximizing all features of its performance simultaneously. The trade-off between false alarms and missed detections (or between false alarms and correct detections) eliminates the possibility that a receiver can ever reach ideal performance. It is impossible to minimize both kinds of error at once. Receivers can only optimize their performance for a particular set of circumstances. They cannot maximize performance globally to reach the ideal of no errors.

Previous thinking about communication has generally not recognized trade-offs for receivers. As a consequence, the performance of receivers has not been a component of models of communication. In the absence of noise they face no insurmountable problems. Instead the focus has been on the more obvious trade-offs faced by signalers.

PERFORMANCE OF RECEIVERS DEPENDS ON SIGNALERS

Signal Detection Theory provides a way to measure the performance of a receiver in the presence of noise and thus to predict its optimal performance, but as a model for the evolution of communication it is missing a piece. The performance of a receiver in noise depends on both the receiver's threshold and also the contrast between signal and noise.

Contrast between signal and signal plus noise, as explained in Chapter 6, differs from the detectability or discriminability of a signal. Contrast depends on features of a signal in a particular environment, detectability on features of the receiver of the signal. The former can be estimated by a third party, perhaps with special equipment and special vantages. The latter is estimated by comparing the receiver's behavior in the presence and absence of a signal. Receivers can do their best to optimize the properties of their receptors, and hence the detectability or discriminability of signals, but they do not control the contrast between the features of signals and the features of noise impinging on their receptors.

A receiver can optimize the location of its threshold for any level of contrast—the difference between the activity of its receptor in the presence of a signal with noise and its activity in the presence of just noise (the ratio of these levels of activity is often called a signal-to-noise ratio). Receivers also have some control over the detectability of a signal in noise by adjusting the sensitivity of their receptors. To some degree this adjustment produces trade-offs similar to those in adjusting the location of a threshold for a decision to respond. Sensitivity to input in general and specificity for a particular signal are not entirely compatible. Furthermore, a receiver's receptors might often have to respond to several kinds of signals. Although receivers might evolve to optimize the properties of their receptors for the particular signals of interest to them and for the particular environments they occupy, nevertheless they cannot completely control the contrast of signals and noise and thus they cannot completely control the detectability of signals in noise.

The signaler, on the other hand, is in a position to alter the amplitude and other properties of a signal and thus its contrast with noise. Yet the contrast between signal and noise for a receiver is also affected by other sources of energy in the background and degradation of the signal during transmission. The signaler thus can influence, but not solely determine, the contrast of signals and noise. Because signaling can evolve, just as receiving can, the contrast of signals and noise can evolve, just as can thresholds for response. Signalers might evolve to optimize the level of contrast between signals and noise, just as receivers evolve to optimize their thresholds for response. If so, the optimal performance of a receiver depends on the evolution of signalers. Consequently, a complete model for the evolution of communication in noise requires not only a model of the receiver's performance but also one for the signaler's performance. Only then is it possible to think about the joint evolution, or coevolution, of interacting receivers and signalers.

Proposals for the behavior of signalers are well developed in studies of animal communication by behavioral ecologists. The approach presented below borrows many of these proposals but, in formalizing them and adapting them to noisy communication, it leads to rather different conclusions. A comparison of the present approach with the previous ones is the topic of Part III. The following sections in this chapter explore the predicament of signalers in noisy communication.

COSTS OF EXAGGERATION

A signaler produces a signal, often called by ethologists a display, by means of some behavior or morphology that broadcasts energy to potential receivers. It might produce patterns of coloration in reflected light, make movements with contrasting parts of its body, emit sounds, leave a confection of scents in a particular place, or touch the receiver. In every case the contrast between the signal and noise reaching the receiver is partly controlled by the signaler. Often the contrast is a simple function of the amplitude of the signal (the intensity of the coloration, the magnitude of the movement, the intensity of the sound, the concentration of molecules). As explained in Chapter 6, in any particular environment and for any distance to the receiver (within limits), the greater the amplitude of the signal at the source (the signaler), the greater its contrast with noise at the receiver. Regardless of the level of background energy or the level of degradation of a signal, a stronger signal at the source arrives at a receiver with greater amplitude than any weaker version of the signal. In the following pages, the amplitude of a signal is called its exaggeration.

Increasing the amplitude (exaggeration) of a signal should normally increase its cost to the signaler. A bigger movement, bolder coloration, louder sound, or higher concentration of molecules all require more energy to produce. Furthermore, they often require more time and often involve risks. Unintended receivers are more likely to detect exaggerated signals for the same reason that intended ones are: they contrast more with the background. Predators and parasites can often locate their prey or hosts by paying attention to communicatory signals intended for mates or rivals. Frog-eating bats locate Túngara Frogs by the mating calls of the males. Parasitic flies locate crickets that serve as hosts for their eggs by listening for the stridulations of the males. Golden Eagles appear to have special tactics when targeting leks of displaying sage-grouse. The risks, like the expenditures of energy and time, must normally rise steadily with increases in exaggeration of signals.

The costs of signals have received a lot of attention in recent decades because costs, it is proposed, are necessary to ensure honesty in signaling. Chapter 12 discusses this proposed relationship between costs of signals and honesty. For the present discussion, all that is needed is the basic premise that evolved signals have costs. Any signal that has advantages for the signaler should evolve some degree of exaggeration to improve the detectability of the signal, and hence responses, by appropriate receivers. It is difficult to imagine a signal with advantageous consequences for the signaler that would not evolve in a way that would entail at least some cost to the signaler. Even signals that originated as inadvertent movements or morphology would tend to evolve some exaggeration to increase their detectability. As noted in Chapter 6, early ethologists called this process ritualization.

The next step is thus to identify an appropriate measure of the cost of a signal. Because the objective is to understand the evolution of signaling and responding, the measure of performance should reflect the consequences of natural selection. As explained in Chapter 7, a reasonable measure of the strength of natural selection on the alleles associated with a phenotypic trait is the probability of survival multiplied by the expected reproduction of individuals with that trait. If we adopt this measure of natural selection on a signal, then the cost of a signal is expressed as a decrease in the signaler's probability of survival or its expected reproduction. The benefit of a signal is an increase in one of these parameters.

For instance, suppose the signal of interest is produced by a male in order to attract a female for mating (other kinds of signals are considered later). A signal advertising for a mate, as just mentioned, has costs in terms of a decrease survival of the signaler. It also has benefits in terms of increased opportunities for reproduction. Whether the cost results from energy and time that could have directly promoted the signaler's survival or from risks of predation or parasitism, the probability of survival normally decreases as the frequency and exaggeration of signals increases. A plot of survival versus exaggeration is thus a line sloping steadily downward as exaggeration increases. The exact shape of this line is not known for any signal. If not linear, it might be concave or convex, but in any case a monotonic decrease is plausible.

Every signaler, on a plot of this sort, has its highest probability of survival when it produces no signal at all, in other words, signals with zero exaggeration. Its probability of survival then decreases monotonically as exaggeration of signals increases. Different signalers probably have different functions for survival as exaggeration increases. They might differ in their survival in the absence of signaling (with zero exaggeration). Such a difference could be called a difference in their *intrinsic survival.* They might also differ in the cost of each incremental increase in exaggeration, in other words the slope of their cost as a function of exaggeration. This incremental cost is analogous to a marginal cost in economics, so a difference between signalers in this respect could be called a difference in their *marginal survival.* There are thus two kinds of costs for exaggeration: an *absolute cost* of exaggeration (the difference between intrinsic survival and survival for a particular level of exaggeration) and a *marginal cost* of exaggeration (the change in survival with an incremental change in exaggeration).

These two kinds of costs suggest two ways that signalers can differ in quality. Two signalers might have the same intrinsic survival but differ in marginal survival, in other words, one might lose more survival for each incremental increase in exaggeration of its signals than does the other. Or they might have the same marginal cost of exaggeration but differ in intrinsic survival, in other words, one might survive better than the other in contexts other than signaling. Of course, they might differ in both respects. To continue the example of males advertising for mates, two males could differ in their intrinsic survival, when not advertising for mates, or in their marginal costs of advertising for mates, or in both ways.

BENEFITS OF SIGNALING

A signaler benefits from the responses elicited from appropriate receivers. In the case of advertising for mates, a signaler would usually benefit from increased opportunities for reproduction. A male's chances for mating depend on females responding to his advertising signals. Because communication is our interest here, the focus is on females' responses to signals from males and not directly to other features of males. For instance, females might have advantages from mating with males of high quality. If they could directly judge the quality of a male, without attending to his advertising signals, they could choose an optimal mate without recourse to communication. If, instead, they could not directly judge of male's quality, they would have to pay attention to his advertisements (his signals). Perhaps, the more exaggerated his signals, the more a female would respond. Our understanding of signal detection from Part I of this book suggests why they are likely to respond more. The greater the exaggeration of signals, the greater their contrast with the environmental background and the greater the separation of signals plus noise and noise alone by the female's receptors. The result is greater discriminability of the male's signals from noise and thus a greater probability of correct detections by the female.

Furthermore, it is easy to see that in this scenario females choosing mates on the basis of their exaggerated signals would mate with high-quality males, because males of high quality inevitably have greater exaggeration of their signals. Figure 8.3 shows survival as a function of exaggeration for two males that differ in quality. The difference in quality can be a difference in intrinsic survival or in marginal survival for exaggeration. The benefit of signaling (the chance of attracting a female) increases with exaggeration, and this increase is the same for both males

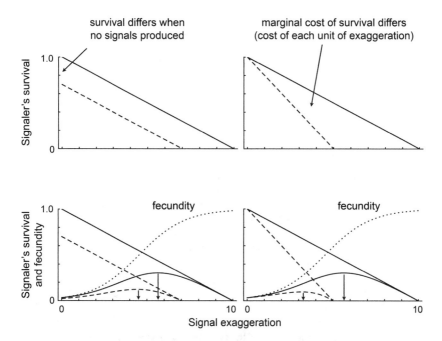

FIGURE 8.3. Optimal exaggeration for males that differ in quality. These plots show
the trade-off faced by signalers when exaggeration of signals is costly. The *vertical
axis* indicates the male's probability of survival, the *horizontal axis* the degree of
exaggeration of his signal (display). Survival is highest when the male produces no
signal (zero exaggeration) and decreases steadily with increasing exaggeration of his
signal. Males might differ in their intrinsic survival (with no signaling) or in their
marginal survival (m, the cost of each additional unit of exaggeration) *(top left and
right)*. All males benefit from signaling in the same way—by increased fecundity as a
result of female's responses. Because females cannot directly determine a male's
quality, they must rely on communication. In this case, they are more likely to mate
with a male that produces a more exaggerated signal (dotted lines in the bottom two
diagrams). As long as this line for any male's fecundity increases steadily
(monotonically) with exaggeration of his signals, its shape does not affect the
conclusions. A sigmoid curve, like the one plotted here, indicates that females are
most discriminating at intermediate levels of males' exaggeration. When the product
of survival × reproduction is calculated for each male *(concave lines)*, it turns out that
high-quality males optimize their signaling at a higher level of exaggeration than do
low-quality males. *Vertical arrows* connect a male's maximal utility and his
corresponding optimal exaggeration. In other words, this system of communication
is honest; both signalers (males) and receivers (females) benefit.

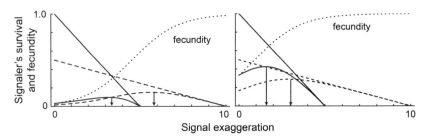

FIGURE 8.4. Optimal exaggeration in more complex situations. In these plots the two measures of male quality conflict: one of the two males has lower costs for each unit of exaggeration (marginal survival, m) but also lower intrinsic survival (in the absence of signaling, i). The *left-hand plot* indicates that the male with lower marginal survival would produce more exaggerated signals at his optimum. Females would benefit by having sons similar to a mate of this sort (her sons' high success in eliciting responses from females would more than compensate for their low intrinsic survival), yet their daughters would also pay the cost of lower intrinsic survival. Females choosing mates with the most exaggerated signals would thus have "sexy" sons but low-quality daughters. When females mostly discriminate against signals with low exaggeration *(right-hand plot),* the relative advantages of the two males do not change although the difference is not so great. Extrapolation of this trend indicates that the difference between the two males converges to 0 only when females fail to make any discrimination at all among males' signals (males' fecundities are constant regardless of any signals they might make).

(provided that females only respond to signals, as just assumed, not directly to the quality of males). Then for every level of exaggeration in males' signals it is possible to calculate each male's survival × reproduction. This product is the measure of natural selection on producing a signal with the particular degree of exaggeration. Each of these products is a convex line that starts at 0 (where the benefit is 0, because no females respond) and ends at 0 (where the cost of exaggeration is fatal—in other words, survival is 0) and in between reaches a maximum. This maximum shows the degree of exaggeration that maximizes natural selection for the respective male. In this plot, natural selection thus maximizes exaggeration at a higher level for the high-quality male than for the low-quality male. Notice that it does not pay for a low-quality male to exaggerate his signals farther than his own optimum. If he did, his decreased survival would then not compensate for increased mating. A high-quality male can pay higher costs for exaggeration and receive higher benefits from females' responses.

It is possible to imagine more complicated situations (see Figure 8.4). To determine which curves might actually apply to animals in their natural habitats would require more careful measurements of costs and benefits of males' displays than are currently available. At least in the simplest situations, however, when females respond to the most exaggerated signals they mate with the highest-quality males.

OPTIMAL SIGNALERS

The optimal level of exaggeration for a signaler depends on its intrinsic survival, its marginal cost of exaggeration, and its expected reproduction as a result of signaling (its benefit). With a few simplifying assumptions, this optimal level can be calculated for any signaler. First, let s = survival of a signaler, i = its intrinsic survival, m = its marginal cost of exaggeration (the decrease in its survival for each unit of exaggeration), and e = the level of exaggeration of its signal. If a male's survival under these conditions is a straight line, it can be expressed by the equation

$$s = i - me.$$

Next, let b = a signaler's expected reproduction (its benefit) from signaling, o = its reproductive offset when it does not signal, and g = its marginal gain from exaggerating its signal. If a female's probability of response increases steadily with the exaggeration of a male's advertising signal, the benefit to a male of a signal can also be expressed as a straight line,

$$b = o + ge.$$

A positive offset indicates that a male has some minimal reproduction even when he does not produce signals. A negative offset assures that a signal must have some minimal exaggeration before it has any effect at all. For any level of exaggeration, a male's survival $\times$ reproduction is

$$sb = (i - me)(o + ge) = io - moe + ige - mge^2.$$

Some calculus locates the level of exaggeration that maximizes $s \times b$. The derivative of $s \times b$ with respect to e is

$$d(sb)/de = -mo + ig - 2mge,$$

so maximal b is attained when

$$ig - mo - 2mge^* = 0 \text{ (the asterisk indicates that } e^* \text{ in this expression is maximal).}$$

By rearranging this expression to

$$e^* = i / 2m - o / 2g,$$

it becomes apparent that optimal exaggeration is higher if a male has either a higher intrinsic survival (i) or a lower marginal cost (m). In other words, a higher-quality male always produces more exaggerated signals than a lower-quality male, provided all males produce their own individually optimal level of exaggeration. High-quality males are thus more likely to attract females.

It is important to assess the assumptions required for these calculations. First, for simplicity, the relationship of survival to exaggeration and the relationship of the signaler's benefit to exaggeration are assumed to be rectilinear. So far as I know, there are no measurements of the costs or benefits of signals as functions of their exaggeration. To determine whether the relationships are straight or curved would require measurements of costs or benefits for at least three levels of exaggeration. A straight line for survival is perhaps not unreasonable. It indicates that each increment of exaggeration reduces survival equally until survival reaches 0. Some level of exaggeration might indeed be fatal for a signaler in many cases.

Whether or not the benefits of signaling increase in a straight line with exaggeration is more questionable. Instead, for many sensory modalities, sensory perception often follows—at least approximately—a logarithmic function. The general relationship is Weber's Law, introduced in Chapter 6. A logarithmic function indicates that, as any parameter of a stimulus increases, a progressively larger absolute change is needed to produce an equal change in a behavioral (or sensory receptor's) response. In other words, the response to a signal would curve convexly as its exaggeration increased. As discussed in Chapter 6, much about Weber's Law remains controversial. It is not clear what mechanism could produce such a general relationship, nor is it clear that responses of sensory receptors or behavioral responses always follow Weber's Law. For the present context, however, an important point is that Weber's Law applies to just-noticeable differences in signals. The difference in a feature of a signal (its intensity, its frequency, its rate) that a human subject can perceive (or that results in a consistent difference in the activity of a receptor) becomes greater as the parameter increases. A just-noticeable difference in loudness (intensity of a sound) or in pitch (frequency of a sound) is greater the higher the intensity or the frequency.

The benefit for a signaler does not depend entirely on just-noticeable differences for a receiver's perception. It depends on the receiver's probability of response, and that probability depends on the receiver's probability of correct detection of the signal. When a signaler exaggerates its signal, a receiver experiences greater contrast between the signal plus noise and noise alone. The signal becomes, depending on the properties of its receptors, more detectable. Assuming that a response is advantageous for the receiver, it then pays for it to adjust its threshold upward. The receiver can thus reduce its chances of false alarm and missed detection while still increasing its chances of correct detection. Thus the probability of a response should increase with exaggeration of a signal, but not as a linear function of exaggeration. This approach based on Signal Detection Theory has the great advantage of explaining why the marginal rate of increase in responses might decrease as the exaggeration of a signal increases.

Signal Detection Theory thus seems to explain incrementally increasing just-noticeable differences as exaggeration of signal parameters rises, just as Weber's Law does. It remains to compare these predictions quantitatively. In the meantime, a convex curvature of the signaler's benefit as a result of receivers' responses should not affect the conclusion about honesty in signaling. A logarithmic (or other monotonic decelerating) function of exaggeration would still ensure that the optimal exaggeration of signals would increase with a signaler's quality.

This approach thus provides a coherent explanation for why high-quality males produce more exaggerated signals and why females respond more to them. It also indicates that the costs of signals should serve to increase their contrast with noise in the environment in which communication occurs. In the previous approach, based on costs ensuring honesty, there is no such constraint on how costs should influence properties of signals. Furthermore, in this previous approach, receivers respond to the costs of signals rather than to their contrast-enhancing exaggeration. The approach here instead explains honesty in signaling as a result of signalers that optimize the exaggeration of their signals and receivers that optimize their responses to those signals.

Continuing to explore of the evolution of communication in noise, the next step is to specify the utility of a signaler's exaggeration. The preceding analysis of the trade-offs faced by signalers leads easily to an expression for a signaler's expected utility from producing a signal with exaggeration e:

$$U_s = p_s sb = p_s (i - me) (o + gp_D(t, e)).$$

The signaler's overall expected utility depends on its utility from producing any one signal, $s \times b$, multiplied by the probability that of producing a signal in any appropriate unit of time, p_s. Furthermore, the payoff from producing each signal depends on the gain provided an appropriate receiver responds, g, times the probability of a response. And the probability of a response is the receiver's probability of a correct detection, p_D, which is a function of its threshold, t, and the exaggeration of the signal, e.

This expected utility for the signaler has the same units of measurement as the receiver's expected utility derived previously: survival $\times$ reproduction. Both of these analyses have tried to capture the features of the participants that are necessary and sufficient to understand the evolution of communication in noise. They are the simplest adequate models for signalers and receivers in noisy conditions. They both express the trade-offs in communication in terms of natural selection, survival $\times$ reproduction of individuals (on the assumption that alleles are associated with, even if they do not determine, the relevant behavior). We are thus in a position to consider combining the receiver's and the signaler's expected utilities to explore their coevolution. In particular, the question facing us is the possibility a joint optimum for both threshold and exaggeration in any particular situation of communication in noise.

THE COORDINATION OF SIGNALING AND RECEIVING

Before proceeding, it is important to consider whether the model developed so far is too simple to capture the essence of communication. One concern might arise because the expressions for the participants' utilities were developed separately for receivers and signalers. Often, however, signalers and receivers are not distinct sets of individuals. Instead, individuals both send and receive signals. Nevertheless, receiving and signaling evolve separately, provided they are influenced by separate sets of genes. Because receiving and sending signals depends on separate structures and separate parts of the nervous system, at least to a large degree, it is not unreasonable to suppose that they also are influenced by different genes.

Some correspondence between the properties of signals and the sensory scope of receivers is of course necessary for any communication. Receivers cannot respond to signals that do not affect their sensory organs. In a diversity of organisms, including insects, frogs, and birds, the features of acoustic signals correspond with the features that stimulate auditory neurons. In comparisons among related species, there is usually a correlation between the dominant frequencies of signals, the best frequencies for auditory neurons, and the preferred frequencies for receivers' responses. In addition to frequency, there is evidence that signals and sensory mechanisms also match in temporal resolution and such features as rise time and tonicity of notes. Nevertheless, there are some conspicuous mismatches reported. Some of these must result from sensory adaptations for detecting predators or other sounds besides those from the receiver's own species. Others currently have no clear explanation.

The close match between the properties of signals and sensors suggest that both rely on a shared neural mechanism—for instance, a timer or oscillator that could assure a matching frequency in both the performance and the reception of signals. Yet even when this possibility seems most likely, the evidence suggests otherwise. For instance, among frogs in general the frequencies of the males' calls correspond, at least approximately, to the best frequencies for auditory neurons (in the amphibian papilla, one of the two groups of auditory neurons in most frogs). In Green

Treefrogs, the males' calls consist of pulses of this carrier frequency at a rate that is preferred by females in approaching a mate. A frog's temperature, which unlike in birds or mammals is close to the ambient temperature, strongly affects this pulse rate. As the temperature increases, the pulse rate of males' calls rises, and so does the females' preferred pulse rate. There is evidently a close coupling of the signaler's and receiver's neural mechanisms for producing and recognizing pulses. As Carl Gerhardt and Karen Mudry have shown, this coupling does not, however, apply to the carrier frequency of the males' calls. The carrier frequency of calls varies only slightly with temperature, but the carrier frequency preferred by females depends strongly on temperature. At 22°C the carrier frequency in production and in preference match, but at 18°C females prefer a carrier frequency almost an octave below the males' calls. Females do not normally mate at 18°C, however, so the disruption of communication is minimal. Nevertheless, the striking difference in the temperature dependence indicates that the males' and females' neural mechanisms are not the same. Investigating abnormal conditions does not reveal an organism's evolutionary adaptations, but it can reveal crucial aspects of its physiology or development.

Another indication that the production and reception of signals might share the same mechanism comes from investigations of hybrids between two closely related species. First-generation hybrids have half of their genes from each parental species. When two species of grasshopper in the Pacific region are hybridized in the laboratory, their male progeny produce signals (rhythmic stridulations of their wings) for attracting females that are intermediate between the two parental species' signals. In addition, their female progeny prefer these intermediate features of signals. Further research, however, has shown that the genes involved are in fact different. Evidently, distinct mechanisms (and genes) for signaling and receiving have evolved because these mechanisms perform more effectively when separated (specialized) than when combined (generalized).

The evolution of coordinated mechanisms in producers and receivers of signals could result from natural selection in an ordinary way. Because signalers do not realize benefits unless appropriate receivers respond, and receivers do not realize benefits unless there are appropriate signals, natural selection would favor matching signalers and receivers. The major difficulty of this sort of coordinated evolution is getting it started. A signaler with a mutation for a new signal is unlikely to encounter appropriate receivers, and a receiver with a mutation for responding to a new signal is unlikely to encounter appropriate signals. This difficulty is eased in the case of signals by males that benefit females seeking mates. In this case coordinated evolution is facilitated by sexual selection. When a receiver mates with a signaler, their progeny carry genes of both parents. As generations pass, genes that influence the production of signals by males become associated with genes that influence the responses to those signals by females. As discussed in more depth in Chapter 13, this genetic correlation of genes for male signals and genes for female preferences can accelerate the spread of both kinds of genes in a population. Genes influencing the males' production and those influencing the females' reception need not be the same, even though their genetic correlation in individuals can promote their mutual spread.

Consequently, there is no reason to think that the coordinated evolution of signals and responses does not involve separate mechanisms influenced, at least to a degree, by separate genes. The approach adopted here assumes that the production and reception of signals coevolve because each is dependent on the other as a result of communication and not because production and reception have a common neural mechanism. Any shared elements of these mechanisms would evolve as a result of convenience, not as a necessary condition for the coevolution of signaling and responding.

MULTIPLE PARTICIPANTS

Another possible concern is the inclusion of just two participants in a model for the evolution of communication. Communication is often more complex. Two individuals in the process of communication might have to contend with others that could interfere. Third parties might take advantage of one or both original parties. For instance, eavesdroppers might respond to signals in ways disadvantageous for the signaler, and deceivers might mimic signals to elicit responses disadvantageous for the receiver. Many good examples of eavesdropping are now documented for animals. The frog-eating bats attracted to the calls of Túngara Frogs are a dramatic case and a model for similar situations. Male birds are known to eavesdrop on the songs of rivals to learn who has lost an encounter and thus might be an easy opponent in the future. Female birds also eavesdrop on the encounters of males to learn which male dominates encounters with rivals and thus might be a more advantageous mate. In these and other cases, the consequences for the signalers can be disastrous, although infrequent. Deception is also well documented in animals. Chickadees, as discussed previously in this chapter, occasionally produce deceptive alarm calls, which elicit escape by larger species, with the result that the chickadee can gain momentary access to food. Monkeys occasionally use alarm calls to interrupt feeding, mating, or attacks by more dominant individuals. Eavesdropping occurs when an inappropriate receiver responds in a way disadvantageous to the signaler; deception occurs when an inappropriate signaler elicits a response disadvantageous for the receiver.

Eavesdropping and deception are prevalent, although infrequent, complications of communication. Both complicate communication in noisy circumstances, because eavesdroppers and deceivers must contend with noise just as primary receivers and signalers must. Many situations could produce natural selection on three or more kinds of participants. Even three would raise a challenge, because a three-way interaction in noise would make the mathematics much more complicated. Three-way optimizations can even become indeterminate. Nevertheless, as a first approximation, the interaction of the primary participants might dominate communication, so that eavesdropping and deception might evolve secondarily. If so, their consequences for communication by the primary participants might be accounted for by adjustments in the payoffs and costs already discussed. Deceivers change the payoffs of false alarms for a receiver, and eavesdroppers change the marginal cost of signaling for a signaler. If these inappropriate interventions in communication occurred infrequently, then it would make sense to focus on the evolution of the primary participants. Adjustments of the payoffs and costs for a receiver and signaler, in this case, might adequately account for the influence of these inappropriate participants, even though they would not include any residual consequences of coevolution between primary and secondary participants. The following analyses assume that the costs of eavesdropping and deception for signalers and receivers, respectively, are adequately included in the marginal cost of signaling and the payoffs for responses. An analysis of three-way coevolution among signalers, receivers, and a secondary participant, all engaged in noisy communication, is left for the future.

THE FUNDAMENTAL PROBLEM OF
NOISY COMMUNICATION

In noisy circumstances, the utility of responding to a signal depends on the receiver's threshold, but also on the exaggeration by the signal, which influences the level of the signal in relation to noise for the receiver. Conversely, the utility of producing a signal depends on the intensity of the signal (its exaggeration), but also on the threshold of the receiver, which influences the probability of a response. In noisy communication, the utility for each participant depends on the performance of the other. Because both signaling and receiving can evolve, what happens if both evolve to maximize their utilities for the two parties respectively? Receivers might evolve adjustments of their thresholds, and signalers adjustments of their exaggeration, in ways that increase the respective utilities of their behavior. Do they reach a mutual optimum? Or is one or the other predisposed to win, so that it maximizes the utility of its own behavior and leaves its partner to make do? If evolution tends to result in optimal behavior for both, does this optimum ensure maximal utility for both? Can evolution produce ideal communication, so receivers make no errors and signalers always evoke appropriate responses? Or must both settle for less?

9

PAYOFFS FOR PARTICIPANTS

TEN PARAMETERS (FOUR OF THEM PAYOFFS)

BEFORE WE CAN FIND the optimal threshold for a receiver, when its expected utility also depends on the exaggeration of the signal, we need values for all the parameters included in the expressions for the expected utilities of receiver and signaler. These equations, as noted in Chapter 8, are

$$U_r = p_S \, (p_D D + (1-p_D) M) + (1-p_S) \, (p_F F + (1-p_F) R),$$

for the expected utility for the receiver's threshold, and

$$U_s = p_S sb = p_S (i - me) \, (o + gp_D(t, e)),$$

for the expected utility for the signaler's exaggeration. There are ten parameters in addition to t, the location of the receiver's threshold, and e, the exaggeration of the signal. Note that t and e do not appear in the equation for U_r explicitly, but they are there nevertheless. The receiver's threshold and the level of contrast between signal and noise, itself affected by the exaggeration of the signal, determine the probabilities of each of the four possible outcomes of a receiver's decision to respond. In a similar way, t appears explicitly in the equation for U_s. The benefit a signaler receives from its signal, b, depends on its benefit if a receiver responds (g, an increase in survival or reproduction) multiplied times the probability that an appropriate receiver does in fact respond (p_D, the probability of a correct detection by the receiver). By now it is clear that the probability of a receiver's response depends on the location of its threshold for response. So we can write the equations above more explicitly as

$$U_r = p_S (p_D(t, e) D + (1 - p_D(t, e)) M) + (1 - p_S) (p_F(t, e) F + (1 - p_F(t, e)) R)$$

and

$$U_s = p_S sb = p_S (i - me) (o + gp_D(t, e)).$$

Other than t and e, the ten parameters for which values are needed are

> p_S, the probability that a signal occurs when the receiver pays attention
> D, the receiver's payoff for a correct detection
> M, the receiver's payoff for a missed detection
> F, the receiver's payoff for a false alarm
> R, the receiver's payoff for a correct rejection
> i, the signaler's intrinsic survival in the absence of signaling
> m, the signaler's marginal survival as exaggeration of its signals increases
> o, the offset or benefit in the absence of a signal
> g, the signaler's marginal gain as exaggeration of its signals increases (gain as a result of a correct detection by an appropriate receiver)
> and, finally, the level of noise.

The level of the signal plus noise results from adding the level of the signal with its variation to the level of the noise along with its variation. As described in Chapter 7, the level of the signal and its variation at the receiver's receptor is determined by the level (exaggeration) of the signal as it is emitted by the signaler and the degradation (attenuation and distortion) of the signal during transmission. The parameters p_D and p_F, the probabilities of a correct detection and a false alarm, are calculated from the level and variation of the noise alone, the level and variation of the signal plus noise, and the location of the receiver's threshold (see Figure 7.3).

So far these ten parameters have remained unspecified. They have been represented by symbols in the equation for U_r, the receiver's utility. The symbols are abstract replacements for actual numbers for the payoffs, the probability of a signal, and the level and variation in noise in a particular situation. In any one situation they are constants that specify the circumstances in which communication occurs. These values presumably change from situation to situation. To replace these symbols with numbers it is thus necessary to consider what is known about some representative situations for communication.

It becomes apparent in the following discussion that there is no situation for which all of these parameters of communication have been measured. Some of the parameters have, as far as I know, never even been considered in studies of natural communication. Nevertheless, communication in animals and humans has received a lot of attention, so plausible hypotheses about relationships of these parameters are not unreasonable.

For a start, it helps to consider two contrasting situations for communication. These can then provide comparisons with other situations. Choosing a mate and reacting to alarms are appropriate for this purpose because, as described in Chapter 7, they have contrasting consequences for the two kinds of errors by a receiver. During mate choice, a false alarm by a receiver is likely to have more negative consequences for the receivers' reproduction × survival than a missed detection; during vigilance for danger, it is a missed alarm that is likely to have more negative consequences.

CHOOSING A MATE

Choice of mates can result in sexual selection, a topic that has stimulated a vast enterprise of its own in the past few decades, with almost no consideration of noise. The preoccupation has been with the costs or tactics of producing signals. Neglect of transmission and interference has separated most studies of mate choice from studies of other forms of communication. Of course, responding preferentially to potential mates that have certain characteristics is a typical example of communication. The choosing individual is a receiver that responds to signals from some potential mates and not others.

That part of sexual selection that results from direct mate choice is a typical case of communication. Males, for instance, produce a signal (such as a display, coloration, or structure, perhaps a long tail) to which females respond by making themselves available for copulation—and for fertilization of their eggs. The females' responses are presumably influenced by the usual forms of noise in communication—degradation of the male's signal during transmission and extraneous energy in the background. The acoustic, visual, olfactory, and even tactile components of a male's display inevitably change during transmission to the female in ways that make the signals contrast less with the background energy. Wind, water, moving vegetation, and signals from other species all contribute to the background energy. In an environment such as a rainforest, the number of species of birds, frogs, and insects attempting to communicate at most times of day or night can reach stupefying levels. On a coral reef, in an analogous way, the kaleidoscopic combinations of colors of fish, corals, and invertebrates create background clutter in visual communication. Degradation of the males' signals and combination with irrelevant energy create a challenge for a female attempting to identify signals from males of her own species.

Furthermore, females often exercise their preference while exposed to many possible conspecific mates. Males often congregate where females mate. Leks, as described in Chapter 13, are an example. Choruses of frogs and insects, discussed in Chapter 5, are often even more extraordinary congregations of displaying males. Experience with choruses of frogs in the tropics, or even the southeastern United States, is enough to question the ability of females to make sense of the almost overwhelming intensity of "noise." Experiments have shown that a female frog in such a chorus might only distinguish three or so males at a time. Any one male's calls attenuate to a level approximately equal to the overall level of background sound within a distance that only includes a few other males. Playbacks of the calls of an individual male in different levels of background noise indicate that females cannot discriminate an individual male's calls from the background noise of a chorus unless the calls are approximately twice as intense as the chorus. A female might thus be in the midst of hundreds of males but only hear three at a time. To make a choice from more than these three she would have to move around on vegetation or in water that can conceal predatory snakes, fish, and even insects attracted to the congregation of frogs. The presence of many conspecific males contributes to noise for a female attempting to identify particular ones.

This form of communication is often noisy in another sense too. Males of their own species can have different advantages for females. Females are known to have higher reproductive success as a result of mating with preferred conspecific males rather than indiscriminately. Evidence often suggests that these preferred males are in better physical and physiological condition than others. Females thus discriminate among high- and low-quality males of their own species as well as between males of their own and of other species. Males of other species are actually the ultimate low-quality males. Noise thus not only consists of signals from other species but also signals from low-quality males of a female's own species.

Errors, as emphasized in Chapter 1, are the key to interpreting communication in noise. The four possible outcomes for a receiver every time a decision is made to respond (or not to respond) include two kinds of errors, as explained in Chaper 5—false alarms (responding when no signal is present) and missed detections (failing to respond when a signal is present). For choice of a high-quality male, a false alarm is thus accepting a low-quality male; a missed detection is passing up a high-quality male. Both have costs for a female. In the first case, she has committed her limited supply of eggs (or some of them) to fertilization by a low-quality male. In the second case, although she retains her eggs unfertilized, she must spend additional time and run the additional risks of further search. The possible outcomes also include two correct responses—correct detections (responding to a signal from an appropriate signaler) and correct rejections (not responding in the absence of such a signal). A correct detection is accepting a high-quality male to fertilize her eggs. A correct rejection is passing up a low-quality male including any males of other species.

A correct detection would usually have advantages for a female. The first principle for the evolution of communication is that it must be advantageous on average for both parties. Often this means that advantages for correct detections must exceed the disadvantages of errors. Even in the case of an arbitrary preference, one with no net advantage or disadvantage for the female so that $U_r = 0$, any disadvantages from costs of searching and the chance of errors would require some exactly balancing advantage from correct detections. Noisy communication makes the evolution of arbitrary preferences much less probable than would communication without noise, because all seven of the parameters for the utility for the female's response must take just the right values to make $U_r = 0$. Such a coincidence of so many parameters seems profoundly unlikely.

A correct rejection is the remaining possible outcome of a female's decision—not responding when there is no advantage to be had. This outcome is like "life without communication." The only possible behavior is continued searching for an optimal mate, with the renewed possibility of all four outcomes the next time she decides to respond or not. Because her future holds some promise and some risks, her survival × reproduction following a correct rejection must have some moderate value. It would presumably be higher than her survival × reproduction following an error in responding (F or M) and lower than her survival × reproduction following a correct detection (D). Because a correct rejection, at least in the short run, is about the same as life without communication, it is often convenient to measure the payoffs for the four outcomes in relation to the payoff for a correct rejection. To do so, we can divide each payoff by the payoff for a correct rejection, R. Thus the adjusted payoff for a correct rejection becomes $R/R = 1.0$. The adjusted payoffs for the other three outcomes then become proportional to the payoff for a correct rejection, D/R, M/R, and F/R. As a result, *adjusted payoffs between 0 and 1 are disadvantageous* in relation to life without communication; all *those above 1 are advantageous*. The remainder of this book dispenses with the qualifiers, "adjusted" and "proportional," in this context and assumes that all values for a receiver's payoffs are proportional to R (or, as mentioned later in this chapter, proportional to the payoff for any outcome that is close to life without communication).

Payoffs relative to "life without communication" allow us to focus on changes in survival × reproduction as a result of a correct response or either of the errors in response.

For mate choice the worst outcome might be absolutely no reproduction and thus a payoff equal to 0. This payoff might apply to a false alarm if mating with a suboptimal mate resulted in no surviving progeny ($F = 0$). A missed detection would usually have less serious consequences because the female would retain her unfertilized eggs. The payoff for a missed detection is not the same as for a correct rejection, however, because the female has lost an advantageous opportunity. She must continue searching, so if the risks or time involved reduce her survival × reproduction even slightly, M would be less than R. It is nevertheless easy to imagine that M would still be considerably higher than F, provided that the disadvantage of further search (a missed detection) is not so great as the disadvantage of committing one's eggs to a suboptimal male (a false alarm).

As a result of these plausible arguments, the payoffs for the four possible outcomes of a female's decision to respond to a male or not (D, R, M, and F) would often have rankings thus—

$$F \ll M < R \ll M$$

—where $\ll$ indicates less than by a relatively large amount and $<$ by a relatively small amount. The adjusted payoffs would have these relationships:

$$0 < F \ll M < R = 1.0 \ll D.$$

It might be clearer to represent these payoffs on a line labeled from 0 to some number greater than 1.0:

F		M R		D	

0		1.0			

Because there are only two points labeled with numbers on this line, the scale of the line can be altered arbitrarily without altering the relationships among the payoffs.

As explained in Chapter 8, this situation might result in the evolution of receivers with "adaptive fastidiousness." They would appear to be choosy in the sense that they would pass by many males that, to an independent observer's eye, produced highly exaggerated signals. In contrast to their high probability of missed detection, they would have a low chance of false alarms.

VIGILANCE FOR DANGER

A contrasting situation in communication occurs when the consequences of a missed detection are more severe than those for a false alarm. This situation might often apply to individuals responding to signals warning of a predator nearby. Many animals have visual or acoustic warning calls. In some species, such as ground squirrels, monkeys, and many birds, there are distinct signals for different kinds of predators, such as terrestrial predators, flying predators, and dangerous snakes. Communication of alarm in this way turns out (whenever carefully studied) to be noisy. Of course, there is the usual extraneous background energy and the degradation of signals during transmission from the signaler to the receiver. Like mate choice, noise also often includes inappropriate signals from other members of the same species or similar ones. Recall once more than the telltale feature of noise is errors in responses by receivers.

For a period of years I spent every winter with students studying the dominance hierarchies of White-throated Sparrows along hedgerows between harvested fields in the University of North Carolina's Mason Farm Biological Reserve. Those of us involved spent hundreds of hours in cramped hides watching the interactions of the small birds visiting feeding tables. Naturally we became familiar with the local predators and with the sparrows' and other small birds' responses to alarm calls. The most impressive predator was the Sharp-shinned Hawk, a small hawk subsisting exclusively on small birds. On one occasion I watched one of these hawks fly at full tilt directly through a rose hedge, without apparently losing a second, to emerge on the other side with a bird in its talons. The hawks routinely patrolled our half-kilometer-long line of traps and feeding platforms. The feeding sparrows and other birds were clearly on edge. Even though attacks were infrequent, the consequences could be severe. The first sign we had that a hawk was nearby was usually the distinctive call that any small bird emits on detecting a flying predator, a thin high-pitched whistle, not very conspicuous but distinctive enough that once we had heard it and seen a hawk several times, our skin tingled whenever we later heard it. A quick look around often revealed the hawk, probably just after it had attempted an attack. This distinctive call elicited panic among feeding birds, as sometimes two or three dozen birds scrambled to cover low in the nearest brambles. About a half minute later they reemerged—at first one or two, then many in a rush to reclaim the best places remaining to resume feeding. Food is important for birds on winter days. The sparrows managed to gain enough fat in a day to support their metabolism during the following cold night without much to spare.

Occasionally the scenario following an alarm call unfolded differently. Following an alarm call out of sight nearby and a panicked stampede to the bushes, almost immediately a Carolina Chickadee would fly straight to the feeding platform, pick up a large seed, and fly off to open it. Sometimes we could confirm that the chickadee had produced the alarm call. It appeared to be an open-and-shut case of deception. The small chickadees often could not feed at our tables because the large sparrows and other species easily displaced them. Whether by innate or learned behavior, the chickadees could use the alarm call for a flying predator to clear a feeding platform and to gain a brief chance to feed. Similar behavior is known for other small birds and also for subordinate individuals in troops of monkeys.

If we take the receiver's perspective, that of a sparrow feeding intently on an exposed platform, a correct detection is an immediate response to a true alarm call. The consequence is reduced exposure to a hunting hawk, which might arrive at full speed within seconds. Correct detections over the course of a winter could potentially raise a bird's survival substantially. Each instance would have an effect. The first principle of communication is thus satisfied: correct detections have payoffs > 1.0. Recall that payoffs are now adjusted to be proportional to the payoff for a correct rejection. A correct rejection, continued feeding in the absence of an alarm signal, is again somewhat like "life without communication," with consequences not much different from those of routine activities.

The two possible errors, as in the case of mate choice, could plausibly have substantial differences in their payoffs, but in the case of an alarm the greater consequence results from a missed detection rather than a false alarm. When nearly every other potential target had also taken cover, to be left alone in the open must involve considerable risk. False alarms include responses to chickadees' ruses. The consequence for a feeding bird is losing a fraction of a minute with access to food. Over the hours even in a short winter day, with the infrequent true or false alarm calls, the lost opportunity to feed is probably nearly inconsequential. In contrast, a missed detection could be disastrous.

Another important parameter in our formula for a receiver's utility is the frequency of signals during the time a receiver pays attention. In the case of sparrows listening for alarm calls, the time attentive must include most of the hours for feeding every day, although the occurrence of true alarm calls probably does not exceed once in several hours or even days. Alarm signals like these are less frequent than some signals for mate choice. As a result, the probability of a signal for attentive individuals is low.

Another kind of noise in alarm signals comes from the possibility that the signaler itself is in error about the referent for the call—the actual presence of a hunting predator. The signaler responds to seeing (or in other cases of alarms, hearing or smelling) a nearby predator. This response to a stimulus is a form of communication. The individual responding to a predator is like an eavesdropper, as discussed in Chapter 8, because its response produces a disadvantage for the predator, so predators should evolve to minimize these responses. Nevertheless, as was mentioned in Chapter 8, eavesdropping shares the characteristics of noisy communication. An individual detecting the presence of a predator faces the same trade-offs in minimizing its errors of false alarm and missed detection. Regardless of the location of its threshold for a response, the individual emitting an alarm sometimes produces false alarms and sometimes misses detections. The signaler, the individual producing the alarm call, is itself a source of errors.

Individuals listening for alarms face this additional layer of error in their own responses of taking cover. An alarm signal, even from an appropriate signaler, might not indicate the actual presence of a predator. The listener might use the procedures discussed in Chapter 6 to reduce its overall level of error—in particular, it might rely on redundancy. If a second signaler emits an alarm, or even if the first emits a second call, the probability of error on the part of the signaler is reduced, although never eliminated completely. Redundancy of this sort takes time, so it involves further trade-offs for a receiver. In the case of alarms, a delay when a predator is actually present could have serious consequences. Once again it seems plausible that individuals attentive for alarms should set low thresholds for responses to effect their escape, and a low threshold should include low reliance on redundancy. Low thresholds would result in more false alarms but fewer missed detections with the possibility of dire consequences.

The problem of errors by signalers must apply to mate choice as well. The physiological mechanisms that couple any one signaler's level of exaggeration to its intrinsic or marginal quality presumably include some variation. In this case, the use of redundancy to minimize errors by choosing individuals would have only slight negative consequences. The payoff for a missed detection in mate choice is expected to be relatively high, not much less than that for a correct rejection. The redundancy of multiple signals from any one preferred male would further minimize the consequences of missed detections. In addition, though, the level of exaggeration in relation to mate quality might vary from individual to individual, as well as from signal to signal of any one individual. Redundancy of signals for mate choice would not help a receiver in this case. This form of variation in signals would raise the level of noise for a choosing receiver.

These plausible conjectures lead to a ranking of the payoffs for the four possible outcomes every time a sparrow pays attention for an alarm call. In this case the payoffs might have the following relationships:

The payoff for a correct rejection, no response when no signal, is as before similar to the payoff (in terms of survival × reproduction) for life without communication. The payoff must be greater for a correct detection (responding to an alarm) than for a correct rejection; otherwise it would not pay to respond to these signals and communication would collapse. D in proportion to R must thus have some value greater than 1.0. The proportional payoff for a missed detection (exposure to a predator) is presumably less than 1.0 and, furthermore, less than the payoff for a false alarm (a few minutes lost for feeding). The ranking of the four payoffs, in particular the payoffs for missed detections and false alarms, differs from their ranking during mate choice.

Attention for alarms should thus result in the evolution of receivers with "adaptive gullibility." They would often succumb to false alarms, when independent observers verified the absence of predator. Chickadees appear to exploit this situation, by occasionally provoking responses to deceptive alarms in order to steal a little food. In contrast, a receiver attentive for alarms should risk relatively few missed detections.

The signaler's utility presents problems of its own. Alarm signals at least in some cases appear to have little cost of energy or time for the signaler. The predator alarm calls of small birds are presumably such a case. Occasional observations of a signaler attacked by a predator suggest that another cost might consist of attracting the predator's attention. On the other hand, it is not clear that predators attack individuals producing alarm calls more than randomly. Alarm signals might also have advantages for signalers. In some cases, alarm calls are produced primarily in the presence of the signaler's genealogical relatives. If so, the spread of alleles associated with producing alarm calls would benefit indirectly from increased survival of the signaler's relatives. In other cases, alarm signals might be directed to the predator itself rather than to other potential targets. Alarm signals might indicate, for instance, that the signaler already has a predator under surveillance or that it is good condition to escape. In either case the signal could notify a predator that further stalking or pursuit of the signaler is unlikely to succeed. For some antelope, conspicuous, stereotyped (ritualized) jumping, called stotting, appears to indicate the individual's physical condition and surveillance of approaching lions or other predators. These signals could also alert other potential prey nearby. The costs and benefits to signalers are thus less obvious in the case of alarm signals than in the case of mate choice.

MIMICRY

Mimicry is a case in which the usual roles of signaler and receiver are less easily recognized. Consider the case of the European Cuckoo, which lays its eggs in the nests of smaller birds. The hosts, if not successful in detecting the cuckoo's egg and ejecting it, are left to incubate the egg and raise the cuckoo's young. Most species exploited by cuckoos in this way can discriminate between their own and a cuckoo's eggs. Cuckoos in turn lay eggs mimicking the patterns of spots and coloration of their hosts.

Evidence suggests that the cuckoos and their hosts are engaged in co-evolution in the sense that the evolution of discrimination by a host species stimulates evolution of egg mimicry by cuckoos, and vice versa. Hosts do not discriminate their own eggs from other species or from experimental models of eggs in areas where there are no cuckoos or other nest parasites. Conversely, cuckoos do not produce mimetic eggs when they parasitize hosts that do not discriminate differences in coloration or markings of eggs. The same applies to mimicry of the hosts' young. It is also clear that mimicry by cuckoos can stimulate the evolution of exaggerated markings on the host's eggs and young. Cuckoos can in turn evolve exaggerated mimicry.

Despite the intricacy and exaggeration of mimicry and discrimination, neither party is perfect. Hosts sometimes fail to detect and thus fail to eject cuckoos' eggs or young; cuckoos do not produce perfect mimicry. Humans can usually detect a cuckoo's egg in a host's nest despite some variation in the patterns on hosts' eggs. For one thing, cuckoo's eggs are usually a bit larger than their smaller hosts' eggs, albeit much smaller than the eggs of other birds the size of cuckoos. Hosts also occasionally eject one of their own eggs, apparently mistakenly confusing it for a cuckoo's. Cuckoos sometimes lay an egg in the nest of an inappropriate host, apparently mistakenly confusing it with a host's.

All the ingredients of signal detection are present in this case, and several studies have used signal detection to explain the behavior of hosts and cuckoos. The potential hosts are usually considered to be the receivers of signals from the eggs of the cuckoos. A correct detection then consists of ejecting a cuckoo's egg (or an experimental model) from its nest; a missed detection is accepting a cuckoo's egg; a false alarm is ejecting one of the host's own eggs; and a correct rejection is taking no action when there is no cuckoo's egg in the nest. In the case of cuckoos, the probability of a signal is often low because only a few nests are parasitized by cuckoos.

Some other brood parasites parasitize large proportions of some of their host species' nests. The North American Brown-headed Cowbird, another obligate brood parasite, is implicated in reducing the nesting success of some of its hosts nearly to the point of extinction of the host species. These interactions that seem out of balance apparently have resulted in part from expanding populations of cowbirds exploiting environments altered by humans. In these cases it seems unlikely that an evolutionary equilibrium has been reached in the interaction of host and parasite.

In the case of the European Cuckoo, a missed detection has catastrophic consequences for the host, which loses an entire season's reproduction. A host, on the other hand, does not work any harder raising a cuckoo chick than it would raising its own brood, so the host might not suffer a decrease in survival as a result of fostering a chick. A false alarm is less severe, because the host only loses one of its own eggs and still has three to five eggs of its own remaining. Again, a correct rejection (*rejecting* the signal from a cuckoo's egg by *not ejecting* an egg when no cuckoo's egg is present) is tantamount to life without communication. A correct detection involves some time and effort not only for the actual ejection but also for the vigilance needed to detect the foreign egg, but it must result in a payoff in terms of survival $\times$ reproduction close to the payoff for a correct rejection. These observations suggest that the rankings of proportional payoffs must often take the form

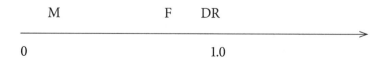

Mimicry to avoid predation likewise implicates signal detection. In the case of Batesian mimicry, a palatable species of prey mimics an unpalatable or toxic species. Predators learn or inherit tendencies to avoid the toxic species and fail to recognize the similar palatable species. As in the case of the cuckoos and their hosts, Batesian mimics and their potential predators show signs of coevolution. Viceroy butterflies mimic their Monarch hosts in coloration, pattern, and in their behavior of flying in the open. Neither party is perfect, however. As astonishing as the mimicry often seems, we can usually distinguish mimic and model once a few defining markings are recognized. In the case of butterflies, birds like flycatchers that catch insects on the wing are frequently involved as predators. The lower survival of experimentally modified butterflies, in comparison to unmodified controls, indicates that mimicry of distasteful models reduces attacks by predators. Nevertheless, bites on the wings of butterflies show that predators attack both mimics and models on occasion.

In Batesian mimicry there are two kinds of communication in play. First, the conspicuous coloration and behavior of an unpalatable or toxic species is an aposematic signal to warn predators to avoid them. Aposematic coloration, like camouflage, is a signal that elicits less, rather than more, response than expected from a receiver. Second, the coloration of the mimic is a strict analogue of camouflage—a signal intended to reduce a predator's response by resembling something unpalatable. The receiver is again a potential predator. The signaler is the mimic, which produces a signal more or less matching the "noise" of the unpalatable model. In this case, a correct detection by a receiver (a potential predator) results in an attack on the palatable mimic. The payoff for the receiver is a meal. A correct rejection (no signal, no response) results in the predator continuing to search elsewhere. As in the previous cases, the payoff is presumably close to "life without communication." A missed detection occurs when the predator fails to recognize a mimic and continues hunting elsewhere, so it too would have a neutral payoff. A false alarm happens when the predator mistakes a model for a mimic and thus attacks unpalatable or toxic prey, often with a payoff demonstrably less than "life without communication." To summarize, for mimicry, the noise is the model, the signal is produced by the mimic, a correct detection is advantageous for a receiver, a missed detection or correct rejection are both neutral, and a false alarm disadvantageous. The signaler benefits from the receiver's missed detections. Exaggeration is progressive approach to perfect mimicry. Presumably the process of acquiring features of the model entails some cost for the signaler (the mimic). Otherwise, there would be no constraint on the evolution of perfection in mimicry. We obtain proportional payoffs for the receiver like these:

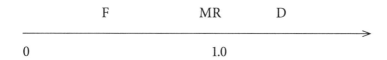

The proportional payoffs for a receiver in the presence of Müllerian mimicry might also fit this pattern. These relative payoffs also resemble those for mate choice and other cases of assessment for acceptance.

ASSESSMENT FOR AVOIDANCE

Advertising signals for defending a resource are expected to deter intrusions by rivals. Singing by birds is known to have this effect, although it can also serve to attract mates. In some species, different songs are apparently used in somewhat different circumstances, some more often directed toward potential mates and some toward rival males. Olfactory marking of territories by mammals often involves odors from urine, feces, or specialized glands. In all of these cases the signaler stands to gain by reducing unnecessary encounters with rivals. For a prospecting or neighboring male seeking an unoccupied territory or a chance to encroach, a correct detection of the signal elicits avoidance. Signalers sometimes indicate their willingness to defend a territory when they are not in condition to do so. A response by a receiver would then be a false alarm—avoiding an area when in fact there was no rival to defend it. A missed detection of an advertising signal could result in an immediate challenge by a vigorous defender. Correct rejection would, in contrast, result in discovery of an opportunity to establish a territory or appropriate a resource. In this case, a false alarm or correct detection would lead to little change in the receiver's survival or reproduction, so these are the neutral payoffs close to "life with communication." The proportional payoffs for such warning advertisements would be:

M   FD   R

$$\xrightarrow{\hspace{10cm}}$$

0   1.0

Other signals of assessment leading to avoidance would also fit this pattern. Such signals could indicate a signaler's ability to defend some resource, with consequences for the four possible outcomes of a response following the pattern just described. Alternatively, such signals might indicate an individual's ability to escape a predator. In contrast, signals of assessment leading to acceptance would have payoffs that fit the pattern for mate choice, as discussed above.

Negotiations involve mutual assessments, either for mutual acceptance or mutual avoidance. Both parties take the roles of receiver and signaler. Nevertheless, overall the interactions are similar to other cases of communication for assessment. Each individual advertises its quality in a way that must balance costs and benefits of exaggeration. Other individuals respond to signals only and have no direct access to opponents' quality. The more exaggerated the signal, the greater the response—for instance, avoiding a challenge to the signaler. An individual of low relative quality would suffer more negative consequences from a missed detection of its opponent's quality. They should set a low threshold for avoidance of opponents. An individual of high quality would suffer less negative consequences from a missed detection of an opponent's quality and set a high threshold for avoidance. As signalers, low-quality individuals might not incur higher marginal costs of signaling. They would also realize less benefit for any level of exaggeration (because of the high thresholds for avoidance by high-quality individuals). Does honest signaling result if we plot survival × reproduction?

DIVERSITY IN PATTERNS OF PAYOFFS

This comparison reveals the diverse relationships that can exist among the payoffs for the four possible outcomes of a receiver's decision to respond or not. In each case payoffs have either negative, approximately neutral, or positive payoffs (see Figure 9.1). In the absence of

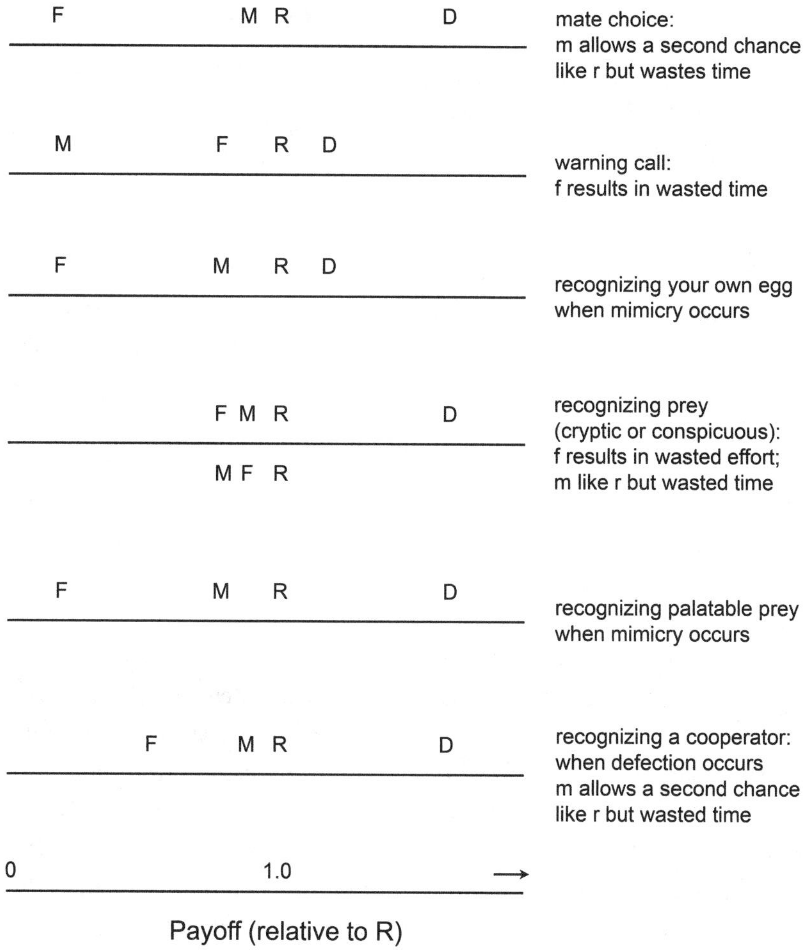

F   M R   D   mate choice:
m allows a second chance
like r but wastes time

M   F   R   D   warning call:
f results in wasted time

F   M   R   D   recognizing your own egg
when mimicry occurs

F M R   D   recognizing prey
(cryptic or conspicuous):
M F R   f results in wasted effort;
m like r but wasted time

F   M R   D   recognizing palatable prey
when mimicry occurs

F   M R   D   recognizing a cooperator:
when defection occurs
m allows a second chance
like r but wasted time

0   1.0   ⟶

Payoff (relative to R)

FIGURE 9.1. Plausible relationships of a receiver's payoffs in six situations. Each time a receiver checks its input, there are four possible outcomes, each of which has a cost or benefit (that is, a payoff) for the receiver: *D, for a correct detection; R, for a correct rejection; F, for a false alarm; and M, for a missed detection* (in the text these payoffs are represented by capital letters for clarity). Even though no one has so far measured all of these payoffs for any example of communication we can surmise their plausible relationships in different situations. This plot compares six situations for communication (plus a variant of one of them), as listed on the right (where d, r, f, and m refer to the four possible outcomes). For each situation the four payoffs are arranged along a scale presented at the bottom, from 0 (survival × reproduction = 0) to some value greater than 1. The payoff for a correct rejection (no response when there is no signal) is set to 1.0. A correct rejection would normally have a payoff nearly the same as the payoff for life without communication. By setting this payoff to 1.0, the other three become proportional changes in reproduction × survival as a result of a receiver's decisions during communication. Notice that the ranking and the relative magnitudes of payoffs for correct detections, false alarms, and missed detections (D, F, and M) vary with the situation. Adapted from Wiley (2013), A receiver-signaler equilibrium.

measurements that would allow greater precision, these three approximate levels for each payoff are all that can be plausibly recognized. The next section in this chapter provides some examples of the measurements available for these payoffs. In no case, so far, have all four been measured.

With four distinct payoffs, each with three possible levels, there are $3^4 = 81$ basic scenarios for the receiver's payoffs during communication in noise. In addition, the level of noise can vary substantially, and so can the probability of a signal when a receiver pays attention. Not all are equally plausible though. Either a correct rejection or a correct detection is always likely to have a payoff close to life without communication and can thus serve to standardize the remaining three payoffs. Erroneous responses, missed detections and false alarms, are likely to have payoffs no greater than life without communication and the remaining correct response is likely to have a payoff no less than this standard. Even with these restrictions there remain 12 possible scenarios by this broad classification for the receiver's payoffs into three categories (low, neutral, and high). Of course accurate measurements could reveal innumerable continuous gradations among these basic options. Accurate measurements would also reduce or eliminate the possibility that two payoffs would be exactly equal. As a result the possibilities might reduce to four basic scenarios for payoffs in noisy communication: the payoff for either error could exceed that for the other and the payoff for either correct response could exceed that for the other.

The five scenarios above include three of these possibilities. The first two, mate choice and alarm signals, exemplify the two situations in which receivers benefit from correct detections, but in one case false alarms have the lowest payoff and in the other missed detections have the lowest. The third and fifth scenarios above exemplify cases in which a correct rejection has a higher payoff than a correct detection—by a marginal amount in the third scenario and by a substantial margin in the fifth. In several cases above, a false alarm seemed likely to have less severe consequences than a missed detection. I have been unable to identify a scenario for communication in which correct rejections would have a payoff greater than or equal to a correct detection but false alarms would have a greater disadvantage than a missed detection.

Although this section has emphasized the patterns of potential pay-offs for a receiver, the expected utility of a receiver's threshold, as formulated earlier, also depends on the probability of a signal and the contrast between noise and signal. The latter determines the probabilities of the receiver's four possible outcomes.

MEASURED BENEFITS FOR RECEIVERS

Before proceeding with this theoretical analysis of coevolution in noisy communication, presenting actual measurements of some of the ten parameters identified above can make the theory more tangible. By now there has been a number of studies, as well as reviews, of the costs and benefits of signaling and of responding to signals. Mate choice has attracted the most attention in this regard and thus is a place to start. A few examples can illustrate possibilities for further investigation. Most important, they indicate that none of the parameters for communication in noise is inherently impossible to measure.

Females can benefit from choosing optimal males for mates either because these males contribute more resources or protection for the female or her offspring or because they contribute good genes that improve the subsequent survival of her offspring of both sexes. These two categories of benefits are often termed *direct* and *indirect* benefits for choosy females, and examples of both have been documented. For instance, female frogs have been shown to benefit by choosing optimal mates, and these benefits can be either direct or indirect. A direct benefit can come from the proportion of a female's eggs that a male fertilizes. A female frog accepts a male by approaching him closely, sometimes even touching him. The male then climbs on her back and grips her by the armpits in a posture called amplexus. Sooner or later this grip results in the female expelling her eggs, which are then fertilized outside her body by sperm expelled by the male on her back. In some species this external fertilization occurs where amplexus occurs. In others the female first carries the male (usually smaller than she is) to a suitable place. One way or the other, a female chooses a place suitable for her eggs to develop. Females often prefer larger males (or sometimes a male of an appropriate size in relation to the female's size). Larger males have the advantage of more sperm and perhaps a better position on the female's back to fertilize her eggs. So a larger male has high quality because a female wastes fewer of her eggs, those that are not fertilized after she expels them. In addition to this advantage of mate choice, females also realize an advantage from good genes of a preferred male. Experiments indicate that when a female's eggs are fertilized by sperm from a male for whom she has shown a preference, the result is more vigorous offspring. In comparison, offspring from eggs fertilized by sperm from a random male grow less rapidly or show other features that indicate lower survival to adulthood. Because these fertilizations are arranged by an experimenter in standardized conditions, the only possible conclusion is that preferred males provide more good genes to improve survival of offspring than do random males. The conditions in a laboratory might not match quantitatively conditions in the field. Nevertheless, there is no clear reason to doubt the greater survival and faster growth rates of progeny that result from female preferences for mates in comparison to those that do not.

High-quality males sometimes are those that provide more nutrients for offspring, a better location for nesting, more protection for offspring, or in other ways contribute to a female's current reproduction. In many insects males present "nuptial gifts," packages of nutrients, to females prior to copulation. Females are more likely to copulate or to copulate longer with males that provide larger gifts. Studies of a number of birds provide other examples. Female Barn Swallows that mate with high-quality males produce young with fewer insect parasites. Overall, more of their young survive to leave the nest. Experiments with cross-fostered chicks indicate that at least some of this effect results from genes that convey inherited resistance to the parasites—in other words, good genes that highly preferred males' convey to their offspring.

Evidence that female Barn Swallows choose males with long tail streamers comes from a tendency of the earliest arriving females in spring to mate with males with the longest streamers. The males, however, have arrived a week or so earlier on the nesting grounds. It is possible that males can judge the best nesting places and compete to defend them. If the males' streamers were important in competition for defense of nesting sites, then females with the first choice of nesting sites would indirectly become the mates of the long-tailed males. Communication (or contests) among males, rather than communication between the sexes, would produce the association between males' traits and their mating success. Females are not passive in this process. They actively select nest sites and thus might determine the conditions for competition among the males. But females would not directly choose males' traits—not in the sense of discriminating the differences among males.

There are other possibilities for good genes from males as well. Greater stability during the process of development would result in more symmetrical sides of the body. There is evidence for some species, including humans, that females prefer more symmetrical mates (those with less "fluctuating asymmetry"), but in other studies this result has been difficult to confirm. There might also be a trade-off between circulating testosterone (which stimulates the development or expression of specifically male behavior and morphology), circulating carotenoid pigments (which are necessary for conspicuous red or yellow coloration), and the strength of immune responses to disease or the intrusion of foreign proteins. The metabolic connections among these three physiological functions in various vertebrate animals are complex. Testosterone is itself a metabolic signal, so its effectiveness depends not only on its concentration (its exaggeration) but also on the threshold for responses in the tissues of the body. The immune system itself participates in noisy communication in the form of responses to diverse signals from endogenous and exogenous proteins. All of these complexities perhaps explain why it has proven difficult to find consistent evidence for good genes for strong immune systems.

MEASURED COSTS FOR RECEIVERS AND SIGNALERS

Choosy females sometimes incur costs that offset any advantage from a high-quality mate. In the investigations of Barn Swallows, an early result was a clear disadvantage for females of a mating with a preferred male. Preferred males with long feathers in their tails bring less food to their mates' nests. Apparently the longer tail is a disadvantage in catching large insects. It might reduce speed or maneuverability, as a result of the extra drag, just enough to make difficult captures less likely. As a result of the reduced help from their mates, females mated to preferred males must work harder to feed their offspring. The availability of insects varies during the nesting season, but the differences in paternal care by preferred and nonpreferred males hold up in comparisons adjusted for the course of the season. Females mated to preferred males nevertheless raise more offspring in a season because they tend to start sooner and, as mentioned above, their offspring receive some resistance to insect parasites.

Males' costs from their preferred morphology or behavior have been estimated in a number of species. Again, the long tails of preferred male Barn Swallows provide the most thoroughly studied example. Anders Møller and his colleagues have demonstrated in two populations and over four years of study that males with long tails survive better over the winter than do males with short tails. The evidence for survival is indirect because it consists of the numbers of males that return the following spring from wintering in Africa. Because males overwhelmingly return to previous sites from year to year, this assumption seems reasonable. Apparently males with longer tails, even though they cannot catch large prey as easily, manage to survive better despite their disadvantage, presumably because they are in better physiological condition. Their better condition allows them to grow longer tails and yet survive better than males in poorer condition.

The pioneering aspect of Møller's study, however, was experimental manipulation of the males' tails. The elongated outer tail feathers can easily be clipped to shorten them. The removed bits can then be glued to other males' tails to lengthen them. Controls for these manipulations consist of clipping and gluing in the same way, without changing the lengths of the birds' tails. Males with extended tail feathers return less frequently the following spring than do controls. Males with shortened tails return most frequently of all. The shortened males have about 7% higher chance of survival and the lengthened ones about 12% lower chance of survival in comparison to the overall mean survival. The controls had about average survival. Evidently, males could not make it to Africa and back as reliably with a lengthened tail, just as they could not catch large insects as easily. A change in tail length of less than 5% (4 millimeters added to or subtracted from an average length of 108 millimeters) resulted in a nearly 20% difference in survival. The marginal cost of tail length in this case is substantial.

It takes a moment of reflection to realize how striking these results are. Although males with naturally long tails survive better over a winter than do males with short tails, all males have lower survival with experimentally lengthened tails, and all have higher survival with experimentally shortened tails. The only possible conclusion is that the length of its tail negatively affects the survival of every male, but some males can produce longer tails and still survive better than can other males. This is exactly the situation predicted by the expression for the cost of exaggeration in Chapter 8.

The picture becomes complicated when the tails of survivors and nonsurvivors are compared. Among males with experimentally lengthened tails, survivors had tails about 10 millimeters longer than did nonsurvivors. The reverse was true for males with shortened tails—survivors had tails about 5 millimeters shorter on average than did nonsurvivors. Among controls, survivors and nonsurvivors did not differ much. In other words, when tails were experimentally lengthened, naturally long-tailed males survived better than those with naturally short tails. When tails were experimentally shortened, naturally short-tailed males survived better. These results, in combination with the previous ones, indicate that the cost of a male's tail is a concave function of its length. Survival must decrease with tail length at first but then increase with length for long tail lengths. As a result, shortening tails would increase the survival of naturally short-tailed males more than it would long-tailed males (so proportionately more short-tailed males would survive). Lengthening tails would conversely decrease survival less for naturally long-tailed males than it would for short-tailed ones (so proportionately more long-tailed males would survive). This conclusion still requires an assumption that survival over the winter is accurately measured by return to previous nesting sites, as mentioned above. Although not emphasized in the original reports, these results provide evidence for a curvilinear relationship between the exaggeration of a signal and the survival of the signaler.

Other situations for which some of the relevant parameters of noisy communication have been estimated include begging by offspring for parental care and mimicry. The latter case is examined further in Chapter 11. First, we continue with an investigation of the fundamental problem of noisy communication. How do signaling and responding to signals coevolve? Is there a joint optimum for signaler and receiver?

10

JOINT OPTIMA FOR SIGNALERS
AND RECEIVERS

INTERDEPENDENCE OF RECEIVER AND SIGNALER

THE PRECEDING CHAPTERS have scouted the terrain around the course of our argument. It is now time to advance to our principal objective—the possibility of a joint optimum in the coevolution of receiver and sender in noisy communication. So far the argument has led to expressions for the utility of a receiver's threshold and the utility of a signaler's exaggeration. Both utilities have been expressed in terms appropriate for an evolutionary analysis, namely survival × reproduction. The receiver's utility depends on the probabilities of the four possible outcomes each time it decides to respond or not. It also depends on the location of its threshold for response, the activation of its sensor that meets its criterion for a decision to respond. It further depends on the activation of its sensor by noise alone and by noise plus a signal. The difference between noise plus signal and noise alone, in combination with the location of its threshold, determine the probabilities of the four possible outcomes for a receiver.

The difference between the levels of noise and noise plus signal in a receiver's sensor is a result of the design of the sensor and the contrast of impinging signal and noise. This contrast in turn depends on the level of the signal broadcast by the signaler—in other words, the exaggeration of the signal. Exaggeration has costs and benefits for a signaler. Greater exaggeration results in higher contrast between a signal and noise. Higher contrast in turn would favor a higher optimal threshold for a receiver, yet at this optimum the receiver could also have a higher probability of responding to a signal. In the cases of mate choice and alarm signals, a greater probability of a response from an appropriate receiver increases the signaler's benefit from producing a signal and thus its utility of exaggeration. As emphasized in Chapter 9, the utility of a receiver's threshold depends on the signaler's exaggeration, and conversely the utility of the signaler's exaggeration depends on an appropriate receiver's threshold.

Increasing exaggeration has diminishing returns (assuming the variance of a signal does not increase as fast as the mean level) for both parties. For a signaler the marginal costs of exaggeration might remain constant or increase with further exaggeration, but the probability of responses by the receiver and hence the signaler's benefit from a signal would increase progressively more slowly. (Figure 7.1 helps in thinking about the probability of responses.) As a result, increasing exaggeration would progressively increase the signaler's costs and decrease its benefits. The receiver faces diminishing returns as well. Raising its threshold reduces the probability of false alarms but raises its probability of missed detections. Because the tails of the distributions of noise alone and of signal plus noise are likely to be concave (with prolonged tails), the probability of missed detection at first decreases faster with an increase in the location of the receiver's threshold than the probability of false alarms increases. Eventually, however, this situation reverses, and the probability of false alarms decreases faster than the probability of missed detections increases. Both signaler and receiver are thus expected to reach optimal performance at intermediate levels of exaggeration and signal. Because the level of the signaler's exaggeration and the location of the receiver's threshold depend on each other, the optimal levels of each are presumably dependent on the other.

A way to investigate this interdependence in adjusting the receiver's and the signaler's optima would be to calculate both the receiver's optima as a general function of the signaler's exaggeration and the signaler's optima as a general function of the receiver's threshold. It would then be possible to explore values of exaggeration of a signal to seek combinations of the signaler's exaggeration and the receiver's threshold at which both parties have optimal performance in relation to each other. If any such a combination is found, it would be a mutual optimum, so neither party could deviate unilaterally without reducing its utility. In mathematics this situation is called a Nash equilibrium.

Chapter 9 described the utilities of the receiver's threshold and the signaler's exaggeration. The first step in finding the optima for each party is to calculate its utility for all values of a receiver's threshold and a signaler's exaggeration. The results of such calculations could be plotted like a topographic map, with utility represented by elevation (on the z axis) and threshold and exaggeration by latitude and longitude (on the y and x axes). Before proceeding to investigate each party's landscape of utility in this way, it is first necessary to consider the mathematical description of noise and other preliminaries.

FORMULATING NOISE

The inherent property of noise is its variability. When subject to noise alone, the activation of a receiver's sensor fluctuates irregularly. A signal reaching a receiver might also vary, so equal signals at the source would arrive at the receiver with some variation and thus would also produce different levels of activation of the receiver's sensor. Furthermore, the signaler itself often introduces some variability in signals, whenever it does not produce exactly the same signal even when conditions, such as its condition or its perception of a predator, are identical. It is this variability that makes it impossible for a receiver to separate noise and signal without errors.

Variation that results from innumerable small deviations from a norm produces a symmetrical distribution of values around a mean. Often called an error function or normal distribution in mathematics and statistics, it is the familiar bell-shaped curve. The normal distribution in mathematics is a symmetrical curve with a maximum at 0 and tails that extend to infinitely high positive values on one side and infinitely low negative values on the other. It represents the probability that a measurable feature takes on any particular value from negative infinity to positive infinity. A defining characteristic of this distribution is the area under it. If the probabilities of all values from negative infinity to positive infinity are added together, the result must equal 1.0. A probability of 1.0 means certainty, so if the feature must have one and only one value at a time, the probabilities of all possible values must sum up to 1.0. The sum of all the probabilities is the area under the normal distribution, or in mathematical terms the integral of the normal distribution from negative to positive infinity. For a normal distribution the highest probability is the mean (average) value. So the most likely value in relation to any other single value is the mean. Yet it is also true that values near the mean are not the most likely in relation to all other values collectively. When noise in communication takes the form of increased variation in the properties of a signal—increased variation around a mean—it can closely fit a normal distribution. Noise from the degradation of signals during transmission or from a degree of unpredictability in the signaler would often take this form.

Noise could often lack symmetry and thus fail to meet the conditions of a normal distribution. Noise from extraneous sources might often take this form. Other sources of noise, such as irregular spontaneous activity of the sensor itself and variation in the production and transmission of signals, might also. Nevertheless, the result of adding numerous sources of such irregular variation would often yield an approximately normal distribution. Consequently, this distribution might adequately describe an aggregate of all sources of noise for any signal in a particular situation.

By aggregating all sources of noise in one normal distribution, it is then possible to set the variance of a signal itself to 0. When such a signal appropriate for a sensor is added to noise, it raises the mean level of activity in the receptor without increasing its variation. In mathematical terms the combined distribution of noise plus signal has a mean that is equal to the sum of the two means and a variance that is equal to the sum of the two variances. In the following analyses, a signal changes the mean level but not the variation in the activity of a receiver's receptor. Signal plus noise has a mean > 0 and variance = 0; noise alone has a mean = 0 and variance > 0.

These distributions require a final adjustment to make them more realistic. The activation of a receiver's receptor can only have values of 0 or higher. To conform to this condition, the following analyses use a truncated and standardized normal distribution for noise—truncated because it only includes values greater than or equal to 0, and standardized because it is adjusted to have the sum of probabilities for all possible values equal to 1.0. When the mode is 0, this probability distribution is sometimes called a half normal distribution. Figure 7.3 illustrates truncated distributions of noise alone and noise plus signal, such as those used for the following analyses.

FORMULATING EXAGGERATION IN SIGNALS

By setting the modal level of activity in the presence of noise alone to 1.0 and holding the variation around this mode constant at a standard deviation equal to 1.0, we can think of all levels of signals in relation to this standard level of noise. If a signal has an exaggeration of 2.0, for instance, it adds that amount to the mean activation of the receiver's sensor by noise alone. It thus produces a modal level of activity in the receiver's sensor equal to 2.0, twice as great as the mode for noise alone. A signal with exaggeration of 4.0 would produce a mean level of activity in the sensor equal to 4.0. By adopting these conventions, the exaggeration of a signal is expressed in units of the standard deviation of noise. A signal with an exaggeration of N units raises the activity in a receiver's sensor by N standard deviations.

This approach focuses on the exaggeration of a signal as it is received (its intensity in relation to the intensity of noise or some other measure of its contrast with noise). As explained in Chapter 6, the exaggeration of signals at the source, where emitted by a signaler, is *qualitatively* proportional to their exaggeration at the receiver. If one of two signals is more intense at the source, then it will also be more intense at the receiver. On the other hand, because the intensity of a sound decreases exponentially with distance as a result of spherical spreading but linearly with distance as a result of scattering and absorption, intensities at the source and at the receiver are not *quantitatively* proportional. A sound twice as loud as another sound at the source is not necessarily twice as loud at the receiver. Because the quantitative relationship between exaggeration of a signal at the source and at the receiver is complex, it is convenient to focus on exaggeration at the receiver for our calculations of signalers' and receivers' optimal behavior.

There are also reasons to expect that exaggeration of signals at the receiver is in some circumstances nearly proportional to their exaggeration at the source. For long-range communication this proportionality should apply, at least approximately. Recall that spherical spreading dominates attenuation at close range between signaler and receiver but that proportional absorption dominates at long range. It is long-range communication that results in high levels of noise external to the signaler and receiver. Short-range communication should also have substantial noise, but the levels of exaggeration are often much lower and spherical spreading is more nearly proportional over short distances. In pursuing our search for joint optima between signaler and receiver in noisy communication, we can accept that exaggeration by a signaler is at least approximately proportional to exaggeration of signals at the receiver. The motivation for accepting an approximate proportionality between the exaggeration of signals at the source and the receiver is the assumption that the costs and benefits for a signaler are proportional to the exaggeration of its signals. As discussed in Chapter 8, however, this latter proportionality is likely to be an approximation also. Costs and benefits of signaling might well have curvilinear relationships with exaggeration. How any disproportionality (nonlinearity) of costs and benefits in relation to exaggeration might combine with disproportionality (nonlinearity) of exaggeration at the source and the receiver is anybody's guess at the moment—there simply are no relevant measurements. As a start, the following analysis continues, for simplicity, to assume proportionality of both sorts. Consequently, the following discussion continues to speak of a signaler's exaggeration interchangeably with exaggeration of signals at a receiver.

FORMULATING OTHER PARAMETERS

The other parameters of communication in noise also require numbers. The proportional relationships of the payoffs for the four possible outcomes for a receiver, as discussed in Chapter 9, differ with the situation for communication. Table 10.1 presents some plausible values for these payoffs that fit the proportional relationships for mate choice. Recall that these values are all proportional to "life without communication" (see Chapter 9). A value of 8.0 for the payoff for a correct detection means that the receiver's survival × reproduction is eight times greater than it would be following a correct rejection (no response when no signal occurs), which is probably close to the receiver's survival × reproduction if no communication occurred. Table 10.1 also includes some possible values for a signaler's costs and benefits, again proportional to a standard situation. In the case of the signaler, a standard reference for survival could be its survival when no signal is produced (exaggeration = 0). By setting a signaler's intrinsic survival (i) equal to 1.0, its survival as a function of exaggeration becomes scaled (proportional) to the actual value of intrinsic survival. A marginal survival of −0.01 means that the signaler's survival decreases by 1% for every unit of exaggeration of a signal. The same principles apply to the signaler's benefit. By setting its benefit (survival or reproduction) to 1.0 when no signal is produced, its benefit from signaling becomes scaled to this minimal value (or offset). A benefit of 2.0 means that the signaler's expected reproduction or survival is doubled every time an appropriate receiver correctly detects (and thus responds to) its signal. Finally, Table 10.1 includes some possible values for the probability of producing a signal in any unit of time. For communication during mate choice, as noted in Chapter 9, the probability of a signal might be higher than during vigilance for alarms. The probability that a signaler produces a signal in a unit of time equals the probability that a signal occurs whenever a receiver checks its sensor, in other words, pays attention. The following discussion thus assumes that receivers pay attention when and where signals might occur.

As emphasized above, the values in the table are just plausible proposals to which we must resort in the absence of measurements. Although all of these parameters can be measured, so far no one has attempted to

Table 10.1   Parameters for the analysis of communication in noise. Included are default values for the parameters of communication in mate choice when they are not otherwise specified in the text. Parameters with no default value are variables in the analyses. The values of parameters for benefits and costs in the absence of communication (R, i, and o) are set to 1.0, so the remaining parameters become simple proportions (multiples) of any actual values for these parameters.

Properties of noise

mean level of noise in the receiver's sensor $= 0$
standard deviation of noise $= 1.0$

Receiver's parameters

U_r	receiver's overall utility
D	payoff for a correct detection $= 2.0$
M	payoff for a missed detection $= 1.0$
F	payoff for a false alarm $= 0.5$
R	payoff for a correct rejection $= 1.0$
p_D	probability of a correct detection
p_M	probability of a missed detection $= 1 - p_D$
p_F	probability of a false alarm
p_R	probability of a correct rejection $= 1 - p_F$
t	location of the receiver's threshold (> 0 for the level of activity in a sensor)
p_s	probability of a signal occurring in any unit of time $= 0.5$ (see also below)

Signaler's parameters

U_s	signaler's overall utility
g	signaler's gain as a result of a correct detection by a receiver $= 2.0$
o	offset (benefit when no signal is produced) $= 1.0$
b	benefit from producing a signal $= o + gp_D$
i	intrinsic survival when no signal is produced $= 1.0$
m	marginal change in survival as a result of increasing the level of a signal $= -0.01$
s	survival as a result of producing a signal $= i + me$
p_s	probability of producing a signal in any unit of time $= 0.5$
e	exaggeration (level or magnitude) of a signal > 0

Alteration of a signal during transmission (not included in the current analysis)

attenuation (reduction of signal level or exaggeration) $= 1.0$
degradation (increase in signal variance) $= 1.0$

do so. Perhaps thinking about the implications of some proposed numbers can stimulate an enterprising colleague to find an example that permits convenient measurement. If enough such measurements are made, surely some will reveal contrasts between situations for noisy communication such as the contrast between mate choice and vigilance proposed here. With these preliminaries settled we can now proceed with an investigation of the landscapes of utilities for a receiver or a signaler separately as well as for both jointly.

CALCULATING A RECEIVER'S OPTIMAL PERFORMANCE

Previous chapters have identified general expressions for the utilities of a receiver and a signaler in noisy conditions. The utility of a receiver's threshold for response, for a specified level of noise, is a function, f_1, of the location of its threshold (the level of activity in its sensor) and the exaggeration of the signal. This general relationship can be written as

$$U_r = f_1(t, e).$$

The utility of a signaler's exaggeration, again for a specified level of noise, is also a function, f_2, of the location of the receiver's threshold and the signaler's exaggeration,

$$U_s = f_2(t, e).$$

As seen in Chapter 8, for the receiver any value of e (exaggeration) is likely to have some value of t (the location of its threshold) that optimizes U_r (its utility). A value of t above this point would result in too many missed detections (and thus lower U_r), and a value below this point would result in too many false alarms and too few correct detections (and thus also lower U_r). Rather than calculate the optimal value of t for just one value of e, it would help to know the optimal value of t for every possible value of exaggeration, e. In other words, it would help to calculate t* (the optimal value of t) as a function of e.

To find this expression for the optimal threshold as a function of exaggeration, we can use simple calculus to find the maximum of a function. We want the value of t that maximizes the value of U_r, which as we have seen depends on both t and e. To find this value of U_r we differentiate the function $U_r = f(t,e)$ with respect to t, set the resulting expression to 0 ($dU_r / dt = 0$), and solve for the first root of this equation to obtain $t^* = f(e)$. By differentiating we find the slope of U_r as a function of t (the rate at which U_r changes as t changes) for any value of e. By setting the expression for the slope at 0, we solve for a value of t that either maximizes or minimizes the value of U_r (when a function reaches a maximum or minimum its slope is horizontal, in other words, it equals 0). By inspecting the shape of U_r (or by differentiating twice to obtain the second derivative, a measure of the acceleration of U_r as t changes), we can determine if the value calculated is in fact a maximum rather than a minimum. If the acceleration is negative at the point where the rate equals 0, the slope is changing from positive to negative (the curve is about to stop rising and start falling), so U_r is concave downward. Negative acceleration with 0 rate means a function has reached a maximum rather than a minimum. As a result, we can find the location of the optimal threshold (t^*) for a response as a function of the exaggeration of a signal for any level of noise.

To put an actual number on this optimal threshold we also need values for the payoffs of the four possible outcomes of a response and for the probability of a signal whenever the receiver checks its receptor (pays attention). If we know these values and the level of noise are set, then with the calculus just summarized we can determine the optimal threshold for any exaggeration of a signal.

It turns out that we need numbers for the payoffs for a more general reason too. In principal it might be possible to solve the differential equation for the general case, so we would have an expression for the optimal threshold with all the relevant parameters retained as variables. This "analytical" solution would be much more satisfying mathematically. We could see at a glance how the optimal threshold would respond to changes in any one variable or any combination of the variables, instead of having to calculate the optimal threshold for particular combinations of parameters. We would have a general solution rather than a specific solution. If we are limited to specific solutions, then general understanding can emerge only by induction after calculating the optimum threshold for many specific cases.

An analytical solution is beyond our grasp, however, because $U_r = f(t, e)$ cannot be differentiated in the general case. U_r varies with the probabilities of the four outcomes and the probability of a signal occurring, and these probabilities depend on the location of the receiver's threshold. If we use normal distributions of probabilities, as explained above, U_r has no simple derivative by differentiation as t varies. Our only recourse is to use numerical rather than analytical methods to differentiate U_r. In essence, we must find the optimal value of t for every value of e by differentiating $U_r = f(t, e)$ by trial and error. To use numerical methods we abandon any prospect of an elegant general solution and resort to induction by number crunching. To do so we must have numbers for all of the "variables" except t and e in the equation for $U_r = f(t,e)$.

As explained below, it turns out that our intuition is correct in expecting a single optimal threshold for any level of exaggeration of a signal. The optimum depends on the values we set for the other parameters of both the receiver's and the signaler's performance, but there is always only one optimum. This result is clear from plots of a receiver's optimal threshold against the exaggeration of a signal. For an even better view of how a receiver's utility varies in relation to its threshold and a signal's exaggeration, we can calculate a three-dimensional plot of the receiver's utility for different combinations of its threshold and the exaggeration of a signal.

A RECEIVER'S ADAPTIVE LANDSCAPE

If we plot the utility of the receiver's threshold for a range of values of t and e we obtain an adaptive landscape for the receiver. Read like a topographic map, our plot shows in relief where the receiver's utility rises and falls as t, the location of its threshold, and e, the exaggeration of the signal, varies (see Figures 10.1 and 10.2). The curving lines connect points with the same elevation (values of the receiver's utility). They show that, over wide areas of this landscape, the receiver's utility changes only slightly as either its threshold or a signal's exaggeration change. In these "plains" and "plateaus" of the plot, changes in threshold and exaggeration almost exactly compensate for each other.

To comprehend this point, it helps to walk through one of these plains in the landscape (Figure 10.1). For instance, in the upper left-hand corner, exaggeration is high but the receiver's threshold is low. Here the threshold is on the left-hand tail of the probability distribution for activity by the receiver's sensor in the presence of a signal. In other words, almost all

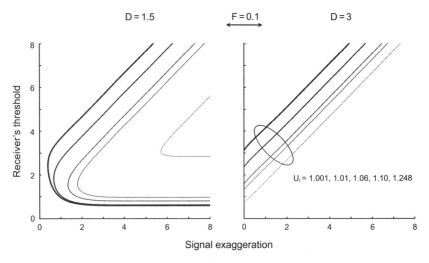

FIGURE 10.1. Landscapes of a receiver's expected utility for combinations of its threshold *(vertical axis)* and exaggeration of a signal *(horizontal axis)*. The lines show different "elevations" of a receiver's utility U_r (heaviest to lightest lines correspond to values of 1.001, 1.01, 1.06, 1.10, and 1.248). Just as on a topographic map, places with closely spaced lines are steep "slopes" where utility changes rapidly; places with widely spaced lines are "plateaus" where utility hardly changes. In these two plots, a false alarm is assumed to have a large disadvantage for a receiver: in both cases the payoff (F) is 0.1, only 10% of life without communication. A correct detection in contrast is advantageous: the payoff (D) is 1.5 *(left plot)* and 3.0 *(right plot)*, 150% and 300% better than life without communication, respectively. All figures in this chapter are adapted from Wiley (2013), A receiver-signaler equilibrium.

signals result in activity above threshold. At points where the threshold is less than 1 (less than one standard deviation above the mean level of noise alone), the threshold is well within the "bell" for activation of the receiver's sensor by noise alone. In other words, in many intervals with noise alone the activation exceeds the threshold and false alarms occur. The receiver thus responds to almost all signals, as well as frequently to noise alone (false alarms), but almost never misses a signal. Because the probability distribution is almost flat far out on its tail, raising its threshold changes its chances of a correct detection or a missed detection only slightly. When the threshold is below 1 (one standard deviation above the mean noise level), the main consequence of raising the threshold is a decrease in false alarms. Above a value of 2, however, the threshold is on the tail of the probability distribution for noise alone, so further in-

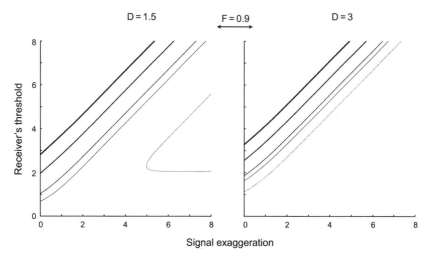

FIGURE 10.2. More landscapes of a receiver's expected utility for combinations of its threshold and exaggeration of a signal (see Figure 10.2). In these plots the payoff for a correction detection (D) is the same as in the corresponding plot in Figure 10.2, but in both of these plots the payoff for a false alarm (F) is 0.9. In other words, a false alarm has a relatively minor disadvantage for a receiver (90% of life without communication).

creases in the threshold make only slight differences in the probability of false alarms.

In other plains or plateaus of the landscape for the receiver's utility, changes in the receiver's threshold or the signal's exaggeration alter the probabilities of the four possible outcomes for a receiver only slightly. Another instructive walk crosses the plain in the lower right-hand corner of the plot. Here the receiver's threshold is low, but the signal is exaggerated to the point that the mean activation of the receiver's sensor is many times higher than 1 (many standard deviations above the mean activation by the signal alone).

The most interesting feature of the landscape, however, is the steep rise and slight ridge that run more or less diagonally across the plot from lower left to upper right. This diagonal lies near the locus of points where the threshold equals the exaggeration of the signal. Near these points the threshold has a large effect on the receiver's utility. Small changes in the location of the threshold in relation to the exaggeration of the signal move the threshold up or over the "bell" of the probability distribution of

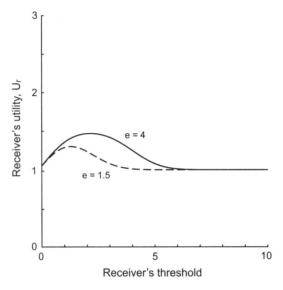

FIGURE 10.3. Two excursions across the landscape of a receiver's utility, each along one "longitude" with signal exaggeration (e) either 2 (solid line) or 4 (dashed line). In these calculations the payoff for a correct detection (D) is set at 3.0 and the payoff for a false alarm (F) at 0.1 (see right-hand plot in Figure 10.2). In each transect the receiver's utility reaches a maximum at a single value for its threshold.

activity by the receiver's sensor in the presence of a signal plus noise. Consequently, small changes in the location of the threshold in relation to exaggeration can produce large changes in the probabilities of correct detections and missed detections. Along this slope, changes in the receiver's threshold produce large changes in the receiver's utility.

At the top of this steep slope there is a gentle ridge, a line of the highest values for the receiver's utility revealed by the contours that loop diagonally downward from the upper right-hand corner and back. "Southeast" of this ridge lies a plateau, nearly (but not quite) flat, as already discussed. If you were to walk entirely across this landscape from top to bottom along a "longitude" of constant signal exaggeration, the elevation of this ridge could be plotted. Figure 10.3 shows two such excursions across the landscape at two different values of exaggeration. In each case, with exaggeration held constant, the receiver's utility rises and then falls to a plain. For higher values of exaggeration (notice that the values chosen for Figure 10.3 all fall in the left half of Figure 10.1), the crest of the ridge comes at a higher value for the threshold and also reaches a

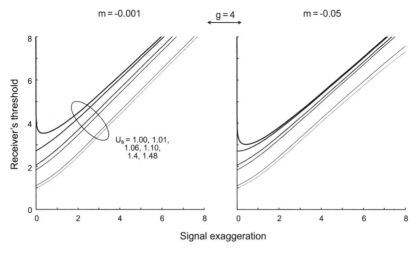

FIGURE 10.4. Landscapes of a signaler's expected utility for combinations of the exaggeration of its signal *(horizontal axis)* and an appropriate receiver's threshold *(vertical axis)*. The lines show different "elevations" of a signaler's utility U_s (heaviest to lightest lines correspond to values of 1.00, 1.01, 1.06, 1.1, 1.4, and 1.48). Just as in the landscape of a receiver's utility, there is a diagonal zone with a steep slope where the signaler's utility changes most. These two plots show calculations for two different levels of a signaler's marginal cost for each unit of exaggeration, either a low cost, $m = -0.001$ *(left)*, or a much higher cost, $m = -0.05$ *(right)*. These costs are respectively 1/10 of 1% and 5% of the signaler's survival for life without communication. In both plots the signaler's benefit (gain) from a response by an appropriate receiver is relatively high, $g = 4$ times life without communication. As usual all parameters not otherwise specified have default values in these calculations (see Table 10.1).

higher value of the receiver's utility. The crest is the receiver's optimal threshold for each case of exaggeration of a signal.

A SIGNALER'S ADAPTIVE LANDSCAPE

The analogous plot for the signaler shows the signaler's utility for different values of its exaggeration and the signaler's threshold for response. Recall that the signaler's utility depends on the receiver's threshold, because the receiver's probability of a correct detection determines the probability that a signal evokes a response from the receiver. The signaler's benefit from a signal depends on a response from an appropriate receiver.

The signaler's landscape, like the receiver's, has large areas of nearly flat topography where the contours are widely spaced (see Figures 10.4

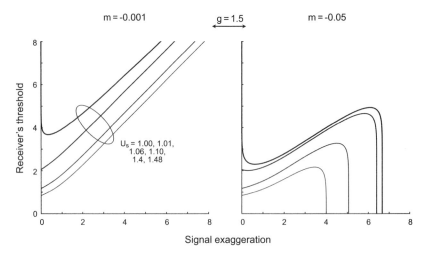

FIGURE 10.5. More landscapes of a signaler's expected utility for combinations of its exaggeration and the threshold of a receiver (see Figure 10.4). In these plots the marginal cost of exaggeration ($m_)$ is the same as in the corresponding plots in Figure 10.2, but in both of these plots the signaler's benefit (gain) from a response is much lower, g = 1.5 times life without communication.

and 10.5). There is also a steep rise along a diagonal from lower left to upper right. The plains result from compensating effects of changes in exaggeration and resulting changes in the receiver's threshold. When exaggeration is high and the receiver's threshold is low, in the lower right-hand part of the plot, the threshold lies on the left-hand tail of the distribution of activity in the receiver's sensor when a signal occurs. Nearly all signals result in a response. In this area neither further exaggeration nor a change in threshold has much influence on the probabilities of correct detection or missed detection. In the upper left-hand corner, in contrast, exaggeration is low but the receiver's threshold is high. Almost no signals elicit responses because the threshold lies far out on the distribution of activity in the receiver's sensor in the presence of a signal, although in this area it is far to the *right* of the mode. The signaler's utility in this area is lower than in the lower right-hand corner of the plot. Yet because the tails of the distributions are relatively flat in both areas, changes in the threshold have little effect on the probabilities of correct or missed detections.

The diagonal slope is not in exactly the same location as the corresponding slope in the receiver's adaptive landscape (compare Figures 10.1

and 10.2 with 10.4 and 10.5). There are three differences that have consequences for the location of a joint optimum for signaler and receiver. First,
in comparison with the receiver's landscape, the top of the slope in the
signaler's landscape lies a little closer to the actual diagonal of the plot.
In other words, it lies closer to the line where exaggeration of the
signal exactly equals the location of the threshold (the threshold thus
corresponds with maximal activity in the receiver's sensor when a
signal occurs). Second, the contour for the signaler's utility turn upward in the lower left-hand corner of the plot, instead of to the right in
the receiver's landscape. Consequently, along the horizontal axis there
is no narrow slope in the signaler's utility. When the receiver has a low
threshold for response and thus responds to even the slightest signal, it
pays for the signaler to produce a minimal signal with hardly any exaggeration. Third, in the upper left-hand corner the signaler's utility is less
than 1.0, the standard for life without communication. In this area the
cost of signaling outweighs its benefit from a response by the receiver. It
does not pay for the signaler to produce any signal at all.

Some of the plots of the signaler's landscape show the contours abruptly
plunging on the right-hand side. At this point, further exaggeration of
a signal results in a sharp decline in the signaler's utility because the marginal cost is no longer compensated by gains in benefits from the receiver's responses. Presumably all of the plots would reach this point
somewhere, but it most of the cases plotted here this upper limit for exaggeration occurs beyond the right-hand edge of the plot.

The signaler's landscape has a diagonal ridge where the signaler's utility
reaches a maximum for any level of exaggeration. By walking across this
landscape (Figure 10.6)—much as we did for the receiver's landscape, but
this time from left to right along a "latitude"—we can appreciate the initial slight dip in the signaler's utility at low levels of exaggeration, a rise
leading to a slight crest, and then a wide plateau with a relatively high
utility for the signaler.

Both of these landscapes thus reveal wide areas where changes in
threshold or exaggeration have little influence of either the receiver's
or the signaler's utility. There is also a band lying close to the diagonal,
where the threshold matches the exaggeration, with a steep rise in
utility and a slight ridge with maximal utility for both parties. In the
bottom left-hand corner (low exaggeration and low thresholds close to

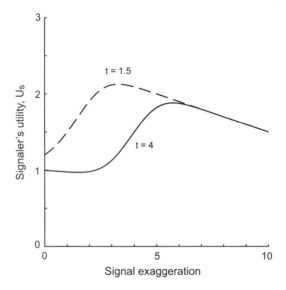

FIGURE 10.6. Two excursions across the landscape of a signaler's utility, each along one "latitude" with the receiver's threshold (t), either 4 (solid line) or 1.5 (dashed line). In these calculations the payoff for a correct detection (D) is set at 3.0 and the payoff for a false alarm (F) at 0.1. The signaler's benefit from a response (g) is 4.0 and its marginal cost of exaggeration (m) is −0.05. In each transect the receiver's utility reaches a maximum at a single value for its threshold.

the mode of activity in a receiver's sensor with noise alone) the utilities of both parties change sharply. Changes also occur rapidly near the horizontal axis for the receiver's utility (where the threshold is less than or near the level of noise) and near the vertical axis for the signaler's utility (where exaggeration produces no benefit). Our calculations have revealed some striking features of communication in noise from the viewpoints of both receiver and signaler.

A RECEIVER'S AND SIGNALER'S JOINT OPTIMUM

Our next step is to calculate the precise locations of the ridges of maximal utility for both receiver and signaler. These lines of maxima describe the optimal performances of the two parties. They indicate the signaler's best level of exaggeration for any level of the receiver's threshold, on the one hand, and the receiver's best threshold for any level of the signaler's exaggeration, on the other. Then we can explore the possibility of an intersection of these lines at a joint optimum. This joint optimum, if it exists, indicates the best that each party can do provided the other party does its best too.

We can find these lines of maximal utility by applying some calculus to the relationship between the exaggeration of a signal and the receiver's maximal utility. A mathematician might say that we compute the partial derivative of the signaler's utility with respect to the exaggeration of the signal (with the receiver's threshold constant). When we compute this partial derivative we obtain the slope of the change in utility at any level of exaggeration (with all other parameters held constant). By setting this slope to 0 we can solve the resulting equation for the level of exaggeration that maximizes (or minimizes) the signaler's utility for any level of the receiver's threshold. A second differentiation can confirm that the solution is a maximum rather than a minimum. Inspection of the landscape for the signaler's utility can also confirm the location of any maxima.

Because the signaler's utility depends on the probability of a response from the receiver, finding the maximal utility as exaggeration changes requires calculation of the probabilities of the receiver's four possible outcomes at every level of exaggeration. These probabilities, as discussed above, depend on the probability distributions of activity in the receiver's sensor in the presence of noise alone and in the presence of a signal plus noise. The differentiation thus requires numerical approximation, a laborious process conveniently handled by programs for mathematical manipulations, such as Wolfram Research's Mathematica.

Once we can find the optimal exaggeration for any one level of the receiver's threshold, repetition of this process can find the optima for every level of the threshold. In practice we can search a range of thresholds, for instance, from 0 to some value many times the standard deviation of the noise. The result is the optimal exaggeration of a signal, e^*, as a function of the location of a receiver's threshold, t, for a particular set of payoffs and frequency of signals (see Figure 10.7, top left).

The same steps can also give us the threshold that maximizes a receiver's utility for any level of exaggeration of a signal. This time we compute the partial derivative of the *receiver's* utility with respect to its threshold, with the exaggeration of the signal (and other parameters) held constant. After setting the derivative equal to 0 and checking for a

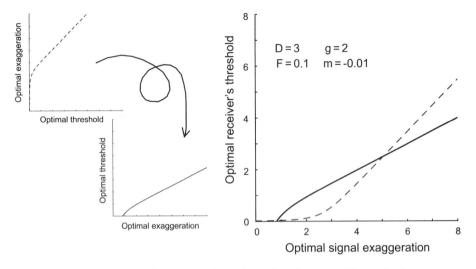

FIGURE 10.7. Each party's optimum depends on the other party. The optimal threshold for a receiver depends on the exaggeration of a signal *(upper left-hand plot)*. Conversely, the optimal level of exaggeration for a signaler depends on the receiver's threshold *(lower left-hand plot)*. In these plots, the receiver's payoff for a correct detection (D) is 3, its payoff for a false alarm (F) 0.1; a signaler's marginal cost for each unit of exaggeration (m) is −0.01, and its benefit from an appropriate receiver's response (g) is 3. The *right-hand plot* is a combination of these two (by switching the axes of the upper left-hand plot), which shows where the receiver's optimal threshold and the signaler's optimal exaggeration coincide. At these joint optima, neither party can unilaterally improve its performance provided the other also does the best it can. In this case there are two such optima. At one of these joint optima, signals are somewhat exaggerated and receivers somewhat selective. At the other joint optimum, communication has almost collapsed. Receivers respond without regard to signals, and signalers make little effort to signal. Further analysis indicates that these marginal optima, when they occur, are not targets for natural selection.

maximum we can find the optimal threshold for any level of exaggeration. Repetition of this process can then give us the optimal threshold for every level of exaggeration. The result is the optimal location of the receiver's threshold, t*, as a function of the exaggeration of a signal, e, for a particular set of payoffs and frequency of signals (see Figure 10.7, bottom left).

The interesting point comes when we superimpose these two plots, e* = f(t) and t* = f(e). To do so, however, we must interchange the axes of one of the plots. For example, exchange the axes for the plot of the optimal exaggeration of a signal: e* = f(t) → t = f(e*). Then superimpose it on the plot of the optimal threshold, t* = f(e). Any point where the two

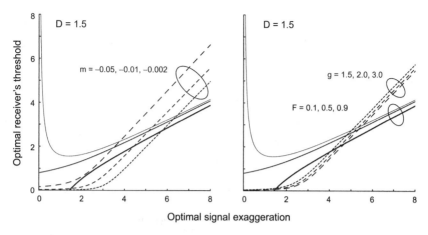

FIGURE 10.8. Overlays of optima for a receiver and for a signaler. These plots reveal how variation in the two parties' parameters affect the possibilities for joint optima. In both plots solid lines indicate possibilities for the receiver's optimal thresholds and dashed lines possibilities for the signaler's optimal exaggeration. In both plots, the receiver's payoff for a correct detection is low, one and a half times life without communication (D = 1.5). The three *solid lines* show the receiver's optimal thresholds for three payoffs for a false alarm (F). The *dashed lines* show the signaler's optimal exaggeration for three levels of the signaler's benefit from a response (g, *right-hand plot*) or for three levels of the signaler's marginal cost of exaggeration (m, *left-hand plot*). Notice that some combinations of signaler's and receiver's lines cross once or twice, and some combinations just approach each other and then separate without crossing. Depending on the payoffs (costs and benefits) for each party, there can be two, one, or no joint optima for noisy communication.

lines cross is the sort of joint optimum we were looking for (see Figure 10.7, right). At a crossing, an exaggeration that is optimal for a signaler occurs at a threshold that is optimal for a receiver. The converse is also true: an optimal threshold for a receiver coincides with an optimal exaggeration for a signaler. At such a point neither party can unilaterally shift its behavior without reducing its utility. Furthermore, if either party did shift, then the other would have to adjust its behavior to regain an optimum for itself, with the result that the first would also have to adjust its behavior, and so forth. Eventually both would move back to the original joint optimum. A crossing is thus a stable combination of a signaler's exaggeration and a receiver's threshold.

Examining these lines of the receiver's and the signaler's maximal utilities (see Figure 10.8), we see that they lie just where we expect—along

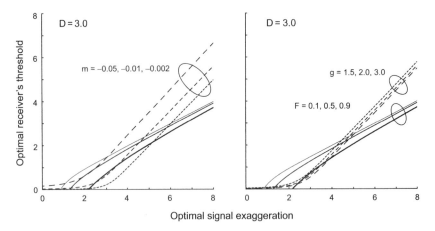

FIGURE 10.9. More overlays of optima for receiver and signaler expand the possibilities for noisy communication. In these two plots the receiver's payoff for a correct detection (D = 3.0) is higher than in Figure 10.8. The *solid lines* still show the receiver's optimal thresholds for the same three levels of its payoff for a false alarm (F), and the *dashed lines* show the signaler's optimal thresholds for the same levels of its benefit from a response (g) and for its marginal cost of exaggeration (m). Changing any one cost or benefit for either party can change the location and number of joint optima. As usual, all parameters not otherwise specified have default values in these calculations (see Table 10.1).

the ridges of the adaptive landscapes in Figures 10.2, 10.3, 10.4, and 10.5. In a general way the lines follow the diagonal from bottom left toward top right. At high levels of exaggeration, the best threshold for a receiver is also high. The converse also holds: at high levels of a receiver's threshold, the signaler's best exaggeration is high. At low levels of either exaggeration or threshold (except very near the origin at the lower left corner of the plot) the best value of the other is also low.

Having computed the exact lines and then superimposed them, we see that the signaler's and receiver's lines of optima do not have exactly the same shape. They have different slopes. As a result, in most cases, they intersect. In some cases they intersect at only one point (Figure 10.9). This point is a unique joint optimum. At this point, neither the receiver nor the signaler can benefit from a unilateral change in behavior. In mathematics, a joint optimum of this sort is called a Nash equilibrium.

In many cases the lines for the signaler's and receiver's optima intersect twice (compare Figures 10.8 and 10.9). One of these intersections is near the bottom of the plot, where the receiver's threshold is close to 0. In other words, this lower intersection occurs at a point where it pays for the receiver to respond almost always—in other words, almost without regard to the presence of a signal. Communication is close to collapse here. It would still pay for signalers to produce weak signals, but responses would only be weakly associated with the occurrence of signals. This extreme form of communication would presumably be difficult for an observer to notice. It might require extensive study just to ascertain that the receiver responded to signals at a rate that was statistically different than random. In these situations with two joint optima, even if natural selection might lead to either of two different outcomes for communication, only one of the possibilities is likely to attract our attention. A joint optimum with the higher levels of threshold and exaggeration is the only one an observer is likely to notice.

Finally, in some cases the lines for the two parties' optima do not intersect at all (Figure 10.9). The receiver's and the signaler's lines converge but then diverge again without crossing. What can we expect in these cases?

THE COURSE OF EVOLUTION

The intersections of the lines of optima indicate points at which the evolution of communication would reach an equilibrium, a joint optimum (a Nash equilibrium) for the two parties. The landscapes of the receiver's and signaler's utilities, however, should tell us more about the evolution of communication than just this point of attraction. Because the utilities for each party were defined in terms of survival × reproduction, or the approximate probability that an allele would pass to the next generation, the steepness of a landscape at any point should indicate approximately the strength of natural selection there. A line tangential to the landscape at a point would approximate a selection gradient there, the approximate "force" of natural selection on a change in the frequency of any alleles associated with the receiver's threshold or the signaler's exaggeration, respectively.

Resorting to some intensive computation once again, we can calculate the slopes at many points on each landscape. Each slope is a vector,

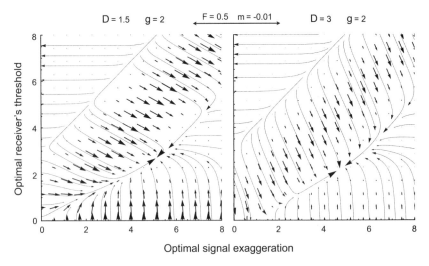

FIGURE 10.10. Paths of evolution for noisy communication. Streamlines calculated from the topography of a receiver's and a signaler's optimal behaviors. Each short arrow indicates the gradient on the landscape, the direction uphill for the receiver and the signaler; the longer the arrow, the steeper the gradient. These arrows (vectors) indicate the strength of natural selection on the receiver and signaler (the length of the arrow) and the direction the natural selection should move a population of signalers and receivers (the direction of the arrow) for the specified combination of parameters (in Table 10.1 if not specified on the plot). The long streamlines combine the shorter arrows to indicate the overall direction of evolution. Notice that the streamlines converge at a single point, a single joint optimum for signaler and receiver, with the corresponding exaggeration and threshold.

an arrow with a direction and magnitude. By combining the vectors at any one point (in other words, by adding the selection gradient for the receiver's threshold and the selection gradient for the signaler's exaggeration), we have a vector that indicates (again approximately) the direction and magnitude of the combined "forces" of natural selection on both parties. This combined gradient indicates the direction in which natural selection "pushes" the joint evolution of receiver and signaler. Furthermore, starting at any one point on the plot, we might follow the vector there to a new point and then add the vector at this second point to the first one. If we continued step by step, adding successive vectors to create an evolutionary streamline, we could follow the probable course of evolution by natural selection through the land-

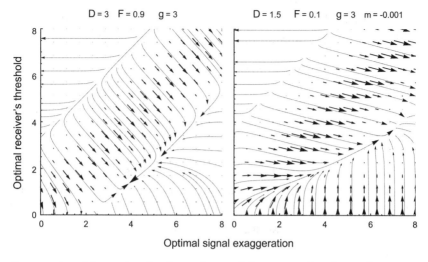

FIGURE 10.11. More paths of evolution for two different sets of conditions for noisy communication. Notice especially the left-hand plot, which corresponds to two lines in the left-hand plot of Figure 10.9 (the upper dashed line and the lower solid line) that converge and diverge but do not actually cross at a joint optimum. Nevertheless, the selection gradients converge on an equilibrial point, a Nash equilibrium where neither party can unilaterally improve. Evolution of communication in noise converges on stable equilibria between signalers and receivers. Notice also that some combinations of thresholds and exaggeration evolve toward a collapse of communication (thresholds or exaggeration = 0). These areas in the plots indicate that communication can only evolve when the properties of receivers and signalers surpass a threshold.

scapes of utilities for a receiver's threshold and a signaler's exaggeration (see Figures 10.10 and 10.11).

Not surprisingly, these paths of evolution lead to the joint optima identified in the previous section. These streamlines from all over the plot show how evolution by natural selection might unfold in these landscapes for threshold and exaggeration. If balls rolled uphill instead of downhill, the lines would show the courses taken by populations rolling through this landscape toward the peaks represented by the joint optima for receivers and signalers.

The streamlines emphasize the importance of the joint optima in three ways. First, in those cases with two joint optima, the one near the bottom edge of the plot where thresholds are close to 0 have small *catchments*, small areas from which evolution by natural selection would reach those low joint optima. In the cases with two optima, the higher joint optima have much larger catchments in the landscape of possible thresholds and exaggeration.

Second, in the situations without any joint optima, because the lines of optimal thresholds and optimal exaggeration do not intersect, there nevertheless exist strong points of attraction close to but somewhat above the places where the two lines of optima reach their closest approach.

Third, the joint optima, including the attraction points, are Nash equilibria—points at which neither party can unilaterally change its behavior without a decrease in its own utility provided the other party also behaves in the best way it can. These equilibria are not Pareto points where both parties each reach their maximum utilities. In the cases of attraction points between lines of optima that converge but do not intersect, it is easy to see that both parties would do better on their lines of optima close by. Nevertheless, neither can move there without a decrease in expected utility provided its partner also does the best it can. Likewise, in cases in which the lines of optima actually intersect, none of the joint equilibria is a Pareto point. Both the receiver's landscape for expected utility and the signaler's landscape rise along the diagonal ridges toward the upper right-hand part of the plots. Each party could separately attain maxima toward (and beyond) the right-hand edge of the plot. Nevertheless, the streamlines emphasize that both parties evolve toward a joint Nash equilibrium where each does as well as it can provided the other does so also.

The sizes of the arrows show that each of the joint equilibrium is approached by selection gradients that are asymmetrical in strength. The plateaus in the landscapes of the receiver's and signaler's expected utilities slope only gradually toward the joint optima. The streamlines and vectors, on the other hand, show that these plateaus are not precisely flat—they slope upward, albeit gradually, toward the joint optima. Their gentle slopes indicate, however, that evolution by natural selection would proceed slowly across these plains or plateaus or even be diverted by other, unknown influences of selection or constraints on genetic changes that would affect the same alleles but for reasons that had nothing to do with communication. These epistatic and pleiotropic effects of genes might, in other words, prevent or retard the eventual attainment of the joint optimum for communication, especially on the plateaus.

COMPARISONS OF MATE CHOICE AND VIGILANCE

The preceding section has developed an analysis of mate choice as an example of communication in noise, and an earlier section identified mate choice as an example of one of at least three scenarios for communication by two parties in noise. Scenarios were divided into those with benefits for the signaler as a result of a correct detection of a signal by the receiver, the usual case for communication, and those with benefits for the signaler as a result of a missed detection, the case of mimicry. When correct detections have benefits for the signaler, then noisy communication can be further subdivided by whether or not false alarms or missed detections have more serious consequences for the receiver. Mate choice fits the scenario with benefits for signalers as a result of correct detections by receivers and, furthermore, fits the scenario for lower payoffs for a false alarm than for a missed detection. A false alarm consists of accepting a low-quality mate; a missed detection involves passing up a high-quality mate. The first might compromise reproductive success, whereas the latter might only require a bit more searching with a chance to find another high-quality mate.

Responses to an alarm signal, like choice of a mate, could have advantages for the signaler. An alarm call might serve to recruit the receiver to help in repulsing a predator, or it might have indirect advantages for spreading the signaler's alleles if it increased the survival of close kin. On the other hand, in contrast to mate choice, a missed detection is likely to have much greater consequences for a receiver than a false alarm. As discussed in Chapter 9, a missed detection would mean exposure to a predator; a false alarm would mean a short interruption of other activities such as feeding. Mate choice and alarm signals could thus be designated: *signaler (CD+) receiver (FA–)* and *signaler (CD+) receiver (MD–)* to indicate the characteristic influences on the expected utilities of the signaler and receiver.

Mate choice and alarm signals probably also differ in the frequency of signals. Females searching for mates often encounter signals from high-quality males at high rates, even if less often than those from low-quality males. On the other hand, an alarm signal might only occur at long intervals although a receiver would have to maintain incessant attention.

Chapter 9 discussed these two situations for noisy communication from the perspectives of the receiver and the signaler separately. This discussion suggested that mate choice should lead to receivers with high thresholds and signalers with high exaggeration. Alarm signals should lead to the converse, low thresholds and low exaggeration. Do the calculations of joint optima for the receiver and signaler in these two situations bear out these predictions?

Alarm calls can be analyzed in the same way that mate choice was. After substituting plausible values for the parameters of the receiver's and signaler's expected utilities, as discussed above, we can calculate the landscapes of each party's utility. The procedures are the same as in the analysis of mate choice. Once again we find plateaus and plains flanking a diagonal ridge of optima for the receiver's utility as a function of the signal's exaggeration. Likewise for the signaler's utility as a function of the receiver's threshold. Next we can plot the lines of optima for each party and, with the same trick we used above, switch variables for one of the plots so that we can overlay the lines of optima for receiver and signaler. An intersection of the two lines identifies a joint optimum. As before, sometimes there are two intersections or a zone of convergence and divergence. We can also compute the selection gradients at many places on the landscapes of utilities and sum them to produce streamlines. Remembering that the utilities for receiver and signaler have the same units of survival × reproduction, the streamlines show the probable courses of evolution by natural selection through the landscape of possible thresholds and possible exaggeration. A point of attraction of the streamlines is a joint equilibrium for the behavior of receivers and signalers, for the selected set of parameters (the payoffs, costs, benefits, and frequency of signals).

To allow comparison of the joint equilibria in different situations, it is convenient to plot just these points for joint optima on the same axes

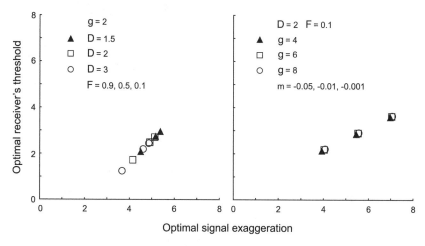

FIGURE 10.12. Joint optima for signaler and receiver during mate choice. Here points indicate the joint optima or stable equilibria for different sets of parameters for signalers and receivers. The lines of optimal thresholds and exaggeration, as well as the vectors and streamlines, are omitted. For conditions with two optima, only the one with greater exaggeration is included (the other invariably is close to a collapse of communication). Each plot compares variation in two parameters. On the *left*, nine points show all possible combinations of three values for the receiver's payoffs for correct detection (responding to an optimal mate) and false alarm (responding to a suboptimal mate). On the *right*, nine points show the combinations of three values for a signaler's benefit from a signal (as a result of a response by a willing mate) and its marginal cost of exaggeration. Other parameters for these calculations have the default values in Table 10.1. As usual, all values are scaled in relation to life without communication. Compare these results with those for warning signals for a predator (Figure 10.13).

of thresholds and exaggeration. Each time we choose a different set of parameters and run through the calculations just summarized, we find a single point to plot, the joint equilibrium under those particular conditions (for reasons already mentioned, it makes sense to include attraction points, regardless of the presence of an actual intersection, and to exclude any optima near the origin). We can thus ask how much a change in one parameter alters the equilibrial point. The right-hand side of Figure 10.12, for instance, compares the equilibrial points for situations that fit the scenario of mate choice, but with different values of the payoff for the signaler's marginal cost of exaggeration—how much each additional unit of exaggeration reduces the signaler's survival. As the cost of exaggeration increases, the signaler's exaggeration (and concomitantly

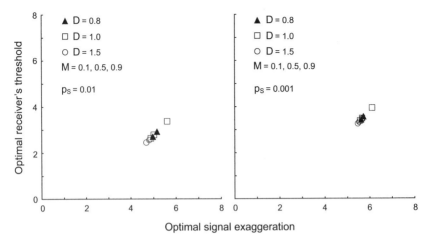

FIGURE 10.13. Joint optima for signaler and receiver during vigilance for warnings of a predator's approach. As in Figure 10.12, each plot compares variation in two parameters. Both plots show nine points for all possible combinations of three values for the receiver's payoffs for correct detection (D, as a result of taking shelter when a predator is near) and for a missed detection (M, as a result of unnecessary exposure to a predator). The two plots show the consequences of different probabilities of a signal (p_s, at least partly as a result of different probabilities of a predator nearby). Values for other parameters in these calculations (see Table 10.1) are $g = 1.5$, $m = -0.01$, and $F = 0.99$, all in line with discussion in the text. Compare these results with those for mate choice (Figure 10.12).

the receiver's threshold) decreases. This result is easily understood. If signals get more expensive (while the benefits of signaling remain the same), signals do better with less. The left-hand side of Figure 10.12 shows a similar result when the receiver's payoff for a false alarm varies. Higher payoffs for a false alarm (closer to the standard 1.0) indicate *less* severe consequences than do lower payoffs (closer to 0). More severe consequences of false alarms result in higher thresholds (and concomitantly higher exaggeration) at the receiver's and signaler's joint equilibrium.

The same sorts of comparisons show similar results for alarm signals. Figure 10.13 indicates that thresholds and exaggeration decrease as the consequences of missed detections become more severe (the receiver's payoff is closer to 0). In addition, the frequency of the alarm signal has an important effect, with higher thresholds and exaggeration when true alarms are less frequent, although all the values for the receiver's payoffs and the signaler's benefits and costs remain constant.

The numbers on the axes of these plots are multiples of the standard deviation of activity in the receiver's sensor in the presence of noise alone (recall also that a signal is assumed not to increase the standard deviation of a sensor's activity). So a value of 6 on the horizontal axis of one of these plots indicates a level of exaggeration six times the standard deviation of noise, which is way out on the tail of the bell-shaped normal distribution of the probabilities of levels of activity in the presence of noise. A change of one unit in exaggeration is equal to one standard deviation of this noise (recall also that the level of exaggeration is the average level of activity in the receiver's sensor in the presence of a signal plus noise). Because the modal level of activity in the presence of noise alone was set at 0, the values of exaggeration on this plot are differences between noise plus signal and noise alone (signal-noise differences, not signal-to-noise ratios). The levels of the receiver's threshold are scaled exactly the same way: a threshold of 6 is a level of activity in the receiver's sensor that is six times the standard deviation of activity with noise alone.

The calculations so far have confirmed that noisy communication can explain the evolution of signals with impressive levels of exaggeration. Noise results in exaggeration. No other features of communication are necessary to explain exaggeration. Noise alone can also explain thresholds for receivers' responses to a signal that are many times higher than the usual variation in receivers' sensors in the absence of the signal. To see whether or not mate choice leads to higher thresholds and exaggeration than alarm signals, we can compare the relevant joint optima in Figures 10.13 and 10.14. With plausible values of the receiver's payoffs and the signaler's costs and benefits, it does appear that mate choice can result in greater exaggeration and higher thresholds, as previously suggested.

The surprise to me is that the difference is not any greater. Levels of exaggeration greater than four times the standard deviation of activity in the presence of noise alone seem impressive enough for mate choice, yet alarm signals seem to reach joint equilibria with levels of exaggeration not much lower. There are several possible solutions to this conundrum. First, of course, is the possibility that the values for the parameters in the expressions for the receiver's and signaler's expected utilities are wrong. After all, they are only plausible hunches, because they have never all been measured for any case of communication. Second, it is also possible that these results are in fact as plausible as the parameters are. Alarm signals might not seem to us as exaggerated as they in fact are. It is my impression that they are—at least sometimes—surprisingly intense. Again, actual measurements are in order. A third possibility is that the probability of a signal is much higher than assumed here. If we consider the noise in this communication to be primarily the deceptive alarm calls of chickadees seeking access to food, the probability of real alarm calls should be expressed as a proportion of the total calls. This probability might reach much higher values, perhaps even higher than the probabilities of signals in mate choice (in the range 0.1 to 0.9, perhaps). Maybe these calculations can spur someone to take on the challenge of making the relevant measurements.

11

EVALUATION AND EXTENSION OF THE MODEL

MATH AND REALITY

THE NEXT STEP is to evaluate the mathematical arguments in the preceding chapter. Aside from the correctness of the math itself, the two crucial issues are the assumptions that precede and the conclusions that immediately follow the math. The assumptions for the preceding calculations include the parameters that determine the properties of receiver, signaler, and noise. As already emphasized, these parameters have been held to the fewest possible. In addition, all are assumed to remain constant or to take the form of simple linear functions. Some alternatives have already been mentioned.

The calculations have also excluded some parameters for complexities of noise. They assume, for instance, that a signal does not increase the variation in the level of activity in a receiver's sensor. Yet it seems likely that signals vary as a result of noise in the signaler and during transmission to a receiver. If so, the standard deviation of noise plus signal would exceed that for noise alone. The distribution of noise (or noise plus signal) might not fit a truncated normal curve, as assumed here. It might, for instance, not have a tail that extends to infinity. If the level of activity in a receiver's sensor had a finite maximum, which seems likely, this level would limit the optimal level of exaggeration in noise. Measurements of noise are the only way to resolve these issues.

A potentially more serious assumption is fixed properties for eavesdropping. As mentioned in Chapter 8, eavesdroppers in communication (such as rivals, parasites, or predators) can have important consequences for the signaler and for the appropriate receiver. Because eavesdroppers are additional evolving organisms, any full analysis of communication should include at least three parties, signaler, receiver, and eavesdropper. In some cases, though, eavesdropping might occur infrequently enough in relation to alternatives for all three parties that assuming a fixed low rate of eavesdropping might not necessarily affect the evolution of the two primary parties, signalers and appropriate receivers. Only relevant measurements can tell.

A fundamental assumption in any analysis of signal detection is the structure of a receiver. As described earlier, a receiver is defined by its three separate operations. First, it must include a mechanism for receiving signals—in other words, a sensor that responds more or less selectively to the presence of signals. Noise then is any source of energy that affects the level of activity in a receiver's sensor other than a signal. It can thus originate, as emphasized in Chapter 1 in the signaler, during transmission of a signal between the signaler and the receiver, or in the sensor itself. Sensors are unlikely ever to achieve absolute selectivity for signals, for the fundamental reason that sensors of any sort are subject to a universal trade-off between sensitivity and selectivity. The less selective a receptor, the more sensitive it is likely to be. This situation is particularly likely in noise. The more sensitive a receptor (the more it responds to low levels of input), the more likely that at some instant irrelevant input can summate with input from a weak signal to put activity of the sensor over a threshold. This trade-off between sensitivity and selectivity is analogous to the trade-off a receiver faces in setting its threshold for response. It is useful, however, to assume that a receptor has a fixed selectivity and that the receiver adjusts its threshold for response.

Second, a receiver must include a mechanism for deciding when a signal has occurred. Engineers call an electrical component with this property a gate, a switch activated by a voltage or current. The preceding analysis has focused on a threshold for response as the simplest mechanism for a decision. A threshold has an all-or-none response—full response if the level of activity is above the threshold and none if the activity is below the threshold. In earlier reviews, I have considered other input-output relationships for decisions in more detail. For instance, a tuning curve (greater output the closer the input comes to a targeted level) results in a trade-off in adjusting the width of tuning that is exactly analogous to the trade-off in adjusting the location of a threshold. It seems likely that even extremely complex criteria for responses, such as those that occur routinely in human cognition, face analogous trade-offs in adjusting the specificity of input that elicits a receiver's response.

Finally, a receiver must include an amplifier. As emphasized earlier, the critical feature of a signal is its limited power. It must provide enough power to influence the activity in a receiver's sensor, but it does not provide all the power for the receiver's response. It is for this reason that a receiver must include an amplifier. An amplifier requires an external source of energy to provide the power for the receiver's response. In organisms the energy comes from the chemical bonds in the food it eats, absorbs, or produces by photosynthesis. In addition, responses need not be immediate or overt. Signals can produce changes in memory or other internal states of an organism, which can in the future affect its overt behavior. A receiver's response to a signal might include the production of another signal that could influence another potential receiver.

These three properties—sensor, criterion, and amplifier—all seem necessary for any receiver, in any form of communication, and furthermore they seem sufficient. Investigating the mechanisms of each of these components in living organisms is the objective for many branches of neuroscience, endocrinology, physiology, and the underlying molecular biology. Yet few investigations so far have explored these mechanisms specifically as adaptations for communication in noise.

The properties of signalers in the preceding calculations are also simplified. One justification for this simplicity is the absence of information. As developed in Chapter 7 and again below, there is almost no information about the benefits or costs of signaling in relation to the exaggeration of signals. For communication in noise, contrast is the crucial feature of signals, not exaggeration in itself. Yet little is known about the relationship between the exaggeration of signals and their contrast with noise. In default, the preceding calculations thus equate exaggeration with contrast of signals. Signal Detection Theory would lead us to expect this equation. As explained in Chapter 6, however, there are other ways a signaler can improve the performance of appropriate receivers. Redundancy and alerting components (or other predictable external associations) also improve performance. Presumably they too can increase the costs of signals. How contrast, redundancy, and predictability interact in affecting the performance of signalers is not known for any case of communication.

In adopting a way to think about the costs and benefits of signals, the preceding analysis has relied on an approximate measure of natural selection, an estimate of the number of progeny contributed by individuals with a particular attribute to the next generation. Chapter 6 discussed the principles of genetics and ontogeny that justify this approach. The result is a procedure for deriving the optimal level of a signaler's exaggeration. As discussed in Chapter 12, this procedure challenges current thinking about the evolution of honesty in communication. The relationship between the cost of signaling and the evolution of honesty needs reconsideration. The approach adopted here makes it clear that there are two ways to measure costs (marginal or absolute costs) and two ways to measure a signaler's quality (intrinsic or marginal survival). As a consequence, there is no simple relationship between cost and honesty of signals. The calculations in Chapter 10 assume linear relationships between cost and exaggeration of signals, but previous discussions of costs imply that these relationships are nonlinear—either convex or concave. The evolution of honesty receives more thorough treatment in Chapter 12.

Before proceeding, a diversion is in order. The calculations thus far have only briefly considered two special cases of communication—mimicry and camouflage. In both of these cases, the signaler benefits by *decreasing* the probability of a response from a receiver rather than increasing it. These cases of communication have also been the only ones to which Signal Detection Theory has been applied previously. Can the approach presented in previous chapters be applied to these special cases? If so, does it yield results similar to previous analyses of these sorts of communication? Affirmative answers would confirm that this approach can apply to the evolution of all forms of noisy communication.

MIMICRY AND CAMOUFLAGE AS SIGNAL DETECTION

Over the past three decades many studies, both theoretical and empirical, have used Signal Detection Theory to interpret the interactions of predators and their prey. Although detection of cryptic (camouflaged) prey was the first problem addressed by Signal Detection Theory, this approach has also been applied to mimicry. As Thomas Getty has pointed out, the evolution of camouflage and the evolution of mimicry have similar mathematical formulations as problems of signal detection. The following sections thus extend the approach in the preceding chapters to the evolution of mimicry and then compare this approach with previous applications of signal detection to the evolution of mimicry and camouflage. Finally, empirical studies of mimicry are consulted for information about the crucial parameters influencing the evolution of mimicry as an example of noisy communication.

Mimicry and camouflage share an important characteristic—the signaler benefits from a signal that reduces the chance of a response from the receiver. Nevertheless, both have the basic features of communication. For mimicry, as for all communication, the signal is a pattern of energy that elicits a response without providing all of the power for the response. Even though the response is a reduction in the frequency of the receiver's action, it is nevertheless a consistent change in the receiver's behavior. If a butterfly in flight did not have colors and patterns resembling those of a toxic species, a predator such as a kingbird would attack it more often. After alighting, if the butterfly did not resemble a leaf, the kingbird would again be more likely to attack. If a European Cuckoo's egg did not resemble a Reed Warbler's, the warbler would more often eject it from its nest. Furthermore, the power for the response comes from the receiver. The receiver makes the decision to respond or not. The mimic has its effect by communicating, not by overpowering the predator or host.

A distinction between mimicry and camouflage is a matter of degree. In both cases the signaler seeks to minimize detection by promoting confusion with objects that a receiver would do better to avoid. For mimicry, a false alarm by a receiver (attacking distasteful or toxic prey or ejecting its own egg) is often strongly disadvantageous. The relative payoff, F in the notation of Chapter 8, is much less than 1. For camouflage, a false alarm (attacking an inedible object) presumably has minor risks, so the relative payoff F must often be only slightly less than 1.

For either mimicry or camouflage to maximize a signaler's expected utility, the signal must have properties that approximate those of noise. For camouflage, noise for the receiver (predator) is the background; for mimicry it is the model. Signals for advertisement, assessment, and warning must also have particular properties to optimize efficacy. In these latter cases, the properties are ones that increase contrast with noise, but in mimicry or camouflage they decrease contrast. Nevertheless, producing a signal with a particular pattern presumably incurs some cost. If there were no cost for mimicry or camouflage, neither direct expenses nor indirect trade-offs, mimicry and camouflage would sooner or later evolve perfection. As in all communication, signals with no costs evolve exaggeration without limit. In a recent review, David Kikuchi and David Pfennig enumerate the various ways that costs and trade-offs might explain the evolution of imperfect mimicry. Even when natural selection for mimics is relaxed (for instance, as a result of the scarcity of models), evidence indicates that slight costs can offset the slight benefits of perfect mimicry. In general, a receiver's decisions influence the evolution of signals only when signals have costs. To explain imperfection, theories of mimicry and camouflage have thus focused on the receivers' limitations.

The signal-detection model of noisy communication developed in previous chapters can accommodate mimicry and camouflage with appropriate modification. In these cases, as in all communication, noise elicits less response from a receiver (predator or host) than does a signal plus noise. Brood parasitism, which occurs when females of a parasitic species lay eggs in the nests of another species, fits this scenario. It is also the best known case of mimicry. As a result, the host species ends up raising young of the parasitic species instead of its own. Many species of cuckoos in Africa, Asia, and Australia are obligately parasitic, as are a few neotropical cuckoos. The remaining cuckoos raise their own young, as do most birds. Numerous studies of the European Cuckoo and its hosts, always smaller species, make it the best known case of brood parasitism or indeed of mimicry in general. Most of the hosts of this cuckoo eject strange eggs from their nests, behavior that is obviously an adaptation to minimize brood parasitism. The eggs of a European Cuckoo thus benefit from a resemblance, in pattern and size, to the eggs of its hosts by reducing the chance that the host will eject them from their nests. The signaler's (cuckoo's) expected utility is increased by confusion of its signal (cuckoo's eggs) with noise (hosts' eggs).

A higher cost for a more exaggerated signal makes sense for brood-parasitic mimicry. A smaller egg might have disadvantages for development of a young cuckoo, which must grow to a much larger size as an adult than do its hosts. Receivers (hosts) probably have payoffs ranked like those for other signals for assessment, such as those for mate choice, as described above. A correct detection of a mimetic cuckoo's egg would provide a chance to eject it from the host's nest. A missed detection by the host would result in raising a young cuckoo instead of its own young. A false alarm, confusing one of its own eggs for a cuckoo's, would lead to ejecting the wrong egg and loss of one of the host's own offspring. A correct rejection, correctly deciding that no cuckoo's egg was present, leads to "life without communication," reproduction in the absence of deceptive cuckoos. Unlike other forms of assessment, however, the signaler benefits from the receiver's *missed detections* rather than its correct detections. As in all other cases of communication, both signaler and receiver must realize a net benefit (a benefit on average) despite instances of erroneous reception and ineffective signals.

Batesian mimicry fits a similar scenario. A palatable mimic produces a signal to match, more or less, an aposematic (warning) signal from an unpalatable model. Usually the mimic and the model, as in the cases of many butterflies and other insects, are separate species. The aposematic signal is the noise for the mimic signal. The mimic benefits by resemblance between its signal and noise (the model). The mimic thus exploits communication between the model and a predator by producing a deceptive signal for the predator. It is a form of evolutionary eavesdropping. This form of mimicry thus starts with a receiver's (predator's) evolved response to the aposematic signal from the model. A mimic then evolves progressively to resemble the model. Increasing precision of mimicry must incur a cost for the signaler, just as other forms of exaggeration have costs. Progressively adopting the features of the model perhaps brings some increased risks or compromises some other features of the mimic. In compensation for these costs, the signaler (mimic) benefits from the receiver's missed detections. The payoffs for a receiver (predator) have the same relative magnitudes as those for assessment for purposes of avoidance. A correct detection by the receiver (predator) results in an attack on the mimic and thus a benefit (meal) for the receiver. A missed detection by a predator is a lost opportunity for a meal. A false alarm leads to an attack on a model (the noise) and a distasteful or even toxic meal. A correct rejection approximates "life without communication"—foraging by the predator in the absence of dissembling prey.

The payoffs for a receiver have the same relationships in brood-parasitic mimicry as in Batesian mimicry. They both have the distinctive feature of mimicry, a high cost for a receiver's false alarm. Camouflage has all the features of mimicry just discussed, but the cost of a false alarm is low. A predator attacking an inedible item presumably incurs little cost other than time wasted. Both mimicry and camouflage share a requirement for all communication, a cost for exaggeration of signals (in mimicry and camouflage for perfecting resemblance with a model). They both differ from other forms of communication in that the signaler incurs a *cost* from a correct detection by a receiver and a *benefit* from a missed detection.

Both signaler and receiver might also face diminishing returns in the approach to perfection in mimicry or camouflage, just as in the other cases of noisy communication investigated above. With increasing resemblance to the model, the relative and absolute costs of perfection continue to rise as the marginal cost remains constant. Yet as the probability density function of the level of signal plus noise flattens, each increment in perfection yields progressively less increment in benefit for the signaler. At the same time, receivers evolve higher thresholds to reduce the probability of false alarms, yet any benefits from correct detections of mimics would decrease. High thresholds might also interfere with recognizing other palatable prey in addition to the mimic. The question raised earlier thus arises in these cases too: Does the evolution of mimicry or camouflage, like that of other forms of communication, lead to a stable equilibrium between signaler and receiver at a point short of perfect communication?

SIGNALER-RECEIVER EQUILIBRIUM IN
THE EVOLUTION OF MIMICRY

To explore the evolution of mimicry requires a different set of parameters for noisy communication. For an example of Batesian mimicry, in which a palatable prey produces signals resembling the aposematic (warning) signals of another species, we can set the pattern of the model to correspond to some high level of activity in the receiver's sensor. A neutral pattern, neither mimicking nor camouflaged, a pattern requiring no special developmental processes, might produce a lower level of activity in the sensor. In Batesian mimicry we assume, as just discussed, that developing a signal with the special features of a model involves costs for the signaler. In return the signaler receives benefits of increased survival from its resemblance to the model.

No measurements of the marginal costs of mimicry exist, just as the marginal costs of signals for advertisement or assessment have yet to be measured. Calculations with some plausible parameters can, however, provide comparisons for future measurements and, in the meantime, can suggest some general features of this sort of communication. The set of ten parameters for Batesian mimicry (with some plausible default values) are

p_S, probability that a receiver encounters a signal $= 0.7$
D, receiver's payoff for a correct detection $= 1.1$
M, receiver's payoff for a missed detection $= 1.0$
F, receiver's payoff for a false alarm $= 0.1$
R, receiver's payoff for a correct rejection $= 1.0$
i, signaler's intrinsic survival in the absence of mimicry $= 1.0$
m, signaler's marginal cost of perfecting mimicry $= 0.01$
o, offset (survival in the absence of mimicry $= i$)
g, signaler's relative survival as a result of a correct detection by a
    predator $= 0.1$
and the level of noise

For Batesian mimicry, we can set the level of the noise (the aposematic pattern) to some relatively high average value, say 10.0, with the standard deviation around this average set to 1.0. The signal can then take mean values between 0 and 10.0 in accordance with how closely it matches the model. Its variation, we can suppose, also has a standard deviation of 1.0. Just as in the calculations for mate choice or warning signals, in Chapter 8, the exaggeration of a signal (its approach to perfect mimicry) is expressed in units of standard deviations of variation in the mimic and the model.

In this case, the receiver (predator) benefits from a correct detection (attacking an edible mimic) and loses from a false alarm (erroneously attacking a toxic model as if it were an edible mimic). Note that a correct detection results from a mimic (signal plus noise) that does not reach the predator's threshold for rejecting a toxic model (noise alone). Correct rejection (refraining from attacking a toxic model) results in further search for the next meal and thus approximates "life without communication." We can set this payoff to 1.0, so the other three payoffs become proportional changes in relation to this standard. A missed detection (failing to attack an edible mimic) has a small cost as a result of a lost opportunity for a meal.

The probabilities of each of these payoffs depend on the separation of noise alone and signal plus noise by the receiver's sensor and on the receiver's threshold for response, just as explained previously for mate choice. By combining these payoffs and their probabilities with the probability of encountering a mimic, we can compute the receiver's (predator's) expected utility for any combination of exaggeration of the signal and threshold for response, again just as before.

For the signaler (Batesian mimic), we can use the approach in Chapter 8 to calculate the signaler's expected utility for any degree of exaggeration—in this case, the degree of perfection in mimicry. The signaler has some intrinsic survival in the absence of mimicry (i). If we set this value to 1.0, then the remaining measures of the signaler's survival are proportions of this standard. The cost of progressive mimicry is expressed as a product of its degree of perfection (exaggeration, e) and its marginal cost of perfection (m). As a result of these costs of signaling, the signaler's (mimic's) survival decreases in proportion to its approach to perfection. For instance, if we set the signaler's marginal cost of perfection to −0.01, it loses 1% of its intrinsic survival for every unit (one standard deviation of the variation in the signal or model) of progression toward perfect mimicry. The benefit for the signaler comes from avoiding attacks by the receiver. An attack is precipitated by a predator's correct detection of a mimic. In this case, the signaler's (mimic's) "gain" following a receiver's correct detection is actually a loss—a reduction in survival (g < 1.0, in the terminology of Chapter 8). In contrast, a missed detection by a receiver (predator) results in its overlooking the mimic. In this case, there is no change in the mimic's survival. The mimic benefits solely by avoiding the loss following a predator's correct detection.

Just as in the previously discussed cases of mate choice and warning (Chapter 8), the signaler's and receiver's expected utilities depend on each other. The expected utility of the signaler's degree of perfection depends on the receiver's threshold for attack, and the expected utility of the receiver's threshold depends on the degree of perfection in mimicry by the signal. The procedure, as before, requires some calculus to find the location of the threshold that maximizes the receiver's utility, if a maximum exists, at every level of a signal's perfection. Conversely, in a similar way we can find the level of perfection in a signal that maximizes the signaler's utility at every level of the receiver's threshold. As before, the expected utilities are expressed as survival × reproduction, the appropriate estimate for the number of progeny in the next generation and thus for the spread of any alleles associated with signaling or responding to these signals.

Provided we can find, in this way, both the receiver's optimal threshold as a function of a signal's perfection and the signaler's optimal perfection as a function of the receiver's threshold, we can then search for points where these two functions coincide (see Figure 11.1). Any such point is one at which neither party can unilaterally alter its behavior (its threshold or signal, respectively) without a decrease in its expected utility. Just as before, this point is a Nash equilibrium, a stable attraction point for evolution.

Batesian mimicry as described in these calculations seems to have more exacting conditions for equilibria than does communication in mate choice or vigilance. The range of values for the parameters that prevent a collapse of communication is narrower. For instance, an

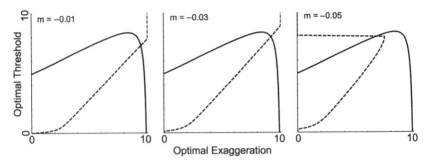

FIGURE 11.1. Overlay of receiver's and signaler's optima for mimicry. Both the receiver's optimal threshold and the signaler's optimal exaggeration vary as a function of the other (see Figures 10.8–10.9 for analogous plots). In these plots the cost of exaggeration increases from left to right (m = −0.01, −0.03, and −0.05, respectively). For Batesian mimicry, exaggeration of signals is progress toward perfection of mimicry. The receiver is a predator that obtains a meal as a result of correctly detecting a palatable mimic. A false alarm leads to a dangerous or toxic encounter with a model. The signaler (mimic) benefits by *decreasing* the probability of the receiver's response (pursuit of the mimic) rather than by increasing it. Because few of the relevant parameters have ever been measured, they are necessarily suppositions. For these calculations, the payoff for a correct detection by a predator (an easy meal, D) is set to 1.1, the payoff for a false alarm (attacking a model, F) is 0.1, the survival of the mimic as a result of a correct detection by a predator (g) is also 0.1, and the probability of encountering a signal (mimic) is 0.7. Perfection of mimicry is assumed to require an exaggeration of 10 (scaled to the standard deviation of noise—variation in the predator's perception of the model). As exaggeration of mimicry increases, the predator's optimal threshold for a response rises (*solid line*) but eventually falls to zero as exaggeration approaches perfection (in the case of mimicry, the mean signal level of the mimic is *lower* than the mean noise level of the model, and a correct detection occurs when the level of signal plus noise is *less than* the receiver's threshold, so when the mimic is almost indistinguishable from the model a predator does best with the lowest threshold possible). Concurrently, as the predator's threshold increases, the mimic's optimal exaggeration rises (*dashed line*) but falls abruptly to zero (*right-hand plot*) when the cost of exaggeration (continuing perfection of mimicry) is too high.

equilibrium for mimicry is especially sensitive to the costs of perfection and the probability of encountering a predator (see Figure 11.1). On the other hand, to my surprise, the cost for the mimic as a result of a predator's correct detection and the cost for the predator of a false alarm have less drastic influence on the location of an equilibrium than they do in the cases of mate choice and warning signals.

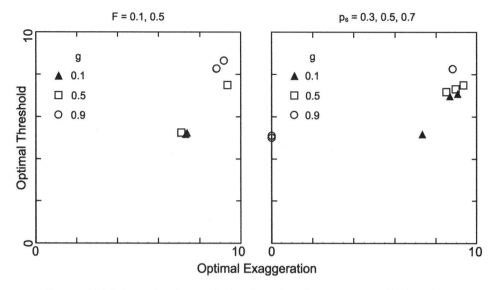

FIGURE 11.2. Joint optima for mimic signaler and predator receiver to illustrate the complex interactions between the parameters of communication in this case. The *left-hand plot* compares all combinations of three levels of a mimic's (signaler's) relative survival when a predator correctly detects it and attacks ($g = 0.1$, 0.5, and 0.9, triangles, squares, and circles, respectively) and two levels of a predator's (receiver's) payoff as a result of a false alarm (mistakenly attacking a toxic model, $F = 0.1$ and 0.5, from left to right, respectively, except right to left for squares). All other parameters are held constant, including the marginal cost of exaggeration ($m = 0.01$). At least for these values, mimicry improves (optimal exaggeration is higher) when attacks by a predator are less damaging for a mimic (g is relatively high). Nevertheless, the joint optimum is strongly affected by an interaction between the mimic's loss and the predator's payoff for a false alarm. The *right-hand plot* compares all combinations of three levels of (once again) a mimic's (signaler's) relative survival when a predator correctly detects it and attacks ($g = 0.1$, 0.5, 0.9, represented by triangles, squares, and circles, respectively) and three levels of the probability that a predator encounters a mimic when checking for prey ($p_S = 0.3$, 0.5, 0.7, from left to right, respectively, except middle–right–left for triangles). Other parameters are held constant (see Figure 11.1). When attacks by a predator are less damaging for a mimic (circles and squares), mimicry (optimal exaggeration) usually increases as the chances of encountering a predator increase. At the lowest chance of surviving an attack (triangles), the pattern is complicated by the collapse of communication as perfection is approached (see Figure 11.1).

The shapes of the curves for optimal thresholds and for optimal levels of perfection in signals make most equilibria cluster in the top right quadrant of a plot (see Figure 11.2). Most equilibria combine high thresholds and high levels of perfection but do not attain complete perfection. As in other forms of noisy communication, evolution does not yield perfection. The trade-offs faced by receivers and signalers result in diminishing returns as perfection is approached, so that the Nash equilibria fall short of perfection.

An anticipated result from this analysis is the influence of the probability of a signaler encountering a receiver (a predator). In populations with fewer encounters with predators, the equilibrium falls farther from perfection and thresholds for attack allow more attacks on less perfect signals (Figure 11.2). Especially when signalers are likely to escape from attacks with little damage, mimicry only evolves when the chances of encountering a predator are high.

PREVIOUS MATHEMATICAL ANALYSES OF CAMOUFLAGE AND MIMICRY

Previous theoretical investigations of the evolution of mimicry and camouflage have included the first application of Signal Detection Theory to animal behavior. In 1983, John Staddon and R. P. Gendron's pioneering study used this theory to explore the evolution of a predator's responses to cryptic (camouflaged) prey. Such a predator must distinguish its prey from other, inedible objects in its environment. The example they propose is a bird searching for stick insects (prey) that resemble ordinary sticks (nonprey). Actually any predator is a receiver that responds to a noisy signal (prey amongst similar less desirable objects) by attacking it. Their concern is to predict when a predator might switch between different types of prey, but their analysis is more widely pertinent to any organism's responses to noisy signals.

They begin in much the same way that the preceding analysis did. They compute a predator's utility as a linear combination of the probabilities and payoffs of the four disjoint outcomes each time a possible prey is encountered. They include the benefits or costs of a correct detection, a missed detection, or a false alarm, and the densities of prey and nonprey (the probability of noise plus a signal or noise alone), much like the analysis in this chapter. Also like the present approach, they assume that one of the four outcomes has little consequence for the predator. They therefore set the relative benefit from not attacking an undesirable object to 0. The present approach set the neutral outcome (the one closest to "life without communication") to 1, rather than 0. Costs then have values between 0 and 1, and benefits have values above 1. This approach treats costs and benefits as proportional changes in survival or reproduction rather than absolute changes. Both approaches reach qualitatively similar conclusions.

To approximate the predator's Receiver Operating Characteristic (ROC; see Chapter 8), Staddon and Gendron use a power function ($p_D = p_F^s$, $0 < s < 1$). The predator's ROC thus varies from perfect detectability ($s = 0$) to no detectability of the prey ($s = 1$). This approximate ROC is not symmetrical and so differs from most experimentally measured ROCs. It also makes the receiver's assessment of its input nonlinear, a possibility that could have consequences for its performance. Nevertheless, by noting that p_D indicates the predator's criterion for attacking, they can then find the predator's best criterion by maximizing its utility with respect to p_D. Differentiation yields this maximum.

When a predator's utility is maximized in this way, p_D (probability of attacking prey) increases, as predicted, with density of prey and decreases with density of nonprey. Also as predicted, p_D increases with the benefit of a correct detection and decreases with the cost of a false alarm or missed detection. The relationships they derive are nearly equivalent to the criteria derived in Chapter 8 for adaptive fastidiousness or adaptive gullibility in receivers.

Staddon and Gendron note that a predator paying extra attention to its sensory input might increase its chances of correctly detecting prey. A predator then faces a trade-off between the time required for this extra attention and its performance each time it decides whether or not to respond. Note that the extra (either prolonged or repeated) attention to sensory input would have consequences similar to those of redundancy in signals (multiple components correlated in time or space). In noise, prolonged attention, like repetition, can improve detectability at the expense of extra time for a decision.

From this early start, predator-prey and parasite-host interactions have continued to provide examples for applications of Signal Detection Theory. An especially thorough application of this theory is Maria Servedio and Russ Lande's (2003) analysis of egg mimicry by hosts of

European Cuckoos, an example introduced in Chapter 9. In Servedio and Lande's analysis, the hosts can evolve discrimination between cuckoos' eggs and their own. In addition, cuckoos' eggs can evolve increased resemblance to a host's eggs, and the host's eggs can evolve decreased similarity to cuckoos' eggs. Servedio and Lande's approach thus allows the noise (hosts' eggs) to evolve as well as the signal (cuckoos' eggs). They also included equations for the population dynamics of both hosts and cuckoos. In doing so they assumed that the number of cuckoos was limited only by the rejection of their eggs by discriminating hosts, and the number of hosts was influenced by the rate of parasitism by cuckoos. Like the analysis in Chapters 8–10, the mathematical equations are complex enough to require a numerical solution.

Servedio and Lande's results indicated that cuckoos and their hosts evolve a stable evolutionary equilibrium, at which hosts have increased discrimination, cuckoos have evolved eggs that resemble the hosts' but not perfectly, and hosts have evolved eggs that reduce this resemblance. Mimicry by cuckoos' eggs evolves much faster and to a greater degree than does either discrimination or the eggs of hosts. The slower evolution of hosts than cuckoos might have resulted from their assumptions about differences in population dynamics. These assumptions might also have produced a second—unstable—equilibrium that could lead to extinction of the hosts when cuckoos were too numerous. The stable equilibrium with a lower rate of parasitism, however, was robust to perturbation of all other variables. This equilibrium short of perfection by either cuckoos or hosts matches the results in this and previous chapters—noisy communication evolves to a stable equilibrium short of perfection for either signalers or receivers.

A stable equilibrium in the evolution of discrimination also emerges from other analyses that incorporate the basics of Signal Detection Theory whenever a response can provide either a benefit or a cost depending on the situation. Some have adopted Staddon and Gendron's method of approximating a predator's (receiver's) ROC with a power function. Most calculations are simplified by exclusion of one or more of the possible outcomes of a predator's decision. The basic issue of discriminating signal from noise applies to all. A stimulus for which a response is beneficial is a signal, one for which a response is costly is noise. Whenever overlap in features of signal and noise prevents absolute discrimination, a receiver faces a trade-off in optimizing its performance. Whenever signalers receive benefits from altering the behavior of receivers but incur costs of producing effective signals, a signaler faces its own trade-off in optimizing its performance. In these models, increasing performance of signalers and receivers faces diminishing returns. In all cases, despite the differences in simplifications and parameters, the result is a stable equilibrium that falls short of perfection. At the evolutionary equilibrium, communication is advantageous on average for signaler and receiver but is not perfect.

EMPIRICAL STUDIES OF THE EVOLUTION OF BATESIAN MIMICRY

The theoretical analyses of mimicry and crypsis as problems in signal detection have identified the important parameters and reached some predictions. Mimicry and crypsis have also attracted many experimental and field studies. These studies have often focused on identifying the species involved as mimics or models, characterizing the relevant behavioral and morphological adaptations, and confirming the toxicity or distaste of models for mimicry. Although mimicry and camouflage often attract attention by their surprising intricacy, many studies have shown that neither mimicry nor camouflage is usually, if ever, perfect. The interest in mimicry and camouflage should produce information relevant to an understanding of its evolution, but information about the pertinent parameters is surprisingly sparse. In particular, the costs and benefits for the signaler and for the four possible outcomes for the receiver (predator), as well as the frequencies of encounter between signalers and receivers, are sparsely documented—if at all.

Mimicry by brood-parasitic European Cuckoos has been studied intensively for decades and provides most of the relevant information about mimicry by brood parasites and counteracting defense by hosts. Nick Davies's 2000 monograph musters the information to examine hypotheses for the evolution of mimicry. It is clear that cuckoos parasitize a minority of hosts' nests, although the rate of parasitism (the probability that an individual of a host species encounters a cuckoo's egg in its nest) varies among host species. Recent evidence shows that cuckoos produce eggs with closer resemblance to a host species when rates of parasitism are higher. Discrimination by hosts, as indicated by the probability of ejecting a cuckoo's egg, also increases when rates of parasitism are higher. In the extreme case, when cuckoos are absent, defensive behavior by potential hosts is also absent. In Iceland there are no cuckoos, but several species there provide hosts for cuckoos elsewhere in Europe. The continental and British populations of these species eject eggs unlike their own, but in Iceland they do not. Hosts thus discriminate against cuckoos' eggs only when they are exposed to them. It is unlikely that this discrimination requires trial-and-error learning, because the opportunities for learning are infrequent (most nests are not parasitized) and discrimination does not improve with age of the host. Evidence thus suggests that continental and British populations of hosts have evolved innate capabilities for recognizing and ejecting eggs that do not match their own, an adaptation that the birds in Iceland have lost.

Recent work has confirmed additional ways that hosts defend themselves against cuckoos and corresponding ways that cuckoos subvert these defenses. A young cuckoo being fed by its hosts learns to match the begging calls of an entire brood of the host's young, yet hosts never provide more food than a brood of their own would require. Hosts harass cuckoos near their nests (a behavior called mobbing). European Cuckoos have evolved a plumage (gray above and barred below) resembling that of a hawk that preys on small birds. This resemblance to a dangerous predator reduces the host's mobbing. To hinder a host's learning to recognize them, cuckoos have also evolved polymorphism. A fraction of cuckoos have brown plumage, rather than gray, so learning the features of a cuckoo presumably becomes more complicated.

These results indicate that cuckoos and their hosts have coevolved counteracting adaptations. This process resembles the coevolution of signaler and receiver in mathematical models. Nevertheless, the empirical evidence for an evolutionary equilibrium between cuckoos and hosts is still uncertain. An examination of birds' eggs in the British Museum, mostly obtained by amateur collectors more than a century ago, has suggested that cuckoos in England previously parasitized a partly different set of hosts than they do now. It thus seems that at least some hosts did not reach an equilibrium with cuckoos. Instead, species no longer parasitized might have "won" an evolutionary "arms race" with cuckoos. These hosts might have evolved such effective defenses that cuckoos parasitizing them became extinct. Yet it is also possible that these lineages of cuckoos died out for reasons not related to signal detection. The environment of England has changed in the past century, possibly in ways that influenced the interactions of cuckoos and hosts. An evolutionary "arms race" in signal detection without an equilibrium, if this possibility were confirmed, is precisely what the mathematical analyses predict should not happen.

Field studies of Batesian mimicry have focused on the advantage mimics realize from avoiding attacks by predators. For instance, among butterflies in the American tropics, several studies have shown that mimetic species are less often attacked by insectivorous birds, such as kingbirds and jacamars, than are palatable nonmimetic species, but the results are surprisingly variable. Experiments have shown that these birds can learn to recognize palatable mimics, in some cases surprisingly quickly. None of these studies, however, provides information on the degree of mimicry in relation to the relative abundances of mimics and models. This information would allow a test of one of the clearest predictions of the mathematical analyses.

Evidence that the accuracy of Batesian mimicry depends at least qualitatively on the presence of the model comes from a series of experiments by David Pfennig and his students, especially George Harper and David Kikuchi. In this case the Scarlet Kingsnake of southeastern North America mimics the venomous Eastern Coral Snake. The kingsnake, however, extends some 300 kilometers farther north than does the coral snake. By placing colored plasticine models resembling snakes in the field, these investigators could determine how often predators (primarily mammals such as opossums, raccoons, and bears) attack these artificial snakes. Particularly revealing were comparisons of areas where kingsnakes and coral snakes occur together (sympatry) and more northern areas where only kingsnakes occur (allopatry). In allopatry, models resembling mimic kingsnakes were attacked more often than were inconspicuous controls, while in sympatry they were attacked less often.

This finding confirms the fundamental prediction that the presence of a dangerous model is necessary for mimics to have an advantage from natural selection. On the other hand, in allopatry imperfect artificial mimics were attacked more often in relation to better artificial mimics than in sympatry. Evidently predators select against mimics in areas where models are absent—and especially against poorer mimics. Paradoxically, selection for perfection in mimicry is stronger in areas without models, although selection against mimics is weaker overall. Why mimics occur at all in areas without models is still unclear. Evidence from comparisons of DNA sequences suggests that movements of kingsnakes carry genes beyond the range of coral snakes. In addition, predators moving from areas with mimics to areas without them might retain their learned avoidance of models and mimics. Or they too might carry genes—in this case for innate avoidance of conspicuous ringed snakes—into areas of allopatry. Despite diffusion of predators' genes or savvy into areas of allopatry, natural selection would not maintain avoidance of mimics there. Attacks on mimics could only persist if migrants from elsewhere brought genes and learning for this behavior.

Quantitative assessment of how often predators encounter either model or mimic presents challenges, because the abundances of these snakes are difficult to assess in the field. Nevertheless, the numbers of museum specimens seem to confirm casual observations that coral snakes are much more frequent in Florida than near the northern limit of their range in North Carolina. In Florida they outnumber kingsnakes, whereas in North Carolina the reverse seems to occur. The resemblance of scarlet kingsnakes to coral snakes is nowhere perfect. Although they have red, yellow, and black rings like coral snakes, the order and widths of the bands differ. Specimens from North Carolina are better mimics of coral snakes, and also less variable, than are specimens from Florida. Furthermore, Scarlet Kingsnakes have greater resemblance to coral snakes in one area where the latter have recently become extinct. Thus the perfection of mimicry increases where models are scarce (and even for a period after they disappear). These results are in line with the predictions of mathematical analyses of the evolution of mimicry based on Signal Detection Theory. A higher probability of encountering a signal rather than noise (a tasty mimic rather than a toxic model) results in a higher optimal threshold for response (more acceptance of imperfect mimics)—more correct detections despite some false alarms.

A further experiment with artificial snakes indicated that predators do not discriminate between the patterns of coral snakes and their mimics. Plasticine models of both were attacked equally often, and both were attacked more often than inconspicuous controls. Perhaps the predators of these snakes do not have the genetic variation that would allow evolution of this discrimination. If so, this constraint on the evolution of discrimination by predators would explain the imperfect mimicry of kingsnakes. Nevertheless, this result is also consistent with the prediction that receivers (predators) and signalers (mimics) evolve to an evolutionary equilibrium without perfect performance by either party.

Possible costs of mimicry in areas where models occur also remain unknown. Kingsnakes might lack genetic variation, for example, for rearranging the order of the colors of their rings. There is no information on genetic variation for further perfection of any mimic. It is also possible that different predators might exert differ sorts of selection on the resemblance of mimics to models. Some predators might not avoid the model, so greater conspicuousness might make a palatable mimic more vulnerable to these predators. The frequencies with which different predators encounter mimics and their capabilities for attacking coral snakes are unknown. Nor are there data on survival of natural mimics with different degrees of resemblance to models. The mathematical analyses of the evolution of mimicry at least serve a purpose in identifying these potentially important parameters.

The evolution of camouflage has also received attention. The classic case is the dark coloration of Peppered Moths that is known as industrial melanism. During daylight, these moths remain motionless on the trunks of trees, where their camouflage protects them from predation by birds. In industrial areas of England during the last two centuries, the pale lichens on trees were killed by air pollution. In these areas, this moth (and other camouflaged insects) evolved darker wings to match the darker trees. The melanistic moths occurred principally in industrial areas, while the widespread paler form occurred elsewhere. Available information shows that in industrial areas the cryptic moths survived predation by birds better than others. In recent decades, abatement of industrial pollution has allowed lichens to return to the industrial areas of England. As a consequence, the dark moths have now disappeared from many areas. Camouflage evolves in response to changes in the background noise.

WHERE DO WE STAND NOW?

Part I of this book presented the properties and prevalence of noise in communication, with a focus on acoustic communication by animals, both nonhuman and to a degree, human. It also introduced adaptations of animals to reduce the effects of noise on the performance of receivers. It became clear that communication in noise is a problem of signal detection. Part II then presented a full analysis of communication as signal detection in noise. A mathematical approach allowed an investigation of the coevolution of signalers and receivers. It had two important results. First, it identified the parameters necessary to understand the evolution of communication in noise, a dozen parameters essential for the most basic understanding of this issue. Second, it reached two general conclusions. (1) Producing and receiving signals reaches an evolutionary equilibrium that falls short of perfect performance by signalers or receivers. Noise is thus inescapable. (2) As a corollary, communication at evolutionary equilibrium is honest on average but with occasional deception or manipulation of both parties. Honesty is the norm.

Earlier mathematical analyses of the evolution of mimicry and camouflage anticipated this analysis of noisy communication. They too adopted, sometimes with shortcuts, signal detection in noise as the paradigm for the evolution of these interactions between predators and prey. None of these earlier analyses were applied to the evolution of noisy communication in general, yet their assumptions and results apply more widely. They identified many of the same parameters essential for an understanding of the evolution of noisy communication, and they all predicted an evolutionary equilibrium between signaler (mimic) and receiver (predator). Empirical studies of mimicry and camouflage have also often had signal detection in mind. As a result, in a few cases they have confirmed some crucial predictions. Yet the most striking realization from these studies is how much is left to learn. Most of the important parameters have not been measured in any case of communication.

Part III of this book takes the two general conclusions of the mathematical analysis for granted: noise is inescapable, and honesty is the norm. The object is to assess how the insights from signal detection alter ways to think about some large issues in biology, including honesty in communication, sexual selection, cooperation, and molecular signaling. Part IV then carries the implications of noisy communication into the realms usually occupied by philosophers. The object is to consider how noisy communication alters ways to think about ourselves as perceivers of the world around us and even as author and readers of this book. Before proceeding, though, it is important to evaluate the two principal conclusions so far.

NOISE IS INESCAPABLE

Despite any uncertainty about the exact results, the calculations in previous chapters lead to a fundamental conclusion about communication: in noisy conditions, communication cannot evolve to escape noise. The earlier applications of Signal Detection Theory to the evolution of mimicry and camouflage lead to the same conclusion—communication evolves to an equilibrium between signaling and responding that falls short of perfection. The trade-offs faced by both receiver and signaler prevent evolution to a point at which receivers no longer make errors and signalers always elicit appropriate responses. There is no escape from noise.

Because both signaling and responding to signals face diminishing returns as exaggeration or thresholds increase, this result seems intuitively plausible. Neither signaling nor responding evolve to the point of eliminating the last degree of noise in communication. The inescapability of noise is a fundamental conclusion of all analyses of the evolution of communication in noise. Noise creates a trap from which communication cannot escape.

Two caveats are in order. First, the analysis in previous chapters does not constitute an analytic proof that the evolution of communication cannot escape from noise. Earlier analyses have produced analytic results, but with the help of additional simplification. If a more complete proof is possible, I must leave it to actual mathematicians.

Second, and perhaps more important, is the question of whether any of the assumptions of the preceding calculations have biased the results in this direction. Does, for instance, the shape of the distribution of noise alter this conclusion? For receivers to respond to signals without error, the upper bound of the distribution of noise in a receiver's sensor would have to be less than the lower bound of noise plus signal. Such a condition might be approached in some situations, particularly in artificial ones, but it seems unlikely that evolved communication would ever achieve it.

The inevitability of noise, in all analyses, derives from the diminishing returns of perfection, for both signaler and receiver. For any level of noise, a receiver's trade-off between the two possible kinds of errors, false alarms and missed detections, results in an optimal threshold or criteria for response with inevitable error. Receivers do not evolve to respond to every occurrence of a signal, nor to ignore all instances of noise alone. For such perfection in performance, signals and noise would have to produce distinct levels of activation of a receiver's sensory input. This separation of signal and noise might occur if signalers evolved sufficient exaggeration of their signals. Signalers, however, also face diminishing returns for perfection. The marginal cost of exaggeration in realistic cases would either remain constant with progressive exaggeration or, although not included in the mathematical models so far, might well accelerate. In contrast, the signaler's marginal benefit from exaggeration must normally decrease. Any benefit depends on the response of an appropriate receiver. As the mean level of a signal plus noise increases in relation to the level of noise alone, the incremental change in a receiver's correct detections decreases. Once a receiver's optimal threshold moves to a point below the peak of the probability density function for signal plus noise, any further increase in the contrast between signal and noise yields progressively less increase in the receiver's correct detections. Rising costs of continued exaggeration combined with falling benefits leave the signaler with diminishing returns. As a result, signals are never likely to reach a level of exaggeration that results in distinct separation of signal and noise for a receiver.

This inescapability of noise as a result of diminishing returns is underscored by realizing that the calculations are scaled to the level of noise alone. In a situation with low levels of external noise, such as intimate communication between two individuals at close range, natural selection leads to an evolutionary equilibrium with low thresholds for receivers and low exaggeration of signals. In a situation with high levels of external noise, such as long-range communication or a dense aggregation of individuals all communicating at once, natural selection leads to high thresholds and high exaggeration. In neither case does communication escape from noise and reach perfection. Intimacies at close range and advertisement in a crowd lead to similar evolutionary equilibria, except everything is scaled up in higher levels of noise.

It is also important to emphasize that, although the joint optima in the above calculations never escape from noise, some of them achieve remarkable exaggeration. Exaggeration that equals ten or even four times the standard deviation of noise does produce communication with very few errors and very high costs of signaling indeed. So the important point is not that evolved equilibria for communication cannot result in remarkable performances by both signaler and receiver but that, no matter how remarkable, these performances never reach perfection.

If so, an understanding of communication is unlikely to progress by assuming that noise is an aberration in communication rather than the norm. Investigators can of course create situations with essentially no noise (with maximal contrast between signal and noise). It might seem that the best way to study animals' responses to each other is to minimize variation in responses. One way to do so is to minimize noise, by excluding extraneous stimulation for the receiver and presenting standardized signals with as little degradation from transmission as possible. Low-noise situations might help in investigating properties of sensory organs and neural control of behavior. They might not, however, reveal why these properties of the nervous system and behavior have evolved.

There are three ways that experiments with natural signals in natural levels of noise might differ from experiments with artificially clean signals. First, by eliminating all, or almost all, noise, animals less often respond in the absence of a signal (a false alarm), so such experiments would show that they respond more reliably (with greater statistical significance) to a signal than they actually do in naturally noisy situations.

Second, in low-noise situations animals respond preferentially to signals that seem excessively exaggerated. This result would occur when animals have evolved a high threshold in order to minimize adverse consequences of false alarms in noisy conditions. This situation occurs whenever animals are trained to discriminate between two similar signals. Training consists of repeatedly rewarding the subject for responses to one stimulus (high payoff for correct detections) and not rewarding, or punishing, the subject for responses to the other (low payoff for false alarms). After such training, subjects can show an unexpected "peak shift" when tested with a wider range of stimulation. Their greatest preference is not for the rewarded stimulus but for a stimulus that is even more removed from the unrewarded stimulus. In other words, they respond best to a "supernormal" stimulus, more exaggerated than the one to which they have been trained to respond. Natural selection can produce a analogous "peak shift" in unlearned behavior. As the preceding calculations have shown, animals are expected to evolve thresholds for responses that balance the advantages of maximizing correct detections and the incompatible advantages of minimizing false alarms. They thus evolve higher thresholds than those expected solely for maximizing correct detections. They might thus have stronger responses to artificial signals that were more exaggerated than those to which they had evolved to respond. Experiments in artificial situations might reveal a peak-shift or supernormal responses rather than adapted responses.

Third, the particular properties of an organism's sense organs and behavioral responses to stimulation might not correspond to the properties of the presumed signal in low-noise situations. For instance, measurements of the audiogram for hearing by Great Tits has revealed that they are most sensitive to acoustic frequencies higher than the frequencies in their own songs and calls. However, in the presence of natural background noise from a European woodland, the frequencies for maximal sensitivity in hearing correspond closely to the frequencies at which their own calls and songs differ most from the background noise. The noise, mostly from wind and movements of vegetation, is strongest at low frequencies, below those of the tits' vocalizations, and decreases steadily at higher frequencies. The tits thus have evolved their most sensitive hearing at frequencies for which their own vocalizations contrast the most with background energy. The real question is thus why the tits' vocalizations have not evolved to match these frequencies that contrast most with noise. The answer, however, is unlikely to come from studying responses to clean signals in low-noise situations. Instead more enlightenment would surely come from further characterization of the noise during natural situations for communication and from experiments comparing responses to their signals under these situations.

The behavior and properties of organisms in situations other than those in which they have evolved do not necessarily have any adaptive explanation. Communication is no exception. To understand the evolution of communication, it is thus important to explore the sources of noise and the errors of receivers in natural situations. For an understanding of the evolution of signals, it is essential to study the noise as well as the signal.

HONESTY IS THE NORM

In all situations of communication considered above, signaling and receiving evolve to a joint optimum at which both parties benefit on average. Signals thus evoke responses from appropriate receivers often enough for signalers to benefit on average. Responses to appropriate signals occur often enough for receivers to benefit on average. At the joint optima, however, the probability of correct detection by a receiver to every signal falls short of certainty. Receivers sometimes make mistakes, false alarms and missed detections, and signalers sometimes fail to elicit appropriate responses. Nevertheless, at the evolved joint optimum honesty is the norm.

Thus honesty is a strong result of the evolution of communication in noise. It assumes, reasonably enough, that signals have costs, but the costs of signaling affect the evolution of communication in a quantitative way. They are no more important for the evolution of a joint optimum than are the features of receivers. Previous arguments for the paramount importance of costs of signaling for honesty are the subject of Chapter 12.

One last point worth emphasizing before proceeding is the broad application of the calculations in the preceding chapters. The scope extends beyond the evolution of communication. Any mathematics is necessarily abstract, a point fully recognized for the first time by René Descartes and his immediate predecessors. Mathematics has indefinite application. In the present case, in addition to communication by organisms constrained by costs, they apply, for instance, to machines designed by engineers sensitive to costs. Engineers face the same sort of diminishing returns for perfection—both rising costs and diminishing returns in the approach to perfection. I am unaware of economic analyses of this sort of situation. It is true that digital communication by machines reaches impressive levels of accuracy these days. Nevertheless, the state of the art is always pushing the limits, as evidenced by the equally impressive fabrications for controlling the temperature of the fastest components of computers. In addition, the compulsion to back up digital files reveals that perfection in engineering, although approachable, is problematic.

PART THREE

ALTERED PERSPECTIVES

THE TWO FUNDAMENTAL CONCLUSIONS, inescapable noise and routine honesty, present some challenges to the way we think about the properties of communication, the evolution of social behavior (which consists mostly of communication), our own perception and cognition, and even long-standing problems of philosophy. To confront these challenges is the aim of the remainder of this book. For a start, Chapters 12–16 explore how the properties of noisy communication can alter our views of the evolution of honesty, mate choice, and cooperation. The scope ranges from complex societies to molecular physiology.

The first topic is the role of costs in the evolution of communication. In Chapters 7–11, costs for both receivers and signalers were essential parameters of the evolution of signals. Yet their contribution to the evolution of honesty was incidental. It seemed obvious that all signals have costs and that costs are essential for some degree of honesty. The principal result of the mathematical analysis was that communication with noise never evolves perfect honesty. Honesty is the expected norm, but not the rule. In contrast, in recent decades costs have been construed as the preeminent explanation for the honesty of signals. As a consequence, the high cost of exaggeration has become the cost of honesty in advertisement. The preceding calculations show instead that honesty and exaggeration in noisy communication evolve despite costs rather than because of them. Higher costs of exaggeration always result in less, not more, exaggeration. Current explanations of dishonesty in communication emphasize the conflicting interests of signaler and receiver. Instead, the preceding calculations have shown that, at an evolutionary equilibrium, communication always benefits both parties on average. Chapter 12 thus explores why noisy communication diverges so fundamentally from previous takes on the evolution of honesty, deception, and exploitation.

The preceding calculations also suggest new approaches for understanding the evolution of exaggerated signals. They have shown that exaggeration can evolve solely as a result of the trade-offs faced by both receiver and signaler during communication in noisy conditions. In developing these consequences of noisy communication, even in the case of mate choice, there was no recourse to arguments based on sexual selection. Yet sexual selection has previously been the predominant explanation for the evolution of exaggerated signals. The task for Chapter 13 is thus examining the relationship between signal detection in noisy situations, on the one hand, and sexual selection, on the other, as explanations for the evolution of exaggeration.

Chapters 14–16 then take up two problems so far largely beyond the scope of signal detection: the evolution of cooperation and the evolution of molecular signaling within an individual organism. Signal Detection Theory has hardly touched these issues. Yet coordination of social interactions is the primary reason organisms communicate. And coordination of the molecular interactions in our bodies is the primary reason for much of our molecular machinery. The complexity of these interactions, both social and molecular, and their importance for survival and reproduction offers a broad scope for noise in the requisite signaling. Excursions into societies, on the one hand, and molecular physiology, on the other, can only indicate this scope. Fully incorporating the consequences of noisy communication in societies or cells will take much more work.

Part IV confronts the ultimate challenges of noise. The course of argument in this book began with Hooded Warblers singing among flowering viburnums in the depths of damp swamps, and it comes around to them again. What is going on in those intent little brains? And, beyond this question, what is going on in our own brains? The path opened by the warblers leads to some fundamental issues of subjectivity, perception, and verification.

12

HONESTY IN COMMUNICATION

COSTS OF SIGNALS IN NOISE

SIGNAL DETECTION THEORY provides a radical reinterpretation of the evolution of honesty in communication. At a joint optimum for signaler and receiver communicating in noise, honesty is the norm. Both receivers and signalers benefit on average by communication. "On average" is a critical qualification, because at the optimum receivers sometimes make errors and signalers sometimes fail to evoke appropriate responses. Because receivers still make errors (false alarms by responding to noise alone or missed detections by not responding to a signal), communication at a joint optimum does not escape from noise. Communication in noise does not evolve to an optimum that avoids noise. Both parties instead evolve to optimize their performance in noise.

Because Signal Detection Theory explains the evolution of honesty in communication without recourse to any special role for their costs, it supersedes the current emphasis on the cost of signals as a preeminent requirement for honesty. It does, however, share with any theory of communication a basic dependence on costs. It seems obvious that signals with no costs whatsoever would have no limits whatsoever for exaggeration or proliferation. One can just imagine what presidential elections in the United States, and presumably in other nations, would be like if advertisement on television and along our highways had no cost. Already judicial rulings that disallow any legal suits by politicians for slander have removed a major potential cost of exaggerated advertising. If television had no cost, all limitations on a deluge of manipulative advertising would vanish. Even habituation by viewers would provoke more exaggerated advertisements. Without costs, it would pay for a signaler to persevere in exaggerating its signals as long as there was any probability greater than zero of just one more response. Thus exaggeration, both in signals for mate choice and in presidential elections, has limits, no matter how high they might be. We can probably all agree that exaggeration entails some cost. It is difficult to imagine a signal, however trivial, that does not involve some cost in time, energy, opportunity, or risk for a signaler. The question is not whether or not signals have costs but how costs influence the optimal form of communication.

HANDICAPS AND OTHER CLASSIFICATIONS OF SIGNALS

The original proposal that honesty in communication requires costs differed greatly from this basic requirement. As Amotz Zahavi originally proposed, honesty requires signals that create handicaps for signalers. As several critics have indicated, his use of the term *handicap* does not correspond to any other use of the word. It is not, for instance, like a handicap in sporting events, such as horse racing or golf. These handicaps are intended to make contestants more evenly matched, so that the one with the most skillful tactics or just the greatest luck at the moment can win. And it certainly is not a handicap that disables an individual. Instead, handicaps in Zahavi's proposal make the contest less evenly matched. They are necessary to create differences among contestants. A structure that interferes with feeding, for instance, reveals that only certain individuals, those in the best condition, are capable of capturing food despite this handicap. A structure that compromises avoidance of predators reveals that only those individuals in the best condition survive. This proposal leads to a corollary emphasized by Zahavi (but less often by other investigators): handicaps should consist of morphology or behavior that are related to the quality advertised. A signal indicating high capability in finding food should be a structure that makes finding food more difficult. A signal indicating high capability in avoiding predators should taunt predators to attack.

Subsequent contributions have identified several kinds of handicaps. These different handicaps are in part defined by when the costs of producing a signal are paid. Costs, for instance, might be paid up front: all individuals, it is proposed, try to produce exaggerated signals, but low-quality individuals die in the process. Or they might be paid as you go: high-quality individuals produce more exaggerated signals each time they signal because low-quality individuals cannot afford to. Or all individuals produce exaggerated signals, but only high-quality ones can display them or maintain them optimally. Another proposed distinction is whether or not a handicap requires an interaction between alleles at two loci (called epistasis). For instance, alleles at one locus might influence the signaler's quality, and alleles elsewhere influence the exaggeration of a signal. When high-quality individuals produce more exaggerated signals than do low-quality individuals, such signals are often called indicators. All signals that are indicators result from an interaction between a signaler's genetic quality and the exaggeration of its signals. They thus all require interaction between at least two genes—in other words, epistasis.

John Maynard Smith and David Harper have proposed that handicaps occur when signals are "excessive," when they are more exaggerated than needed "to avoid ambiguity." They were not thinking of noisy communication when they proposed this definition of a handicap. Claude Shannon, the key figure in the development of information theory as introduced in Chapter 1, on the other hand, saw that ambiguity is one of two distinct forms of noise. *Ambiguity* occurs when different instances of the same signal elicit different responses—for instance, when a word evokes different definitions for different receivers or at different times. The converse, *equivocation,* occurs when two different signals elicit the same response, when saying something two different ways evokes the same definition. Ambiguity and equivocation are thus one way to classify noise in two mutually exclusive but exhaustively enumerated categories. In the full version of Signal Detection Theory, ambiguity is aligned with missed detections (a signal elicits either of two responses, correct detections or missed detections), and equivocation with false alarms (a signal and irrelevant stimulation—in other words, noise—both elicit the same response). As explained in our development of signal detection, there is no separation between exaggeration that reduces ambiguity (or equivocation, for that matter) and exaggeration that does not. All exaggeration of a signal results in a progressively lower probability of missed detection, for any threshold adopted by a receiver. All exaggeration reaches an equilibrium with the receiver's threshold. All exaggeration also results in progressively lower probability of false alarm. There is thus no distinction between "necessary" and "excessive" exaggeration. Instead, exaggeration becomes progressively more costly until the additional gain from avoiding errors no longer compensates for the costs of further exaggeration.

There have been other proposals for classifying signals, partly congruent with classifications of handicaps. An aim has been to distinguish signals that do require handicaps for honesty from those that do not. Some signals, it is argued, do not require costs (or at least large costs) for honesty. For instance, *indices* are signals physically associated with the feature of a signaler that is of interest to a receiver. A deep voice, for instance, is a physical consequence of a larger vibrating structure, as discussed in Chapter 2. It seems like a less "cheatable" indication of overall size than alternatives. It is true that deep voices are signals for mate choice and aggression in a variety of species in which the size of a mate or a rival makes a difference to the choosing individual. Many species have evolved special, even grotesque, structures to produce deep voices (see Chapter 2). Presumably it is more difficult—that is, it costs more—for a smaller individual than for a larger one to produce these structures.

Just the opposite of indexical signals are arbitrary ones, also sometimes called conventional signals. The exaggeration or structure of such signals has no apparent relationship with a referent (in this case, an attribute of the signaler's state of interest to the receiver). The evolution of communication always involves frequency-dependent interactions between signaling and receiving and thus the chance that signals evolve as arbitrary fads. Possibilities include most human words and also many sounds and visual displays in communication between nonhuman animals. Arbitrary signals, it is sometimes suggested, only evolve by cooperation, in other words, when signalers and receivers share the same interests in communication. They thus differ from indexical signals that require handicaps to assure honesty when signalers and receivers do not always share common interests.

Both of these distinctions—between arbitrary and indexical signals and between communication with shared and conflicting interests—become less clear for communication in noise. It is, for instance, difficult to imagine completely arbitrary or completely indicating signals. As noted in Chapter 5, in the presence of noise, signals evolve adaptations for efficient transmission, for increasing contrast with background energy and other signals, and for density of information. Even when signals have some relationship with the signaler's other attributes, such as a deep voice with size, the relationship is never beyond the possibility of adjustment. Furthermore, features such as low frequencies and high intensities also affect the active space of signals, as also noted in Chapter 5. If arbitrariness is viewed as a continuum between extremes of adaptations and fads, it is not clear that any signal is completely arbitrary nor completely adapted.

Furthermore, the distinction between shared and conflicting interests in communication is also less clear in the light of the discussion of noise. Preceding sections have shown that all communication is cooperative. Both signalers and receivers must benefit on average from communication or the evolution of communication collapses. Signalers and receivers need not of course have all interests in common, but they must overall have a mutual interest in the production and reception of each kind of signal.

The evolution of indicator and arbitrary signals is explained by the same balance between costs and benefits of signaling presented in Chapter 8. All signals evolve in the presence of noise by the same principles, as a result of a positive balance of benefits and costs. On the one hand, noise unifies the study of signals by eliminating the need for separate categories of signals based on the relationship of the features of a signal to its meaning. On the other hand, noise splinters the study of signals by requiring investigation of each signal's features in its appropriate context. A signal's features must be understood in relation to both signalers' and receivers' trade-offs and in relation to encoding information and counteracting noise.

PROOFS OF HANDICAPS EXAMINED

A crucial step in the wide acceptance of Zahavi's proposal that handicaps explain honesty came with Alan Grafen's twin papers in 1990. These papers aimed to prove mathematically the correctness of Zahavi's proposal that honesty in communication occurs if and only if signals have costs. Previous articles had purported to demonstrate mathematically that it was in fact wrong. In the following year another influential paper came to a conclusion similar to Grafen's. In this case, John Maynard Smith applied game theory to a particular case of communication, which he dubbed the Sir Philip Sidney Game. Like Grafen's papers, this analysis claimed to have vindicated Zahavi's proposal. Another prominent paper that year by Yoh Iwasa, Andrew Pomiankowski, and Sean Nee reached a somewhat different conclusion—that only a particular kind of handicap could result in the evolution of honest communication.

The following decades have seen nearly universal acceptance of handicaps as a prerequisite for the evolution of honest signaling. The definition of handicaps has tended to drift toward *net costs* of signaling, without the original emphasis on *waste* and *meaning*. The formulation of costs of signaling in Chapter 11 is entirely consonant with this trend. Nevertheless, the papers of 1990 and 1991 make an explicit case for the importance of handicaps and, as a consequence, have had a pervasive influence on thinking about the evolution of communication. Because this present book is so different in advocating an entirely different approach to studying the evolution of signals and an entirely different conclusion about the evolutionary equilibrium for communication, it is imperative to be clear about the conclusions of these papers. A close examination of each of them is in order, but nonetheless impractical here. Instead, this section considers in more general terms the issues they raise.

All of these papers show that costs are necessary for honesty in communication. This point is the one that we can all agree on at the outset.

None of these papers demonstrates how much the costs must be, however, nor what the nature of the costs must be. These mathematical analyses thus have little relation to the original proposals about waste and meaning.

GRAFEN'S VIEW OF THE EVOLUTION OF COMMUNICATION

Grafen's first paper, "Biological Signals as Handicaps," sets out to show that communication is evolutionarily stable only when signaling and preferences for responding meet some specified conditions. He uses advertisement by males and choice by females as an important example of communication. Nevertheless, Grafen's goal is to prove a general theorem: if and only if certain conditions are fulfilled does honest communication evolve. His derivations are more abstract than Maynard Smith's and often difficult to comprehend, but both come to nearly the same conclusions.

In the end, Grafen proves his theorem. If signals incur costs in a reasonable way, then honest evolution can evolve (with the possible exception of cases in which no communication evolves). Furthermore, the converse is also true. If communication is honest, signals must have costs of a reasonable sort. Signals with costs and receivers with preferences for those signals imply that signals are honest and, conversely, honest signals imply that signals have costs and receivers prefer them. For signalers, the costs of signals are compensated by the benefits from the receivers' responses. All of these conclusions match those reached for communication in noise.

On the other hand, it is unclear that these conclusions justify the original verbal arguments about handicaps. In particular, proposals about waste as a necessary part of honest signals are dubious. Proposals that the meaning of a signal is necessarily related to its costs are also. Nevertheless, Grafen easily equates his costs with handicaps. In addition, the conclusions he draws share the original exclusive emphasis on the costs of signals. He wraps up his discussion with the exhortation, "A major facet in the study of signals should be the fitness cost imposed by them," and he mentions no other facets of signals that merit study. To a degree, this approach has dominated the study of communication for two decades.

Just as in the Sir Philip Sydney Game, a conclusion that communication is honest if and only if signals have nonzero costs provides little help in understanding communication. As already emphasized, it is difficult to identify or even to imagine signals that have no costs of production or performance. Furthermore, cost-free signals are not a helpful alternative, as they would have no limits whatsoever on proliferation and exaggeration.

Perhaps the most important, but overlooked, conclusion reached by Grafen is that—at equilibrium and on average—signals must be honest. Otherwise, responses to them would not evolve. If signals were not honest on average, natural selection would not produce responses to those signals. If appropriate receivers did not respond to signals, natural selection would not produce costly signaling. Honesty on average is introduced as an assumption of his models, but it really is a conclusion based on an understanding, at the most basic level, of the evolution of communication by natural selection.

The conclusion that receivers must realize benefits (at equilibrium and on average) restores some balance in thinking about evolution of communication. The focus on costs of signals is incomplete because it leaves the receivers out. When we consider how receivers might behave, new questions and new parameters become relevant for the evolution of communication. As we have seen, the costs of signals remain important, but they are just costs, not handicaps. For noisy communication, costs of signals are not even the majority of the parameters that must be investigated if we are to understand the evolution of communication.

EPISTASIS IN THE DEVELOPMENT OF SIGNALS

While Grafen's and Maynard Smith's papers make convincing cases for an intuitively obvious point, Iwasa and colleagues make a case that would be interesting if it weren't so peculiar. The first two papers prove that honesty in communication requires signals with nonzero cost. The latter claims to show that one particular kind of costs for signals, "pure epistasis handicaps," cannot result in honest communication, although "conditional" and "revealing" handicaps can.

The distinction between these two classes of handicaps is based on diagrams of the causal paths in signaling. Once again the model of communication is choice of a mate based on signals produced by potential suitors. For convenience, as in most cases, females are the recipients and males the signalers. By responding to a male's signal a female chooses him as a mate. Males differ in *quality*, some feature that affects the females' own reproductive success or the survival of her offspring. When she makes her choice, however, the only information she has is the male's signal. She has no way of directly assessing the male's quality. This is the same situation that Grafen analyzed and that Part II of this book has analyzed. What conditions make the males' signals honest indicators of their quality?

Iwasa and colleagues identify four causal "nodes" in this situation:

1. A male's "potential" signal (or trait, t) is presumably his allele influencing the production of a signal, perhaps a long tail or a complex song.
2. A male's general viability (v) is presumably an allele that influences survival—in other words, his general quality, which can be inherited by any of his offspring.
3. A male's "realized" signal (s) is the magnitude of the signal he actually produces.
4. A female's preference (p) is the strength of the female's response to a male's signal.

The authors then suggest that in most handicaps a male's realized signal is influenced concurrently by both his general viability and his potential signal (v and t together influence s). In other words, expression of alleles for viability interacts with expression of alleles for the signal to produce the actual signal. A male with alleles for the signal but low viability produces less of a signal than a comparable male with high viability. And a male with alleles for some level of viability but low potential for signaling produces less signal than a comparable male with high potential for signaling. This pattern of causation is genetic epistasis in development, the interaction of alleles at different genes in the development of a trait.

A *pure epistasis handicap*, on the other hand, is supposed to have a different pattern of causality. General viability is supposed to influence the potential trait directly, and the latter is then the sole influence on the realized trait (v influences t, which in turn influences s). If we translate these terms into their genetic and developmental equivalents, as just summarized, the result is a particular mechanism for the development of a signal: expression of alleles for general viability affect the expression of the alleles for signaling, which then is the sole influence on the development of the signal. In this case there is no direct influence of alleles for general viability on the male's actual (realized) signal. So the association between general viability and the actual signal must always be less than the association between the alleles for the potential signal and the actual signal. Consequently, this pure epistasis, as the authors call it, allows signaling to evolve under final control of the gene for signaling and consequently with no assurance of honesty in signaling a male's general viability.

On reflection, this result is not as relevant as it seems. Although we know of many cases in which the expression of one gene alters the expression of another, this process is hardly a criterion for epistasis in development. Epistasis occurs whenever a trait develops as a result of an interaction of the effects of two genes, but usually we do not know the exact mechanism of the interaction. Most students of communication would assume that genes for general viability (such as resistance to diseases or agility in avoiding predators) affect the expression of an exaggerated signal because a sicker or weaker individual cannot muster the resources to signal as much as a healthier or stronger one can. These cases—Iwasa and colleagues' *conditional* or *revealing* handicaps—are good examples of epistasis. By excluding one particular mechanism of development, one not usually thought to apply to signals anyway, we have not learned much about the evolution of communication.

CONCLUSION

Despite the wide influence of the papers discussed in this chapter, they contribute little to understanding the evolution of communication. The most fundamental insight is that both signalers and receivers must benefit, on average, in all systems of communication. The conclusions about costs of signaling are intuitive. They have no relation to the distinctive features of the original handicaps, and they provide no explanation for the nature or magnitude of exaggeration in signals. The trade-offs between costs and benefits are not clearly formulated for either signalers or receivers.

In contrast, as Chapter 13 emphasizes, signal detection in noise makes precise predictions, for specific situations, about how costly signals should be (as a result of trade-offs for signalers and receivers communicating in noise) and also about the form that exaggeration should take (as a result of its influence on signal detection in noise). The role of costs in communication is thus both less mysterious and more interesting than previously imagined.

13

SEXUAL SELECTION AS COMMUNICATION

COMPLEMENTARY THEORIES

FOR DECADES, SEXUAL SELECTION has been the primary explanation for the evolution of the spectacular displays of many animals.
Spectacular displays are examples of exaggerated signals, the evolution
of which is also explained, as shown in Part II, by noisy communication. These two explanations for the evolution of exaggerated signals, as
a result of sexual selection or noisy communication, do not however
conflict. They are complementary perspectives on the evolution of mate
choice and exaggerated signals. Both are needed for a complete understanding of exaggerated signals.

Sexual selection as currently understood is a special case of natural
selection. As explained in this chapter, its special features apply only to
certain kinds of alleles, those that influence an individual's access to
mates. Like other mechanisms of evolution it is best described mathematically by the spread of alleles in populations. In this procedure, each
genotype (allele or combination of alleles) is assigned a phenotype. By
specifying the frequency of each genotype in one generation, the selection on each genotype (the proportion surviving to reproduce), the associations among all genotypes for mating, and the proportionate pro-

duction of offspring for each of these associations (plus, in some cases, the proportionate immigration and emigration of each genotype and any random variation in each of these parameters) it is possible to calculate or to simulate the frequencies of all alleles in the next generation. Generations are usually assumed not to overlap, so none of the parameters depends on age. The consistent use of proportions (relative frequencies) for all parameters is equivalent to assuming that evolution does not change the overall size of a population and that none of the parameters changes with the density of a population. The exact calculation of the dynamics (rates of change) of the relative frequencies of alleles in a population is the strength of this approach. The abstract associations of alleles with phenotypes is the weakness.

The analysis of signal detection in this book uses a different approach, a mathematical description of the spread of phenotypes in populations rather than the spread of alleles. Each phenotype is assumed to be associated with one or more alleles in unspecified ways. By specifying how phenotypes interact with each other and the payoffs for each kind of interaction in terms of survival × reproduction, it is possible to calculate the number of offspring of each phenotype in the following generation. These calculations are a form of game theory in which individuals choose their tactics in interactions and receive payoffs (numbers of offspring in the following generation) that depend on the tactics both they and their opponents choose. Natural selection moves a population toward a composition of phenotypes that maximizes the number of offspring in successive generations. The maximum might only be a local one, valid only within a limited range of phenotypes. Unlike the previous approach, there are no limitations on growth of populations but, like that approach, none of the parameters depends on the density of the population. The calculations in preceding chapters have not included any differences in the age-dependent development of phenotypes (although this option could be included). The exact calculation of the proportions of phenotypes in a population at equilibrium is the strength of this approach. The lack of any way to calculate the dynamics of the spread of alleles is the weakness.

These two approaches are complementary. The evolution of exaggeration in signals reveals this complementarity especially clearly.

EVOLUTION OF EXAGGERATION

Charles Darwin first proposed sexual selection as an explanation for differences between the sexes that do not seem directly related to survival or to reproduction. These differences include morphological structures other than those directly necessary for production and fertilization of gametes or for maturation of eggs or embryos, such as gonads, genitalia, oviducts, and uteruses. The differences also include behavior associated with mating, often competitiveness in one sex and coyness in the other, usually males and females respectively. Although Darwin included sexual selection in *The Origin of Species by Means of Natural Selection* (1859), he described it in much more detail over a decade later in the course of his thorough case for the evolution of humans from ancestral apes in *The Descent of Man and Selection in Relation to Sex* (1871). He identified two mechanisms for sexual selection. First, he proposed that males might engage in physical contests with each other to gain access to females for mating. This situation would favor the evolution of various forms of "weapons" that increase the chances of success in interactions with other males, such as antlers, horns, enlarged canine teeth, or larger overall size. Alternatively, females might prefer males with certain features, such as elaborate songs, coloration, or elongated tails or other plumes. This situation would result in higher success in mating by males that developed these features more prominently. Darwin proposed that females' refined tastes for beauty might result in the evolution of the spectacular plumage or songs of some birds. Thus, in its first introduction, sexual selection was presented as an evolutionary process peculiar to mate choice. It took several generations, however, for biologists to express clearly why the evolution of mate choice had special properties.

For the first few decades following *The Descent of Man,* a primary objection to Darwin's proposal focused on the possibility of females' preferences. As astonishing as it seems today, many biologists were unwilling to accept that females of species other than humans had the cognitive abilities to appreciate beauty in their mates. In the early decades of the twentieth century, Julian Huxley and R. A. Fisher laid this doubt to rest by pointing out that genetic tendencies of females to respond to certain kinds of stimulation rather than others were enough to result in sexual selection.

Fisher went on to describe the process of sexual selection in a little more detail than Darwin had. Despite being one of the great mathematicians of the century, his two descriptions of sexual selection (1915, 1930) included no math. His exclusively verbal descriptions suggested that sexual selection would produce accelerating (snowballing) evolution of the male trait preferred by females and that the process would eventually stop when the decreased survival of males as a result of expressing the trait balanced their increased success in mating.

Discussions of Fisher's process of sexual selection have often been muddled, in part because it is not clear what Fisher actually had in mind. The problem that he—along with Darwin previously and others afterward—sought to explain is the apparently profligate exaggeration of male ornamentation, such as peacocks' trains, birds' songs, and a host of other examples of extraordinary coloration, structures, and behavior in one sex. This exaggeration of ornamentation suggests accelerated evolution as well. In addition, closely related species often differ dramatically in sex-related ornamentation, a phenomenon noted by Darwin and many others. It thus appears that sex-related ornaments can evolve both rapidly and divergently.

TWO MECHANISMS FOR THE EVOLUTION OF EXAGGERATION

It turns out that there are now two well-documented mechanisms for accelerated evolution of signals, and it is not clear which one Fisher had in mind. The first is *frequency-dependent selection*. It should apply to all forms of communication, including mate choice. Frequency dependence arises because the spread of alleles for signaling and responding depends on how many individuals in a population carry these alleles. Alleles for signals and for responses spread slowly in a population when only a few individuals carry them. The reason is that signalers rarely evoke responses, and receivers rarely encounter signals. Mutations for new signals or responses encounter even greater problems. It is difficult for alleles associated with a new signal (or a new response) to begin to spread in a population because initially there are no receivers to respond to them (or appropriate signals to respond to). At intermediate frequencies, alleles associated with advantageous signaling and responding increase progressively faster. Every new receiver for a signal increases the benefits of signaling, and every new signaler increases the benefits of responding. The spread of alleles associated with signals and responses is self-reinforcing, at least at first. The more they spread, the more individuals carrying these alleles benefit. The strength of natural selection on these alleles is thus frequency-dependent.

This sort of process was demonstrated by Peter O'Donald in the first quantitative model of evolution by sexual selection. His simulation calculated changes in allele frequencies in a population as a result of advantageous mate choice, but he did not track genotypes. Consequently, O'Donald's model could not include genetic correlation (discussed below). He obtained strongly S-shaped curves for the spread of alleles for signals in a population. In other words, spread of a rare signal accelerated dramatically at least at first, but spread of a signal decelerated as it approached fixation. O'Donald pointed out that his model predicted the accelerating evolution of male ornaments expected from sexual selection. In fact, however, this sort of frequency-dependent accelerating selection would apply to any signaling system and not just to communication about choice of mates.

O'Donald's simulation indicated that the spread of alleles eventually decreases as fixation of the alleles in the population approaches. As the number of individuals with genes for signaling approach saturation, the advantage of responding to signals decreases. If most of the males are signalers, then there is a good chance of mating with one just by choosing at random. As the advantage of responding decreases, the alleles for responding spread more slowly. In turn, the advantage of signaling decreases, so alleles for signaling also spread more slowly. The initial accelerating and final decelerating evolution of signals and responses, as the alleles spread in a population, resembles the evolution of any advantageous allele. Initial acceleration (as genetic variation increases) and final deceleration (as variation decreases) characterize the spread of an allele in a finite population, even when the advantage conferred by the allele (the strength of natural selection on the allele) remains constant. The advantages of alleles for signaling and responding depend, however, on how many complementary individuals there are in the population. So alleles for communication are subject to frequency-dependent changes in the strength of natural selection. Their spread is at first explosive and later damped, in comparison with alleles not associated with communication. Accelerating evolution of signals by frequency-dependent natural selection thus applies to all forms of communication.

The second process for accelerating evolution of signals and responses requires *genetic correlation* and applies only to attraction and choice of mates. It was explained almost simultaneously by two theorists around 1980, Russell Lande and Mark Kirkpatrick. A genetic correlation arises when individuals with a particular allele at one locus in their genome also tend to have a particular allele at another locus. For instance, individuals with allele A (as opposed to allele a) at one locus would also tend statistically to have allele B (as opposed to allele b) at another locus somewhere else in the genome. One way this correlation between alleles at two loci can occur is by linkage of the two loci on the same chromosome. Eventually crossing over between chromosomes tends to eliminate this sort of genetic correlation. For historical reasons, however, this form of genetic correlation attracted attention as soon as chromosomes were discovered early in the twentieth century. It was called linkage disequilibrium and to this day population geneticists use this term for any genetic correlation although in fact linkage disequilibrium is a special case of genetic correlation.

In Lande's and Kirkpatrick's models, genetic correlation arises in a completely different manner. Whenever males with a particular allele at one locus tend to mate with females with a particular allele at another locus, their progeny tend to inherit both alleles. The progeny thus contribute to a genetic correlation between the alleles at these two loci in the population. For instance, consider males with an allele T that results in an ornamental trait (hence the T, as opposed to males with t that retain dull traits). Further consider females with an allele P that results in mating with males with the ornamental trait (as opposed to females with p that do not care and thus mate at random). In this system, females carrying P tend to mate with males carrying T, while females with p mate with any male they encounter. The progeny of the P females and T males tend to have both alleles. At the trait locus, they have T; at the preference locus, they have P. Of course, we assume that the trait locus is expressed only in males and the preference locus only in females, but the progeny, both males and females, still have both of their parents' alleles. They contribute to a genetic correlation between the T and P alleles.

Now imagine what happens as this process continues. Lande and Kirkpatrick showed mathematically, in two rather different ways, what happens. As generations pass, the number of individuals with both T and P tend to increase in the population, because T males have more possible mates than do the t males as a result of the P females that mate with them. The preferences of the P females thus produce an advantage (a special case of natural selection) for the T allele. The crucial point, however, never explicitly acknowledged before Lande and Kirkpatrick, is that the P females also produce selection for the P allele. When a P female mates, her mate not only carries the T allele but also has a greater than random chance of carrying the P allele, as a result of the genetic correlation between these alleles. This case is thus an unusual situation in which an allele produces selection that spreads itself. The result is self-promoting spread of the P allele. Furthermore, the effect of the P allele in spreading itself increases as the P allele spreads in the population. The result is an accelerating increase in its frequency. In this case, evolution is explosive in the strict sense. This mechanism explains accelerating evolution of male ornaments, just as does frequency-dependent selection.

In species with reversed sex roles, competitive females and choosy males evolve by spread of T alleles in females and P alleles in males. Just as in the usual arrangement of sex roles, the P alleles tend to spread themselves. Or, in another way to look at it, the P alleles in one sex produce selection for T alleles in the other sex, and the P alleles come along for the ride, indirectly, as a result of the genetic correlation of the trait and the preference loci.

This accelerating evolution as a result of genetic correlation differs in an important way from evolution by frequency-dependent selection. Genetic correlation can result in the spread of arbitrary male ornaments, but frequency-dependent selection cannot. An arbitrary ornament is one that has costs for the males that develop them but has no consequences for the females that prefer them. Females do not have to realize a benefit from mating with ornamented males (although subsequent work has shown that they cannot incur net costs). In these models the costs to males of developing ornaments are compensated fully by the advantage they gain in mating. In contrast, Alan Grafen's models discussed in Chapter 12, like the frequency-dependent models, require that females with preferences receive some benefit from mating with males with ornaments, as might happen if those males had higher quality as providers of parental care, assistance for mates, or good genes for progeny.

Another difference between the two processes concerns the direction of evolution. Genetic correlation can result in the evolution of male ornaments of any sort. In particular, there is nothing that favors the evolution of bigger or brighter traits in males rather than smaller and duller traits. Any arbitrary female preference and its corresponding male trait can spread as a result of a genetic correlation between them. In contrast, models of frequency dependence and Grafen's models explicitly include both costs to signalers and benefits for receivers. Recent work has shown that benefits for receivers can affect the kinds of signals likely to evolve.

In both Lande's and Kirkpatrick's models, genetic correlation can produce explosive evolution of both ornament and preference. The eventual result can be either a stable equilibrium with both T and t alleles and P and p alleles persisting in the population or fixation of the T and P alleles. What happens depends on the starting conditions (the initial frequencies of the alleles). Both models include a threshold that must be crossed before the self-promoting or explosive evolution begins. In Kirkpatrick's model the relevant alleles must reach a minimal frequency in the population. In Lande's model the genetic correlation between the alleles must reach a minimal level. In small populations, these minimal criteria might arise by chance and thus explosive evolution would also occur by chance. These models thus raise the prospect that some traits of males and the preferences for these traits might evolve in a completely random way, with no advantages for the females or males, solely by the self-promoting evolution that results from the genetic correlation.

FISHERIAN EVOLUTION

For our present purposes, there are several important points to make. First, it is unclear whether Fisher had in mind genetic correlation (then and now known as linkage disequilibrium) or simply frequency-dependent selection when he proposed a mechanism for sexual selection. His brief descriptions are not clear about this detail. In letters published posthumously (and over a decade after Lande's and Kirkpatrick's papers), Fisher's response to a critic mentioned correlated evolution, but even this letter is not explicit about the importance of genetic correlation as opposed to phenotypic coevolution. It thus might be more appropriate to call the results of genetic correlation a Lande-Kirkpatrick mechanism, since they were the first to make it clear. Nevertheless, *Fisherian* is a well-established term in this context.

More important is some clarification about the role of genetic correlation in sexual selection. Mathematical models can make assumptions that either include or exclude genetic correlation, but genetics in the real world does not have this flexibility. If individuals with one set of alleles mate with individuals with another set of alleles, the result is bound to be genetic correlation. Consequently, a Fisherian process (in other words accelerating evolution as a result of genetic correlation) is not one among several alternatives for the evolution of male traits with benefits for females. Whether or not female choice has benefits for females, if male ornaments and female preferences are associated with at least somewhat different sets of alleles, then genetic correlation occurs and evolution, once past a threshold, is self-promoting.

Sexual selection is thus a special case of the evolution of signals and responses. It occurs only when the signaler and receiver mate with each other. So it applies only when signals by some individuals of one sex increase their chances of mating with individuals of the other sex. All such cases produce genetic correlation between alleles for signals and responses and result in self-promoting evolution. No other examples of communication have these consequences.

The term *Fisherian process* is sometimes thought to apply only to the evolution of arbitrary traits—those with no consequences for females. This use of the term is misleading. The Fisherian process (accelerating evolution by genetic correlation) occurs in all cases of mate choice, whenever signal and response have some association with alleles. As a result, it is not possible to separate sexual selection as a result of advantages for females (often called a good genes mechanism) from sexual selection of arbitrary traits (often misleadingly called a Fisherian mechanism) by measuring the genetic correlation between signal and preference. Genetic correlation always results from sexual selection of any sort.

It is hard to devise a test for the evolution of arbitrary ornaments. Genetic correlation is not diagnostic for this sort of evolution. The only recourse is to exclude all possible advantages for females. The evolution of arbitrary traits as a result of genetic correlation thus has the same status as a null hypothesis. It is what can happen randomly when nothing else happens. It is a case of the evolution of neutral traits.

Returning to Grafen, we can now make two further observations about his models. First, although he provides a demonstration that sexual selection can theoretically occur without genetic correlation, we already know another way for this situation to occur—by frequency-dependent selection. In fact, frequency dependence possibly underlies the results of his model. Furthermore, even though it might be an interesting conceptual exercise to imagine mate choice without genetic correlation, in reality it never occurs (at least if we exclude situations in which either the male ornament or the female preference have zero heritability). Grafen's model of sexual selection thus nicely confirms his more general results for the evolution of communication, but it does not illuminate the special case of sexual selection.

INDIRECT MATE CHOICE

Although arbitrary mate choice remains a possibility, evidence increasingly suggests that individuals choose mates for advantages that they can provide. Most research on mate choice investigates signals that serve to identify these advantageous mates. Communication is assumed to be a requirement for optimal mate choice. Nevertheless, as my former student Joe Poston and I have suggested there is an alternative for mate choice that involves no communication between potential mates at all. When faced with enough uncertainty in the association of males' signals and their quality—in other words, faced with especially noisy communication—females might do best to avoid communication altogether.

For instance, females might set conditions for competition between males and then mate with whichever individual prevails in this competition. Unlike direct mate choice, which requires direct assessment of the signals from potential mates, indirect mate choice does not require receivers to attend to signals at all. In many situations, errors from indirect mate choice seem less likely than those from direct mate choice.

During direct mate choice, the choosing individuals rely on signals associated with the quality of potential mates. These signals must honestly indicate the signaler's good genes (those that confer higher expected survival on progeny) or favorable phenotype (absence of disease or parasites or ability to provide more resources or assistance for offspring). As discussed in Chapters 10 and 11, such communication, even when honest on average, is nevertheless inevitably noisy. If the frequency or consequences of error are too great, it might not pay for receivers to rely on signals at all. Particularly when mates are chosen quickly, costs of prolonged searches are high, or the neural mechanisms required for complex sequential comparisons are inadequate, frequent errors are likely.

Indirect mate choice can provide a thorough assessment of potential mates without the need for complex neural mechanisms. A female's behavior, for instance, might set conditions for prolonged competition between males, so that a male's success required continual survival, health, or command of resources. If such a female mated indiscriminately with any male succeeding in this competition, she would have indirectly chosen a thoroughly tested, high-quality mate.

Poston and I suggested that indirect mate choice could take a variety of forms. One possibility would require breeding females to choose particular locations that provide the best resources for reproduction (for instance, secure nest sites or abundant food for offspring) and to mate with any males that can defend such locations. If both males and females could identify these locations, it would then pay for males to compete with each other for control of these locations. Successful defense might require repeated interactions among rivals over periods of days, months, or even years. Males defending these locations are thus likely to have been repeatedly tested for their quality, so they would have high genotypic or phenotypic quality with a high probability. Females might arrive at these locations and choose to mate within minutes. By relying on indirect mate choice (mating with whichever male defends the optimal location) rather than direct mate choice (mating with the male with the brightest plumage), females could have shifted their decision to a mechanism with lower chance of error.

This process requires only that both males and females agree on the optimal locations. Many male hummingbirds defend patches of flowers with abundant nectar, where females of the same species both feed and copulate with the defending male. Male Red-winged Blackbirds provide another well-studied example. They defend small territories in marshes, especially in locations with secure nest sites and abundant food. Experiments indicate that females choose a location rather than a male when they decide where to nest. The most successful males defend areas large enough for two or more females' nests and then sire most of those females' young.

A special case of choosing a location, rather than a mate, could explain the social interactions of some lekking species. In these species, males congregate in small areas where they defend tiny territories or at least establish dominance relationships. The area where males congregate is called a lek, from Scandinavian terms for "play" or, by extension, "courtship." Males defend their positions on a lek for prolonged periods, despite the absence of food or shelter. Females visit these leks solely for copulation following a few visits apparently to examine the situation. In all lekking species, females then leave to incubate their eggs and raise their young without help from a male. In tropical forests, female manakins visit nearby leks several times as they lay eggs in a new nest. With prolonged nesting seasons and high predation on nests, females nest repeatedly through most of the year. Males remain in their territories all day long, throughout most of the year, except for brief absences to feed.

The lekking species of grouse of Eurasia and North America, in contrast, have restricted laying seasons lasting only a few weeks each spring. Females visit leks only during these weeks, each female for just a few visits. Males, on the other hand, occupy their places on leks every

FIGURE 13.1. Females around a central male on a lek of Greater Sage-Grouse in western Wyoming. A neighboring male on the left is almost encroaching on the small territory defended by the male to the right of center. In the distance the white chests of other males can be seen against the sagebrush at a second mating center on this large lek. On this morning at the peak of the brief period in April when most matings occur, females jostle to hold positions in a small area—in this case, mostly in one male's territory. He might copulate more than 20 times on such a morning, while other nearby males never copulate. The females seem not to pay attention to the nearby males; in fact, they often avoid them. The question is, are they carefully choosing one male by his superior attributes, or have they gathered at a traditional site to mate with whichever male can defend it? This is the alternative between direct and indirect mate choice.

morning and evening for weeks before any females arrive and for weeks afterward, even during autumn. Only during the exigencies of molting during summer and the challenges of survival during midwinter do males fail to attend their leks. My observations on leks of Greater Sage-Grouse in western North America suggested that females pay little attention to males on leks they visit. Indeed they tend to avoid them. Instead they congregate with other females near the center of a lek, where after a few visits they copulate with the males able to defend parts of this central location. Those males perform a large proportion of all matings and many males at a lek never mate at all in any one season (see Figure 13.1).

There are two plausible explanations for the interactions between males and females in these central locations. One possibility is that females rely on direct mate choice. In this case, females visit leks in order to compare the traits of males. A lek in this case is like a smorgasbord of males (to use another appropriate Scandinavian term) for direct mate choice by females. Several investigators have reported that features of plumage or behavior correlate with a male's success in mating, but it has proven difficult to find consistent evidence of these relationships. Nevertheless, most females prefer one or more particular individual males for copulation. Perhaps a tendency for females to copy each other's preferences can explain this high concordance in choice. Copying might also improve the chance of identifying an optimal mate despite their noisy signals.

A second possibility is that females rely on indirect mate choice. In this case, females visit traditional locations and then copulate with whichever male they find there. A high concordance in females' choices of males requires that they copy each other's choices of a location, rather than choices of signals from potential mates. Female sage-grouse in their first nesting season arrive on leks nearly a week later on average than older females, so young females might copy the older females' remembered preferences for locations. Whenever many females choose to mate at a traditional location, learned by both males and females, then males do best when they compete for control of this location. For some lekking birds there is evidence that success in defending a central location requires a prolonged period of time. In their first breeding season males often establish positions at the edge of a lek. Depending on their luck, their survival, their strength, their endurance, and perhaps their savvy in managing relationships with rivals, they gradually move their positions toward the center of the lek. Each spring the males in control of the preferred locations have managed to maintain consistent competitiveness for a period of months or years. Their location might thus be a better predictor of their overall competitiveness than is their immediate behavior or appearance.

Females mate each spring after several visits to leks and thus only a few hours of experience with displaying males. They might have greater success in choosing a male with proven ability to survive and to compete by choosing a traditional location than by choosing a male with the longest tail or loudest sound. Even if a female could make this direct discrimination among males accurately, she would be choosing a male by qualities that only reflected the male's recent achievement rather than a lifetime of achievement. It thus seems possible that females could choose a mate with genes associated with survival and health more accurately by indirect choice of a traditional location than by direct choice of a particular feature of his phenotype.

There are other possibilities for indirect mate choice also. Females' behavior also determines how "defensible" they are. When female mammals cluster in herds and mate with whichever male can defend the herd against his rivals, they again exert an indirect mate choice. By adjusting their tendencies to congregate or to drift apart, they can determine how much activity a male must exert to maintain defense of a group of females. Females could also behave evasively. A female that fled from every approaching male but then mated with whichever male had the strength and savvy to catch her would also have an indirect mate choice for males with these qualities rather than for less-qualified, males.

Females could stimulate competition simply by advertising their fertility or even by attracting the attention of males. Such behavior or morphology of females would be another form of indirect mate choice if an attractive female evoked competition among males to guard her from rivals. By mating with any male that could guard her consistently, she would exercise indirect mate choice for competitive males. If an ability to guard a female in a large social group required development of high dominance in the group, a female's indirect mate choice would result in matings with males tested by competition with rivals over periods of months or years.

In all of these cases, indirect mate choice has the potential to avoid the problems of error in direct mate choice. As a form of communication, the latter has all of the problems of noise in brief interactions. The former has the double advantage of requiring no communication and subjecting potential mates to prolonged tests. Females do not have to rely on spur-of-the-moment decisions.

On the other hand, direct and indirect mate choice both lead to sexual selection and the consequences of accelerating evolution as a result of genetic correlation between alleles influencing the phenotypes of mates. In direct choice, as described above, a preference by females for males with a particular phenotypic trait results in an advantage for males with that trait and a genetic correlation between the alleles associated with the preference and those associated with the trait. As a result the preference tends to spread itself. Furthermore, as the trait escalates so does the preference. In indirect choice, the same processes occur. Female behavior that results in mating with males with particular attributes creates an advantage for males with those attributes. A tendency of females to mate in a traditional location generates an advantage for males that compete successfully for a position there. Whatever traits of males contribute to success in that competition would tend to spread in a population. These traits might include such conspicuous features of morphology as large size, spectacular plumage, or special weapons such as antlers, but might also include less obvious traits such as savvy, endurance, or overall health. Furthermore, any alleles associated with the females' tendency to mate in a traditional location would eventually have a genetic correlation with any alleles associated with these traits of males. The alleles influencing female behavior would eventually reinforce their own spread in the population. The males' traits and the females' behavior would both escalate. Indirect mate choice thus leads to sexual selection despite the absence of communication.

This coevolution of female preferences and male behavior has an interesting consequence for indirect mate choice. As just mentioned, advertisement of a female's potential fertility is one possibility for indirect mate choice by setting conditions for competition between males. In this case the indirect choice is a result of both females' tendencies to copulate with whichever male can guard her and also her traits that advertise immanent fertility. These traits are partly behavior but also morphology, such as the estrus perineal swellings in primates that live in troops with multiple males as well as, perhaps, features of the morphology of human females, such as youthful skin, breasts, and waists. Because indirect mate choice involves the same processes of sexual selection as direct mate choice, the males' traits and the corresponding females' traits coevolve. They evolve exaggeration together. Indirect mate choice thus could result in exaggerated evolution of female signals as well as male traits. Direct mate choice would result in exaggerated evolution of female preferences, as well as male traits, but female preferences are less obvious to an observer in the field than are female signals that serve to serve to advertise fertility.

Indirect mate choice changes the role of sexual selection in the coevolution of the sexes. Because it includes so many possibilities for a female's behavior and morphology to influence her set of potential mates, the scope for sexual selection in the evolution of organisms with separate sexes is greatly expanded. Any aspect of a female's behavior or morphology that reduces the set of males with which she might mate initiates the process of sexual selection with its special dynamics.

Indirect mate choice has received almost no attention from behavioral ecologists interested in the evolution of mating systems of animals. The original reports describing the mathematics of sexual selection were all phrased in terms of female "preferences." This term seems to suggest a direct discrimination by females, but in fact any behavior or morphology of females that restrict their matings to a subset of available males follows the same mathematical dynamics. In addition, Darwin's original description of sexual selection identified two components: female preferences and male-male competition. He presented them as two potentially unrelated mechanisms. Male-male competition in Darwin's connotations led to a few males able to gain access to females. He never suggested that females actually control the conditions for males' access to them. Nevertheless, female behavior is undoubtedly the crucial factor in male competition for mates in all species. There are very few cases in which males physically force females to copulate, and some of those are questionable. In lekking birds, in particular, males do not force copulations. Instead, female behavior sets the conditions for male competition for mates. Darwin's two mechanisms are actually examples of indirect and direct mate choice. These two forms of mate choice share the same evolutionary dynamics. Perhaps their most important distinction is their consequences for error by females. Indirect mate choice offers the possibility for reducing the errors that attend noisy communication in direct mate choice.

EXAGGERATED SIGNALS

Either direct or indirect mate choice could result in the evolution of exaggerated signals. In the case of direct choice, exaggerated signals could promote responses by females. In the case of indirect choice, exaggerated signals could discourage rivals. In one case, females judge the quality of individual males, in the other males judge each others' qualities. Both the attraction of mates and the repulsion of rivals often involves long-range communication, at least at the early stages of either process.

If the males on a lek were a "smorgasbord" for female direct choice, they might all benefit by attracting more females to the lek from a distance. As shown in Chapter 8, each male is expected to display with an exaggeration correlated with his quality, but every male would benefit from an overall increase in the probability of attracting a mate. If, on the other hand, males on a lek were defending their access to a site preferred for mating by females, they could again benefit by attracting more females to the lek. The advantage would be greatest if nearby males were likely to succeed to the preferred spot when the current defender dies. Both direct and indirect choice thus are likely to result in exaggerated signals. In both cases, females should evolve high thresholds for response. Furthermore, both cases often include long-range communication with high levels of extraneous noise.

The evolution of the spectacular traits and behavior of animals in conjunction with mating needs two kinds of explanation. The dynamics are explained by sexual selection with genetic correlation, and the direction of evolution is explained by signal detection in noisy communication. Exaggerated signals in other contexts could also have accelerating dynamics, as a result of frequency-dependent selection. In all contexts, whether mate choice or not, the direction of evolution results from signal detection. Evolution should proceed toward signals with greater contrast, redundancy, and predictability. As explained in Chapter 6, these three features of signals improve the performance of receivers and thus have advantages for signalers. They also have disadvantages that increase the marginal costs of exaggeration. A complete explanation for the evolution of exaggerated signals thus requires a combination of sexual selection, on the one hand, and signal detection in noise, on the other. So far only part of this program has been broached. Most research has focused on direct choice by females and absolute costs of signaling by males. Much less has considered indirect choice by females, the performance of receivers, and the marginal costs for signalers.

14

COOPERATION BY COMMUNICATION

RECOGNITION IN COMPLEX SOCIETIES

CHAPTERS 12 AND 13 have explored two fundamental features of communication. First, communication in noise explains some of the salient features of communication, especially the exaggeration of signals and the inevitability of errors. Exaggeration should accentuate not just increased intensity or magnitude of signals in general but specifically features that counteract noise, including contrast, redundancy, and alerting components. Second, communication in noise is inescapable. As long as costs and benefits influence the evolution or design of communication, the conflicting errors of receivers assure that both receivers and signalers face diminishing returns in approaching perfection in communication. An equilibrium is reached that perpetuates noise. Although honesty is the norm, noise is inescapable, and perfection in communication is unattainable.

This equilibrium is the goal toward which natural selection tends to move populations of signalers and receivers. In Chapter 7 we described natural selection as a difference in the spread of alleles in a population as a result of a difference in the numbers of progeny left to future generations by phenotypes associated with these alleles. This paraphrase of natural selection needs an important modification when the behavior of an individual influences the survival or reproduction of another individual. In this case it is necessary to take into account the changes in the survival and reproduction of both individuals as a result of their interaction.

A behavioral interaction between two individuals can be either beneficial or costly for one or both individuals. When these costs and benefits include changes in the survival or reproduction of the parties involved, they affect the spread of alleles associated with the individuals' behavior. From this point of view, behavioral interactions fall into two categories: cooperation, when one individual's actions increase the survival or reproduction of another, and antagonism, when an individual's actions decrease the survival or reproduction of another. In each case the interaction might also decrease or increase the actor's survival and reproduction.

The consequences of an interaction for the spread of alleles can depend on which individuals in a population are involved. Information about the previous behavior of an individual might predict the consequences of interacting with it. Furthermore, an actor's genealogical relatedness to a recipient makes a difference for the spread of alleles associated with the actor's behavior. Consequently, the way individuals recognize and categorize other members of their species affects the costs and benefits of their interactions and the spread of their genes. Often investigations of the evolution of social behavior, either in theory or in observation, take for granted individuals' capabilities to categorize each other appropriately—in ways that maximize the spread of relevant genes.

Recognition is a special case of categorization, and both are forms of communication. An individual responds toward others based on signals received from them. This communication, as we have seen, is bound to be noisy. Yet studies of social behavior rarely consider the consequences of mistakes of recognition. There is so far little that can be said, either theoretically or experimentally, about noisy recognition of social partners. Yet to begin thinking about the implications of noise for recognition and thus for the evolution of social behavior, it is important to consider what we do know about recognition and about its imperative in complex societies.

RECOGNITION IN NOISE

Categorization of other individuals is perhaps the fundamental cognitive process in social behavior. In its minimal form, individuals might accomplish no more than classifying other organisms into just two categories: members of the same species and all others. Such a minimal classification would support a congregation of individuals of the same species. Some such congregations, such as schools of fish or flocks of roosting birds, achieve striking coordinations of spacing and movement when individuals respond to each other based on their current spatial relationships. There is no need in these cases for individuals to distinguish between other members of the congregation in any way. Everyone elicits the same response in relation to its current relative position. As soon as individuals differentiate two or more categories of conspecifics (individuals of the same species), social behavior acquires rudiments of the complexity in the most elaborate societies.

When an individual adjusts its responses to other individuals (or sets of other individuals) it recognizes them. Recognition of individuals is thus is a form of communication in which receivers discriminate (respond differently to) signals associated with sets of other individuals. Noisy communication therefore affects the properties of recognition in the same sorts of ways that it affects other forms of communication. In the first place, inevitable errors in noisy communication lead to inescapable uncertainty in the recognition of other individuals. Noisy communication also forces a receiver to categorize its sensory input, including features of other individuals, in a way that communication without noise would not.

Categorization results from the dual nature of error in noisy communication. As already emphasized, a receiver cannot simultaneously minimize both types of error. An alternative to noisy communication might instead include error as variation around a mean response. This sort of error does not affect the evolution of communication much. The association of a particular signal with a particular response acquires some uncertainty, so the benefits of responding and signaling vary around their means. Yet the mean values still govern the direction of evolution by natural selection on responses and signals. Furthermore, there is no restriction on minimizing error. In contrast, in noisy communication the two sources of errors are not independent of each other and cannot be simultaneously minimized. Error cannot be minimized to an arbitrarily small degree. The two possible errors each time a receiver checks its sensors are a bifurcation in a receiver's subsequent experience. Noisy communication thus assures that responses to external stimulation, including signals from other individuals, categorize stimulation and signals. Recognition of sets of conspecifics as a result of noisy communication thus has some features it would not have in the absence of noise.

In a paper in 2013, I emphasized that categorization of conspecifics varies in both specificity and multiplicity. *Specificity* is a term for how narrow the sets of other individuals are. A set might, for instance, include a single other individual, the most specific form of recognition, or sets might include two or many more individuals, such as sets of an individual's parents, siblings, brood mates, rivals defeated or submitted, mates, or potential mates. Possible sets of individuals recognized also include those associated with previous events or locations or just familiar individuals without further distinctions. *Multiplicity in recognition* is a term for how many sets of other individuals there are. Binary recognition is the minimal case, in which only two sets of other individuals are differentiated. A frequent case of binary recognition is mate versus any other individual. Other cases include offspring as a group versus all others or parents versus all others. Another widespread form of binary recognition distinguishes between familiar individuals and strangers, those not encountered previously (or not encountered more often than some criterion). In some cases, such as recognition of strangers, the boundary between two categories might not be an abrupt transition but a more graded distinction. Greater multiplicity in recognition requires discrimination of more than two, sometimes many more, categories of individuals.

Nevertheless, we know only a little about the limits of multiplicity in recognition, even for humans. There are no estimates of the number of individuals, or sets of individuals, that a person recognizes. Much of the research on face recognition by humans focuses on the psychological mechanisms for identifying of individuals or sets of individuals. Are individuals or sets of individuals recognized by particular features or by relationships among features? Sets might in theory be recognized by a common feature (or a common relationship among features) shared by all members of the set. The problem is similar to the recognition of sets of objects in general. As Ludwig Wittgenstein pointed out in another context, sets of exemplars might also be recognized by "family resemblances," overlapping but incompletely shared sets of features. Such a set, for instance, includes all individuals with features A, B, or C ("or" in this context is the inclusive "or" in logic, commonly expressed as "and/or"). In this case there is no single feature shared by all individuals in a set but instead a complex of overlapping partially shared features, a family resemblance.

My review emphasized that humans also recognize other individuals with a diversity of levels of specificity. Some individuals we recognize as individuals. We respond to them in a different way than we would respond to any other individual, for instance by calling an individual by name or discussing the individual's garden, performance on a task, or seat in a classroom. Others we might recognize as a member of a set of individuals. Some we recognize from class, our neighborhood, or a picture in a newspaper but cannot associate them with anything more specific. Some we recognize in a familiar context but fail to recognize in a novel context. Some we only recognize as familiar—someone noticed previously but without further associations.

As recognition of other individuals varies in specificity and multiplicity, it is not surprising that the neurological mechanisms of recognition also vary in complexity. Some forms of specific recognition require no more than habituation to familiar stimulation. Habituation of responses to a repeated or continual stimulus occurs in all organisms with nervous systems. It provides the only mechanism needed to recognize familiar as opposed to strange individuals. Other cases require associative learning and complex cognition.

RECOGNITION OF RIVALS

Study of the social behavior of animals usually focuses on aggression, cooperation, mating, and parental care. Each of these poles of social behavior inevitably involves recognition of other individuals. Surprisingly little is known in any of these contexts, however, about specificity, multiplicity, or error in recognition.

Consider, for instance, aggression. In all animals, aggressive behavior is usually directed more toward some individuals than others. Often it is more intense between individuals of the same sex than between those of different sexes. Often there are differences in aggression toward young and adult individuals. Frequently, aggression is less severe or frequent toward familiar rivals than toward strangers. These cases require binary recognition of two sets of rivals. In the last case, habituation alone can provide the mechanism for this binary recognition.

In a few nonhuman animals we have evidence for more complex recognition and more complex differentiation of aggressive responses. In these cases an individual's aggression is directed toward multiple specific individuals each in specific locations. Consider the case of songbirds in which males defend territories during the breeding season, such as the Hooded Warblers introduced in Chapter 1. It is often observed that aggression between two neighboring males at the beginning of the season, when territories have recently been claimed, is prolonged and intense. Within a week or so, however, neighbors rarely engage in aggressive encounters and instead respond to neighbors' songs by singing in return from a distance. Males for the most part sing only within their own territories and rarely confront each other at their mutual boundaries. Experiments with playbacks of individual males' song confirm that each male recognizes its neighbors by the individual idiosyncrasies of their songs.

The first such experiments compared a male's responses to a neighbor's song and a stranger's song. They thus only provided evidence that these territorial males discriminated familiar songs from songs never heard previously. Subsequent experiments, following the lead of Bruce Falls and his student John Brooks, showed that territorial male songbirds differentiated the songs of each neighbor depending on where the song was played. My wife, Minna, and I found similar results, for territorial group-living birds, Stripe-backed Wrens, in Venezuela. Subsequently, my student Renee Godard thoroughly studied this capability in Hooded Warblers inhabiting forests of the Mason Farm Biological Reserve near Chapel Hill, North Carolina.

Godard presented recordings of Hooded Warblers' songs to a territorial male from locations just inside that male's boundary (it takes a couple of hours a day for several days to determine the extent of a male's territory by mapping the locations at which it sings). For instance, a neighboring male's songs could be presented either near the correct boundary (one shared by that neighbor and the subject of the experiment) or near an incorrect boundary (a boundary on the other side of the subject's territory across from the neighbor's usual location). The subjects of these experiments responded more strongly to a neighbor's song in the incorrect position than in the correct position. It was as if a neighbor singing out of place had trespassed on the subject's territory and thus had evoked an antagonistic response. Godard also performed the reciprocal experiments: the song of a second neighbor was presented near its mutual boundary (on the opposite side of the subject's territory from the first neighbor) and then across the subject's territory at the incorrect boundary for that neighbor (although the correct boundary for the first neighbor). The experiments with the two neighbors thus used the same two locations (within a few meters) for presenting recordings of their songs. Each of these locations was near the boundary for one of the neighbors but opposite the boundary for the other. The results of repeated experiments of this sort confirmed that each territorial male recognizes each of its neighbors' songs (not just all of its neighbors collectively) and associates each with the correct side of his own territory. Thus when any neighbor's song is presented out of place, the subject responds as if a serious trespass had abrogated any mutual agreement on a boundary and thus required a vigorous aggressive response.

The Hooded Warblers exemplify a general principle: recognition occurs when it has an advantage for the individuals involved. This principle suggests a hypothesis that the neurological mechanisms required evolve only when they provide an advantage for the individual and then only with just enough complexity. When specificity is not needed to obtain the benefits, individual specificity of recognition does not evolve. This situation might apply to recognition of parents or young by birds. There are many experiments that demonstrate that young birds recognize their parents and that parents recognize their young, yet there is no evidence that parents recognize each of multiple offspring individually, nor that young differentiate their two parents. In these cases it is difficult to imagine an advantage for individually specific recognition. "Family resemblance" is apparently sufficient to obtain the benefits. So far evidence suggests that specificity in recognition in these cases has evolved no farther. Natural selection produces complexity only when it reaps benefits for reproduction or survival.

Primates and, of course, humans have still greater capabilities for specificity and multiplicity in individual recognition, although as already noted we still do not know the scope of these capabilities in any case. In primates, as in humans, recognition plays complicated roles in both aggression and cooperation.

RECOGNITION IN COOPERATION

The relative importance of aggression and cooperation in animal and human societies has excited much discussion, perhaps especially in the past century or so. Theories of human history and social organization have in some cases emphasized the importance of conflict between classes or other sets of individuals. Human tendencies to dichotomize others have often been linked to allocations of power in societies, allocations which have sometimes been rationalized by putative analogies with natural selection. The past century has been punctuated by human conflicts more catastrophic than any in previous history. Perhaps partly in reaction there has been complementary emphasis on human cooperation as the characteristic feature of human societies. As a result, there has been a concerted effort to show that cooperation is as compatible with natural selection as aggression is.

Much of this discussion of human social behavior has been confounded by insupportable assumptions about human development or the "natural" condition of humans. The issue is once again the roles of genes and environment in the development of behavior. Chapter 7 of this book has already emphasized that the development of behavior, like all other features of phenotypes, of humans as well as other organisms, results from interactions of genes and environment. Human behavior, whether aggressive or cooperative, thus follows the four rules listed earlier: no feature of human behavior is determined by genes, none is determined by experience (environment), all features of human behavior are influenced by genes, and all are influenced by experience (environment).

Beyond these ontogenetic issues there remains the question of how natural selection affects the frequencies of the alleles of genes influencing behavior in a population of organisms, including a population of humans. This question is easiest to understand when focused on aggression. Ever since Darwin's generation, it has been obvious how tendencies for aggression might evolve as adaptations of social behavior. After all, natural selection results in the accumulation of alleles that promote the survival × reproduction of individuals. To the extent that access to resources influences either survival or reproduction, alleles associated with behavior that increases an individual's control of resources would spread in a population. Expressed in the terms of ecology and evolution, competition for limited resources results in natural selection for control of those resources. The behavior expected from such a process is not necessarily objectionable. Many people would probably agree that assertiveness in claiming a fair share of resources is acceptable behavior—at least as long as it does not compromise another individual's fair share. Economic competitiveness is also widely valued in a society. Of course assertiveness and competitiveness, although influenced by genes that accumulate in a population by natural selection, is also influenced by experience.

Cooperation, on the other hand, has no such facile explanation. Particularly when cooperation involves behavior that has some disadvantage to a helping individual (despite some advantage to a recipient), it has no simplistic explanation in terms of the spread of alleles associated with helping. Instead alleles associated with helping should spread less than those associated with receiving help. Behavior that might solicit help from others but not return it would spread in a population, if all else were equal, and would eventually eliminate alternative alleles associated with helping. The problem of understanding the evolution of cooperation is thus the problem of how alleles that promote helping others at a cost to the helper can spread in a population.

One prominent theory to explain how such alleles might spread in a population is group selection. Until about 50 years ago, especially during the wars of the 1940s, group selection provided the accepted explanation for the evolution of cooperation and helping. If groups that included alleles associated with helping (and thus cooperation) were more successful in competition with other groups with less cooperation, then the cooperative groups would prevail while the less cooperative ones would dwindle and eventually vanish. During the 1960s, however, a number of biologists pointed out that group selection needed more careful thought. The basic problem is the interplay between any advantage of cooperative groups in competition between groups and any disadvantage of helping individuals within their own groups. Could the spread of cooperation by competition between groups outrace the diminution of cooperation by competition within groups?

There are a several points to clarify here. First of all, in cases of cooperation between two individuals, the problem is only interesting if helping involves some disadvantage for the helper. If a helper benefits individually as much as the recipient, then it is easy to imagine that alleles associated with this sort of helping could spread in a population. One way that a helper might benefit is through reciprocity. The recipient of help might return the favor so that both parties benefit. The helper's immediate cost of helping is then compensated by its eventual benefit from reciprocity. The possibility of reciprocity does not, however, provide a simple explanation for the evolution of helping and cooperation. The problem is that most reciprocity involves some contingency. Because a recipient of help might or might not reciprocate at some later time, any contingency whatsoever in the reciprocation of help would create an opening for alleles associated with cheating. Cheating, in this context, is receiving help but not returning it. Those alleles would thus spread and eventually supplant any alleles associated with helping. When helping has an immediate disadvantage, despite the possibility of eventual reciprocity, the spread of alleles for helping within groups remains problematic.

Cases of obligate immediate reciprocity, such as algae or cyanobacteria and fungi in lichens or algal zooanthellae in corals, usually involve different disparate species in a highly evolved relationship. The discussion in Chapter 15 returns to such advanced cases. Cooperation between members of the same species, on the other hand, invariably involves some contingency in reciprocation, no matter how brief.

By-product mutualism is a term often applied to another special case of cooperation. For instance, trees in temperate climates drop their leaves in autumn and earthworms feed on the decaying tissues. The trees thus provide food for earthworms, which in turn macerate and digest the leaves into nutrients for absorption by the same trees the following spring. In this case there are no costs of helping; both earthworms and trees do no more than promote their own immediate advantages. It pays for trees to drop their leaves before winter, and it pays for earthworms to digest them. That each promotes the well-being of the other is an interesting consequence, but not one that requires any special evolutionary explanation. The evolution of cooperation becomes problematic only when helpers incur costs of helping.

As a corollary, in cases of cooperation within groups, the only interesting situation arises when an advantage for groups depends on behavior with disadvantages for individuals within groups. The problem of group selection vanishes if individuals have no option other than cooperation in groups. Ostracism of an individual, for instance, might preclude any chance for reproduction or survival. In this case, participation in a group raises no evolutionary problems, even if the individual is at a disadvantage in comparison to other members of the group, provided all individuals in a group realize some non-zero chance for reproduction or survival. In general, the possibility for reproduction and survival without membership in a group sets the lower limit for expected reproduction and survival of individuals within a group. Above this lower limit, the spread of an individual's alleles is inevitably associated with the spread and multiplication of the group of which the individual is a member. Nevertheless, the spread of alleles within a group remains an issue. If any alleles influence the conformity of individuals to a group, then those associated with usurping resources within a group spread and alleles associated with sharing or cooperating within a group dwindle. Regardless of whether individuals can survive or reproduce without a group, a crucial issue in the evolution of cooperation in a group is the spread of alleles within a group.

A frequent question is, why don't helping individuals just identify themselves? Then they could easily direct their help solely to other helpers and thus assure reciprocity. Identifying oneself as a helper is a form of communication. If helpers produced a signal indicating that they had alleles for helping, then other such individuals could identify them and direct their helping toward those who would return the favor. Individuals not producing this signal would receive no help from others. The problem is that if the genes associated with the signal were not the same as those associated with helping, then individuals that had the alleles for the signal would benefit from receiving help regardless of whether or not they had the alleles for helping also. Such individuals are yet another case of cheating and the associated alleles could easily spread in a population.

Signals that identify a cooperator were dubbed "green beards" by Richard Dawkins in his 1976 best seller *The Selfish Gene*. If all prosocial, cooperative humans grew green beards, one could easily identify them and reserve cooperative action just for them. As Dawkins immediately recognized, the problem is cheaters—individuals with genes for green beards but not for cooperation. In recent decades, however, a possibility for green-beard cooperation has been proposed for microorganisms. It has become clear that cooperation is a prominent feature of many bacteria and protozoans, even viruses. The medical consequences are alarming, because resistance to antibiotics (as well as other adverse environments) results in part from cooperation by microbes in forming resistant films or foams. The molecules that attach cells to each other in these films seemed to be candidates for green beards that would explain the evolution of this form of cooperation. Molecules on the cells' surfaces match each other, like with like, so the signal, the response, and the cooperation all result from a single protein and thus a single gene. The problem of contingency between the signal and the cooperative behavior seem circumvented. Yet it did not take long to realize that cheating is, in fact, the norm among bacteria and protozoans. The molecular mechanisms have been identified in some cases—alleles of the relevant proteins that alter how much benefit a particular cell can realize in proportion to its cost. Some alleles are more selfish than others, and it is clear that selfish cheating reduces the benefit others receive from cooperation. The good news is that such cheaters might provide medical scientists a way to combat the antibacterial resistance that results from bacterial cooperation. Anything that maintains the tenuous ascendancy of humans over bacteria is welcome!

Other possibilities for the evolution of helping require recognition of partners and thus can only apply to complex societies. As we have seen, the spread of cooperation requires that cooperating individuals interact with each other preferentially, in one way or another. Perhaps the most obvious way to meet this requirement is for individuals to keep track of partners that have actually helped them previously and then help only them in return. This strategy amounts to the tactic Tit for Tat—you cooperate with me and I'll cooperate with you. For this strategy to work, of course, someone has to try cooperating first. More important, individuals must interact repeatedly, identify each other, and remember their partners' behavior in the past. Chapter 15 returns to this possibility in more detail. Even more complex recognition and memory is required for indirect reciprocity. In this case, individuals behave cooperatively toward any other individual previously observed or reputed to also behave cooperatively. Even in groups with only a few individuals, the score keeping becomes a challenge. Not often considered in the theoretical treatments of such indirect reciprocity is the possibility that selfish individuals might manage to appear to cooperate more than they do or to cooperate only conditionally in ways that minimize their costs and maximize their benefits. If so, the possibilities for errors in scoring "cooperativeness," not to mention errors in the recognition of possible partners, need more attention.

Despite the complexities that arise, recognition of partners can promote the evolution of helping in every case. Even when recognition is not necessary for the evolution of cooperation, such as when interactions are restricted to local neighborhoods, recognition nevertheless always allows preferential treatment of cooperative partners and thus promotes the spread of cooperation. This conclusion is easier to see in the context of a general principle for the evolution of cooperation.

Alleles associated with helping, rather than those for cheating, can spread in a population provided helping individuals interact with each other sufficiently more often than by chance. If helping and cheating individuals interact with each other at random, then alleles for helping cannot spread in a population. Only if helping individuals interact preferentially with each other (rather than with cheaters) can alleles associated with helping have a chance. This principle can be encapsulated in a general mathematical expression for natural selection with positive interactions, an expression developed by George Price and now recognized as the most general formulation of the process of natural selection. It clarifies the basic condition for the spread of alleles influencing cooperation in a population—cooperative individuals must interact with each other more than expected by chance. In its generality, however, it loses contact with the mechanisms of social interactions.

One simple possibility applies to sessile organisms, those that lack the ability to move around. If such a population includes a random distribution of individuals with alleles for helping and those with alleles for selfish cheating, just by chance there might occur clusters of neighboring individuals with the alleles for helping (and other clusters with alleles for cheating and other areas with other random combinations of the two). Because individuals cannot move, cheaters cannot take advantage of the clusters of helpers. Those clusters thus prosper in comparison with clusters of cheaters or indeed areas of interspersed helpers and cheaters. Provided the density of helpers reaches some threshold by chance, these fortunate clusters can spread and supplant all cheaters. The key is the lack of any possibility for movement, which preserves clusters of helping individuals, so within such clusters helpers are more likely to interact with each other than they are in the population at large.

Even if individuals can move, small partially isolated groups might form at random. Particularly if these groups were small, some of them would by chance consist primarily or even entirely of helpers. If these groups of helpers were effectively isolated for sufficient periods of time from random immigrants (which could include cheaters), they would again prosper while other groups of cheaters or of mixed composition would dwindle. If the rates of group turnover in relation to the rates of individual turnover within groups were high enough, alleles for cooperation could spread in a population. The conditions must be just right.

KIN SELECTION PROMOTES COOPERATION

Kin selection offers another possibility for the spread of alleles for helping in special conditions. Kin selection has been controversial from its introduction in the 1960s. In recent years Martin Nowak and Edward Wilson at Harvard University have challenged the logic of kin selection, but they have also elicited some equally adamant rebuttals. The following discussion presents one view of these issues and, most important in the present context, indicates the crucial influence of noisy recognition on each.

Kin selection, first worked out mathematically by William Hamilton as a graduate student in the early 1960s, almost cost him his doctoral degree when examiners at the University of London doubted it had sufficient significance. He eventually prevailed and, perhaps more important, prevailed over Edward Wilson's doubts. His results also caught the attention of Robert Trivers, a graduate student in the late 1960s at Harvard. Trivers's five papers in the early 1970s provided a context for kin selection in the social behavior of a wide variety of animals. Wilson's trend-setting treatise in 1975, *Sociobiology,* is to some degree an extended discussion of the implications of Trivers's papers and what has come to be called Hamilton's rule. Nevertheless, Wilson has recently changed his mind under the influence of Martin Nowak and disparaged kin selection for logical and biological reasons. Before we get to that development, first consider some basics of kin selection.

Individuals are more likely to share alleles with close genealogical relatives than with others, because of their descent from a recent common ancestor. Provided individuals interact more often than randomly with kin, then individuals with alleles for helping interact with each other more often than by chance. Helping individuals thus tend to provide help for other potential helpers. This situation fits the basic condition for the spread of alleles associated with helping as noted above. In this case, rather than immobility or small isolated populations, the key is interactions between genealogical relatives. It requires an ability to recognize kin or some other mechanism that promotes interactions between kin.

There are many cases in which individuals preferentially help their close relatives. Many of the hundred or more species of birds that are cooperative breeders have, for this reason, received extended study. These birds live in stable groups in which some individuals produce most of the offspring but all help to raise them. These groups turn out, with few exceptions, to consist of close relatives. Many mammals are also known to help their relatives. Some, such as social canids like Gray Wolves and African Hunting Dogs, engage in cooperative breeding not unlike that of birds. In ground squirrels in western North America, females tend to settle as adults near their female kin, while males tend to disperse farther. Individuals call to alert others of the presence of predators only when the nearby receivers are likely to be kin, so males are less likely to alert others than are females. The calls bring some risk to the signaler, so they qualify as help with a disadvantage for the helper. At least in some species of ground squirrels, it is now known, individuals do not actually recognize their genetic similarities to each other. Instead they follow a rule of thumb: "Treat individuals as kin if raised in the same nest with me." In natural circumstances, such individuals almost invariably are genealogical relatives. Experiments by biologists that cross-foster juvenile squirrels among nests reveal that the ground squirrels do not recognize genetic similarity as such and instead follow the rule of thumb.

Kin selection is thus a form of natural selection in which the spread of alleles in a population results from differences in survival or reproduction of genealogical relatives whose interactions are associated with those alleles. It is that subclass of natural selection that results from interactions of relatives. A great advantage of kin selection as an explanation of cooperation is that it makes mathematically precise but simple predictions about when cooperation should occur. This precision comes from the prediction that every one of an individual's alleles has a precise probability that an identical copy is present in a close genealogical relative. Because two genealogical relatives share a recent common ancestor, they are also likely to share copies of each of their alleles—including of course any alleles influencing a tendency to provide help. For instance, in diploid organisms, those with complementary sets of chromosomes from each parent (all birds and mammals and most other vertebrates and invertebrates), each individual leaves half of its chromosomes to each offspring; the offspring inherits half of its chromosomes from its father and half from its mother. So any allele in a parent has a 0.5 chance of occurrence in each offspring. We say that the *coefficient of genealogical relatedness* of an individual with its offspring is 0.5. The same is true for the relatedness of an individual with either one of its parents: any allele in an individual comes either from its mother or its father, half from each, so any allele chosen at random has a 0.5 probability of coming from its mother and a 0.5 probability of coming from its father. The relatedness (to substitute a shorter, albeit less precise, term) of an individual to a sibling is a bit more complicated to calculate, but the principles are the same: an individual has a coefficient of genealogical relatedness with its sibling also equal to 0.5. Its relatedness to a half sibling is 0.25 (1/2) and to a cousin 0.125 (1/8). Because of their recent common ancestors, a close genealogical relative usually has a higher chance of sharing an individual's allele of any gene than does a random individual in their population. In human populations the probability that an individual shares a allele with a randomly chosen member of the population is about 0.1 (averaged over many polymorphic genes), only slightly less than the probability that a cousin has that allele.

Recall that behavior (like all phenotypes) is influenced (not determined) by alleles, so it is plausible to assume that helping is associated with one or more alleles. If helpers directed their behavior toward random individuals in a population, they would help others with the same alleles for helping as well as others without these alleles in proportion to their representation in the population. Their helping would thus not change the frequency of the allele in the next generation. If instead they directed help toward genealogical relatives, who are more likely to have the same alleles for helping than random individuals are, then helpers would spread any allele for helping. Genealogical relatives do not always share the same allele—only as often as the coefficient of genealogical relatedness—but for close relatives it is more often than between random individuals.

Suppose, for a simple example, that helping results in a decrease in reproduction by the helper (say, two fewer offspring) but an increase in reproduction of the recipient (say, five more offspring). Furthermore, suppose that the recipient is a full sibling of the helper, with a 50% chance of having the same allele for helping and a 50% chance of leaving that allele to each of its offspring. Then the number of offspring in the population with the allele for helping is, on average, 0.25 more than the number there would have been if no helping had occurred. With helping, the additional number of offspring with the allele equals the benefit to the recipient (five additional offspring), times the chance it carries the allele for helping, times the chance that each of its offspring carry this allele, or $5 \times 0.5 \times 0.5 = 1.25$. Without helping, the additional number of such offspring equals the cost to the helper (two additional offspring), times the chance that any offspring carries its allele for helping, or $2 \times 0.5 = 1.0$. By taking the additional number of alleles with helping and subtracting the additional number without helping, $1.25 - 1.00$, we obtain the expected increase in the number of alleles for helping in the next generation, 0.25. It all depends on the decrease in helper's offspring as a result of helping (the helper's cost), the increase in the recipient's offspring (the recipient's benefit), and the probability that any allele in the helper allele is also present in the recipient (the genealogical relatedness of the helper to the recipient's offspring and its own offspring).

Hamilton's rule for the evolution of helping by kin selection is $Br - C > 0$. The cost to the actor (C) must be less than the benefit to the recipient (B) devalued by the coefficient of genealogical relatedness (r). Somewhat earlier J. B. S. Haldane had encapsulated this rule in the aphorism that he would sacrifice his life only to save the lives of more than *two brothers* or more than *eight cousins!*

There are some crucial points to notice in the preceding argument. First, the cost and the benefit do not include the entire reproduction or survival of the actor and recipient, only the *change* in these measures as a result of their interaction. Otherwise, double accounting can obfuscate calculations when several individuals help the same recipient or when the recipient has some success on its own. Second, it is essential to account for all alleles at one point in time. The example above did the accounting of alleles in one generation by comparing the offspring of the actor and the offspring of the recipient. Combining a cost in terms of the survival of a helper, for instance, with a benefit in terms of the offspring of a recipient is comparing consequences in different generations, a comparison of apples and oranges. Wilson's 2012 book pointedly criticizes kin selection for making these mistakes in accounting. It is true that such mistakes have been made, but it is not true that the problem has not been recognized. Among investigators of kin selection in natural populations, the accounting is often done correctly. Chapter 15 works out another example of kin selection in a more complex situation. Finally, it is important to note that Hamilton's rule predicts individuals' decisions to help a relative or not; further refinements are necessary to compute the actual change in frequencies of alleles associated with helping.

Despite the initial enthusiasm of the 1970s, studies of kin selection in animal societies have often shown that it does not alone provide a sufficient explanation for the maintenance of helping. Chapter 15 confirms this point in a specific case. Although there are some exceptions, it is safe to conclude that Hamilton's rule provides no sweeping explanation for helping. His rule did not match the success of Isaac Newton's three rules as explanations for how the world works. Nevertheless, it is important to realize that kin selection always contributes to the evolution of helping. Kin selection always affects the spread of alleles in populations whenever close genealogical kin interact in ways that affect each other's survival and reproduction. The only issue in the ongoing debates is whether or not the effect is large enough to provide a sufficient explanation for the observed occurrences of helping.

Another issue is that the most complex societies often deviate from the expectations of kin selection more than the simplest societies. For instance, most anthropologists argue that kin selection is not a sufficient explanation of human cooperation. Despite the importance of kinship for cooperation in most human societies, it cannot explain everything. In addition to humans, social insects—perhaps especially those with the most complex societies and, most notably, many ants—also provide examples of cooperation that are not explained by kin selection.

It is partly for this reason that Wilson and his colleagues have in recent years so steadfastly promoted the idea that complex societies are "superorganisms." The interdependence of individuals is so great that competition between groups, such as colonies of ants or tribes of humans, overwhelms any consideration of the maintenance of helping within colonies. To elucidate these issues and the influence of noisy communication, Chapter 15 considers an example of cooperation by birds. The species, South American wrens, are ones my colleagues and I have studied. Fortuitously they provide just the sort of case to illustrate some of the complexities in the evolution of extreme cooperation.

15

COMPLEX SOCIETIES

COMPLEX COOPERATION

Stripe-backed Wrens provide a case of extreme cooperation and illustrate the limitations imposed by noisy communication. Many other examples of complex social behavior have received study in nonhuman animals. Nevertheless, these wrens raise, in a particularly clear way, the issues of noisy recognition and second-order cooperation. First, this chapter examines the possibility of kin selection. Second, it considers the possibility of queuing for favorable social positions. Queuing, it develops, is a form of second-order cooperation with similarities to indirect reciprocity. And third, this chapter broaches the possibility that these wrens are on the cusp of becoming a superorganism. This possibility could lead to escalating complexities of recognition and thus further forms of inevitably noisy communication.

Stripe-backed Wrens (see Figure 15.1) occupy savannas (grassy areas with scattered groves of trees) in the llanos of central Venezuela and eastern Colombia. They live year-round in stable groups that average 5.5 individuals but can include as many as 14. Each group defends its own exclusive territory. A small fraction of territories is occupied by pairs, just a male and female, an arrangement widespread among other spe-

FIGURE 15.1. Members of a group of Stripe-backed Wrens meet beside their nest. Two of them are producing a coordinated duet. On many occasions, other members of the group join in to produce a chorus of raucous chattering.

cies of songbirds. These wrens are in many ways typical cooperative breeders.

Studying a population of these wrens was a long-term effort begun by my wife Minna and me and then continued by my postdoctoral associate Kerry Rabenold. He subsequently moved to Purdue University and assembled his own set of associates, including Patricia Parker, Walter Piper, and Steve Zack. Altogether eleven investigators at Purdue and the University of North Carolina contributed to the effort.

A great advantage of this species was the tendency of each group to roost every night in the same nest. As a result, we could capture all individuals in each group of wrens relatively easily as they emerged at dawn. By placing different combinations of colored plastic bands on their legs, we identified each individual and followed its behavior throughout the remainder of its life. For nearly 15 years we could identify every individual in a core area that included about 20 groups, and in some years

Rabenold's team had almost every individual marked in a much larger area with some 60 groups. Annual survival of adults over one year of age was 63% during the 10-year period when the demography of the central population was thoroughly studied. Survival did not depend on the size of the group and, for breeding individuals, did not differ with sex.

All young individuals, with a few exceptions mentioned below, joined the group in which they were raised. All adult members of a group participated in defending their group's territory during encounters with neighbors. They also all participated in loud vocalizations produced by coordinated efforts of two or more individuals. We confirmed that each group produced a repertoire of about 10 unique rhythmic patterns of these duets or choruses. As mentioned previously, playbacks showed that each group recognized the patterns of each of its neighboring groups. In addition, individuals produced quieter nasal calls, of which each group again had a repertoire of about 10 distinct patterns. Jordan Price showed that all males in a group shared one set of patterns and all females a second set. These wrens, as far as known, have an unprecedented complexity of vocalizations for group-living birds.

During the four- to eight-month rainy season, each group built a second nest in which one female laid and incubated a clutch of eggs. The breeding female associated predominantly (but not exclusively) with one male. Prior to and during egg laying, copulations were mostly surreptitious, presumably inside the breeding nest, which the breeding female and her partner entered together for a few minutes every morning shortly after the group had emerged from its roosting nest. Eventually, genetic testing confirmed that all eggs were produced by the presumed breeding female and almost all were fertilized by her partner. The exceptions, about 10% of eggs overall, occurred mostly in groups with a stepmother, as explained below.

Every group with more than two adults thus included individuals in two roles. There were the two breeding members; the remainder usually did not breed. The latter nevertheless provided much of the food for the nestlings once the eggs hatched and for fledglings for as long as several months after they left the nest. They also helped to defend the breeding nest from predators, including snakes and raptors, and from brood parasites—female Shiny Cowbirds that would lay their eggs in the wrens' nests for the wrens to raise, usually at the expense of their own young.

These helpers, Rabenold confirmed, increased the number of young produced by a group. There was no evidence that helpers increased the overall amount of food brought to a nest (nor did breeding females in larger groups lay more eggs in anticipation of more help). Instead, feeding by all members of a group reduced the average number of feeding trips by each individual. A significant influence of helpers on reproduction came in fewer losses to predators and cowbirds. The helpers apparently contributed to vigilance near nests and to repelling predators and parasites before it was too late. The most striking effect of helpers, however, was a group's ability to raise two broods in one season. A group with at least two helpers could divide their efforts when the first brood fledged. The helpers would take the primary role in feeding the fledglings for over a month while the breeding female produced and incubated a second clutch of eggs and the breeding male guarded the nest. By the time the second brood hatched, the first fledglings needed less attention, so all adult members of the group could then feed the new hatchlings and still have time to keep an eye on the older progeny. Breeding pairs without helpers and even groups with only one helper never successfully raised two broods of young in one season, whereas larger groups managed this feat about one year in four.

KIN SELECTION AND COOPERATIVE BREEDING

Kin selection, as so often the case for cooperative breeders, is not a sufficient explanation for the behavior of Stripe-backed Wrens. Members of groups are closely related, but the costs and benefits of helping, as described below, do not fully satisfy Hamilton's rule. Helpers in a group are usually the progeny of the breeding pair and thus usually each others' siblings. If the breeding female has died and been replaced, then helpers are half siblings of the new offspring. The slightly increased genealogical relatedness of wrens in adjacent groups, a result of the short dispersal of most breeding individuals, means that mates are slightly more closely related than distant individuals and that siblings are also slightly more related than expected if the population at large mated randomly. Nevertheless, even if helpers raise full siblings, the benefit to the parents does not compensate for the loss of the helpers' own reproduction.

To see this point we can rephrase Hamilton's rule, as explained in Chapter 14, for the case of an individual helping its parents to raise full siblings instead of raising its own offspring alone with a mate (recall that the coefficient of genealogical relatedness of an individual to its full sibling is 0.5 and to its own offspring also 0.5):

(gain in helper's siblings) $\times$ 0.5 > (loss in helper's own offspring) $\times$ 0.5.

This condition for the spread of helping states that the number of alleles added to the population by helping to raise siblings must exceed the number of alleles added to the population by reproducing with a partner. It compares the expected numbers of alleles in the population in the next cohort of young as a result of two alternative possibilities for an adult wren's behavior, either reproducing with a partner or helping parents to reproduce. The expectations for both alternatives are calculated at the same point in time—the end of the current breeding season. Both progeny and siblings, the two alternative results of the wren's behavior, are appropriately devalued by their coefficient of genealogical relatedness to the adult helper. The gain of siblings includes only the difference a particular helper makes.

There is a possibility that the second helper in a group increases the production of new offspring disproportionately in comparison with the addition of the first or of any further helpers. If we use observations from 20 groups studied exhaustively over a 10-year period, the gain of siblings by the second helper in a group is the difference between the progeny produced by a group with two helpers (2.4, on average) and the progeny produced with one helper (0.4). Nesting with a mate alone or with a mate and one helper have statistically indistinguishable results, 0.4 independent young per year. In this case, Hamilton's condition for the spread of alleles associated with helping is met:

$$(2.4 - 0.4) \times 0.5 > 0.4 \times 0.5$$

or

$$1.0 > 0.2.$$

Assisting parents *as a second helper* leaves more copies of any associated alleles in the next cohort of young than finding a mate and producing progeny.

Some mental arithmetic quickly reveals that this result does not apply to an adult wren helping in any role other than as a second helper. Becoming the only helper for parents (to make a group of three) does not significantly change the parents' reproductive success. Becoming the third, fourth, or fifth helper in a group increases the reproductive success of the breeders by only 0.1 young in each case.

This advantage of the second helper is lost if we calculate the gain of siblings for a helper by pooling all observations from six populations, each studied less exhaustively over more than 20 years. If we calculate the marginal increase in the breeders' reproductive success over the entire range of sizes of groups, the average increment in young each year is 0.5 per helper. Even if helpers raise full siblings, this figure is not enough to satisfy Hamilton's condition for the spread of alleles for helping.

Furthermore, if the female breeder in a group is replaced before the following breeding season, the remaining helpers are now only half siblings of the young they help to produce. In these stepmother groups, one of the male helpers often fertilizes some of the female's eggs, as does her principal mate. This individual, usually the next male helper in the line of succession, thus includes some of his own progeny in the group's success. Most of the helpers raise half siblings with a coefficient of genealogical relatedness of 0.25, still enough to satisfy the condition above for second helpers in a group. Helping to raise half siblings is even less likely to explain helping by other helpers or an average helper.

Nevertheless, these helpers show no change in their contributions to feeding the young, defending the territory, or protecting the nest. Furthermore, in two unusual cases, an entire brood of orphaned fledglings was adopted by a neighboring group. This situation occurred only when all adults in a group died in a short period of time and left their fledglings still too young to fend for themselves. In hundreds of group-years this circumstance only happened twice. Nevertheless, in both cases, a neighboring group adopted the dependent fledglings. The adopting individuals were avid in recruiting these new members of the group, and the adopted individuals subsequently helped like the other members. These rare adoptions perhaps do not require an adaptive explanation.

Instead, a more parsimonious explanation requires no more than a rule of thumb that is adaptive in more ordinary situations. Nevertheless, these two cases emphasize the strength of Stripe-backed Wrens' tendencies (or motivation, if you wish) to join a large group when young and to recruit new members of a group once adult. The fact that adoption is so rare also emphasizes one feature of their behavior that is so obviously explained by kin selection that we tend to overlook it in our attention to the problematic features—when young wrens join a group, they almost invariably join their natal group, the one in which their genealogical relatedness with other members is highest.

Kin selection therefore explains some but not all features of these wrens' cooperative behavior. Interactions with close genealogical relatives significantly alters reproductive success, but the effects are not strong enough to provide a sufficient explanation for their behavior. Kin selection clearly influences the spread of alleles in their populations, but not enough to account for their strong tendencies to help. Rabenold and I realized, however, that another process could explain their behavior. This possibility surfaced only as information accumulated on how wrens become breeding members of groups.

SUCCESSION TO BREEDING STATUS

Because so many individuals were followed throughout their lives, we could determine in quantitative detail how individuals became breeding members of a group. Almost all individuals remained in their natal groups until a vacancy, following the disappearance of a breeding bird, became available in their group or a nearby group. In no case did we ever observe a nonbreeding individual challenge a breeder and aggressively usurp its position. Instead, nonbreeders always waited for a vacancy following the disappearance of a breeder. No contests preceded these disappearances, and the individuals involved, with one exception, never showed up elsewhere, so all evidence suggested that individuals remained breeders until they died from predation or disease.

A male usually stayed in his natal group in the role of a helper until he succeeded to a breeding position in his natal group (see Figure 15.2, *left*).

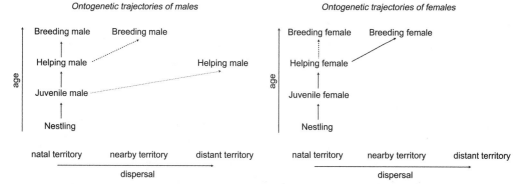

FIGURE 15.2. How Stripe-backed Wrens obtain breeding positions. Ontogenetic trajectories (the sequence of social positions) from birth to breeding positions follow different courses for males and females. Males *(left)* are most likely to reproduce (if they survive) in the group in which they were born, although some of them disperse to a neighboring group and occasionally one moves a long distance. In contrast, females *(right)* almost never reproduce in their natal group. If they survive they eventually get a breeding position in an adjacent group, often after intense competition with other females from near and far.

If a group included several nonbreeding males, the one taking a vacant position for a breeder was always the oldest. Even if there were two males of the same age, no obvious aggression occurred. They had apparently long ago settled their relationship by subtle interactions or extremely infrequent overt aggression. On occasion a waiting male left his natal group to take a vacancy in a nearby group after the death of the breeding male there, provided that group included no nonbreeding male in waiting.

In contrast, females almost always had to fight hard for their positions as breeders. They always left their natal group to take a vacancy after the disappearance of a breeding female in a nearby group (see Figure 15.2, *left*). When a female breeder disappeared, waiting females in other groups from as far away as a kilometer arrived to compete for the open position. The fighting among these females was outrageous, with repeated pursuits, repeated stabbing with beaks, and raucous vocalizations, from just after dawn until just before dusk. Sometimes these competitions were settled in a day or so, but sometimes they lasted as long as a month. Once one female had achieved ascendancy, the others all returned to their natal territories, where they resumed their status as non-breeding helpers until the next opening arose.

There were two remarkable patterns in these contests. First, when an opening arose nearby, often all of the waiting females in a group left to join the fray. Thus two to four females from the same group would fight for an opening, along with females from other groups. The females from any one group were usually sisters, or at least half sisters. If one of them succeeded in claiming the vacancy, it was always the oldest (or if two sisters had the same age, their precedence had long ago been resolved just as in the case of same-age males). Once an older sister had her new breeding status settled, the younger sisters returned to their natal territory to resume waiting. Sisters, we carefully noticed, never fought as a team on behalf of their oldest sibling. They also never fought against each other. Having a sibling or two fighting her rivals would presumably have helped the older sister to prevail in these contests. The younger sisters would, of course, benefit the next time a nearby vacancy arose, when one of them would be the oldest. Just as in some human societies, younger sisters only had a chance for "marriage" after their older sisters had successfully found mates. In these brawls, of course, quick recognition of contestants was crucial. Perhaps nothing more than discrimination of familiar and strange rivals would suffice, but—as described Chapter 14—we had demonstrated that these wrens could recognize the vocalizations from each of the neighboring groups.

Second, there was a strong pattern for the successful female in one of these contests to come from a territory immediately adjacent to the vacancy. In most cases, a female dispersed from her natal territory to take a breeding position in an adjacent territory (and the second most frequent situation was dispersal to a territory only one removed). As in many other cooperatively breeding birds, dispersal from place of birth to place of reproduction was surprisingly short for most individuals. Perhaps females from next door had an advantage because they knew the terrain in neighboring territories. Yet their interactions with the neighboring helpers and with the widowed breeding male were so persistent that it seemed more likely that the next-door females' advantage came from knowing their neighbors as individuals.

QUEUING FOR ADVANTAGEOUS SOCIAL POSITIONS

If the members of a group form a linear queue for breeding positions, then it might pay to wait in a queue for a favorable position rather than to strike out alone in an unfavorable position. Waiting for a favorable position in a group is the same argument as waiting for an opening in suitable habitat. It focuses however on a favorable social position rather than a favorable ecological one. It is not the availability of food (or shelter from predators) that is crucial but the availability of helpers to collect the food (or to warn of predators). The demographic question is the same in both cases: Does it pay to wait? Waiting brings the risk of dying at any time but the possibility of greater success eventually. Because we knew the probabilities of survival (and thus mortality) for wrens in all social positions and the expected reproductive success in groups of all sizes, we could calculate the trade-offs. We found that, for Stripe-backed Wrens in our population, it would pay for a young female to join a group and to wait for an opportunity to obtain a breeding position provided she did not have to wait longer than four years on average. It paid for a young male to join a group and to wait for an opportunity no matter how long it took. The difference between the sexes comes from the lower survival of female helpers, presumably a result of the costs of searching and competing for vacancies in nearby groups. Males usually succeeded to vacant positions in their natal groups, but females invariably became breeders in another group, usually a neighboring group. Females in this population first obtained breeding positions at an average age of 2.1, males at an average age of 2.8. Our demographic calculations thus revealed that young wrens leave more progeny in their lifetimes by waiting for favorable social positions in groups than by striking out to breed with a mate alone.

If the helpers in a group are in line for breeding opportunities and if the number of helpers increases the reproductive success of the breeding pair, then it is easy to understand why all members of a group should accept some costs involved in recruiting new members. New members joining the end of the queue would (1) eventually serve as helpers once those ahead of them in the queue become breeders (or, in the case of females, teammates in competitions for breeding positions nearby) and

(2) help to recruit further members to join the end of the queue. By helping those ahead in the queue, each member would later receive help from those behind in the queue.

There are some prerequisites for queuing. First, the demographic conditions must remain constant from generation to generation (or fluctuate unpredictably around a steady mean). Furthermore, the nonbreeding members of groups must form a queue, a stable linear order for succession. It is as if each takes the next number in line as they join their natal group. Queuing is a form of cooperation. Each individual behaves as if it agrees not to challenge those ahead. If so, each individual is relieved from challenges by those behind it in the queue. The alternative would be a scramble for every vacancy of a breeding position, so that the individual in best shape or with luck at the moment would prevail. Females face this sort of competition for vacancies. In all of our years of observation, there was never an exception to the rule that the oldest male or female helper in a group takes an available vacancy (or among same-age individuals, the one that revealed subtle signs of precedence). Queuing is a cooperative agreement among rivals to reduce aggression and to accept a mutual relationship in order to release time and effort for each individual's own advantage.

COMPLEX RECOGNITION IN THE SERVICE OF COOPERATION

An analogous situation arises in the case of a "dear enemy" relationship between territorial individuals. When settling on territories at the beginning of the breeding season, Hooded Warblers and other territorial birds, following several days of chases and physical interactions, quickly accept the status quo. Individuals with neighboring territories accede to a mutual boundary and thereafter refrain from trespassing—at least when singing to advertise their territories. Each individual confines its singing within its settled boundaries and accepts the presence of its neighbors and thus the density of conspecifics in the vicinity. In return, each individual has time and energy to devote to its mate and offspring (and in many cases also to surreptitious interactions with neighbors' mates).

In a few species for which measurements are available, territorial males with former neighbors have higher reproductive success than those with new neighbors (even when variation in the males' ages is removed statistically). Apparently having boundaries already in place at the beginning of the breeding season promotes prompt attraction of a mate and optimal timing of nesting.

The "dear enemy" cooperation between territorial Hooded Warblers in North Carolina, a student of mine discovered, depends on Tit-for-Tat reciprocity between neighbors. Renee Godard, whom we met earlier in this book when discussing these warblers' abilities to recognize each of their neighbors, conducted an experiment that confirmed their Tit-for-Tat reciprocity. As she had shown previously, Hooded Warblers not only recognize each of the neighbors but also respond aggressively when any one of them appears to sing out of place.

The question now is, do Hooded Warblers retaliate against such uncooperative neighbors? Godard first presented a territorial warbler with songs of a neighbor near its correct boundary. This step confirmed her previous result—warblers respond weakly if at all to a neighbors' songs in the expected place. Then she staged a trespass by the neighbor deep into the subject's territory. This trespass consisted of recordings of the neighbors' songs presented at two locations near the center of the subject's territory. Godard took care to present these recordings only briefly. If the subject approached, she promptly terminated the recording. This precaution assured that the subject would have no opportunity to explore the area near the recording and thus potentially to discover that only a loudspeaker was present. Finally, she played the neighbor's song once again near the correct boundary. This fourth playback allowed Godard to determine whether or not the subject's response had changed following the apparent intrusion by the neighbor. Indeed, the subject now responded to the neighbor's song, even though it was in the correct place, with a level of aggression appropriate for a strange rival that required full repulsion.

No conclusions were drawn, however, until after a second series of playbacks. Godard had to show that the change in the subject's behavior was indeed a response to a specific neighbor. An alternative possibility was that any intrusion into a male's territory would aggravate his aggressiveness in general, not just toward the guilty party. So Godard repeated the same series of four playbacks to each subject, but with the songs of a stranger presented near the center of the subject's territory. This stranger had been recorded far enough away that the subject was unlikely ever to have heard him. After this staged intrusion by a stranger, the presentation of the neighbor's songs near the correct boundary evoked no more aggression than the minimal response before the intrusion. Godard could then conclude that the increase in her subjects' aggressiveness was specific to the trespassing individual and not the result of a generalized increase in aggressiveness. She had demonstrated that male Hooded Warblers not only recognized each of the neighbors' songs but also used this ability to police their relationships with them. Any infringement of the mutual boundary was met with renewed aggression directed specifically toward the uncooperative rival.

For Stripe-backed Wrens we could confirm that individuals recognize neighboring group's vocalizations. Yet we had no opportunity to confirm that the queuing of helpers within a group fit the expectations of Tit-for-Tat cooperation. If Hooded Warblers are capable of such behavior, it does not seem like a stretch to suppose that Stripe-backed Wrens might also. In the field, without experimentation, all seem to live in peace with each other or at least they revealed only inconspicuous forms of antagonism. Their adherence to an invariant rule for succession, however, must require cooperation and thus some form of continually monitoring partners' compliance with the rules. Low levels of aggression must result from trade-offs between the potential gains from self-promotion and those for mutual restraint.

COOPERATION WITH INEQUALITY

The helpers in a group of Stripe-backed Wrens do not contribute equally to caring for the young. I could never interest a student in examining this perplexity thoroughly. In other cooperative species, unequal contributions have by now been carefully investigated. Females in prides of lions are the classic example. Not only do they not all contribute equally in potentially dangerous interactions with other prides, in the sense that some are slower to approach the rivals, but those cooperating less in this way do not always compensate in some other way, such as nursing the young of other females while they hunt. Whether Stripe-backed Wrens show similar patterns of inequality among partners is not yet known.

The issues are whether or not cooperation in one sphere of activity requires cooperation in all others and whether or not all individuals in a cooperative group must cooperate. Territorial birds seem not to co-operate in all spheres of activity, as already suggested. "Dear enemies" respect mutual boundaries for singing, thus presumably also their catch-ment areas for attracting mates. Nevertheless, studies have repeatedly shown that trespassing in silence occurs frequently when foraging for food and when interacting with neighboring females. Even if territorial behavior does not eliminate trespassing by neighbors, it could still be mutually beneficial by limiting the density of individuals in the vicinity. The possibility that territoriality limits density, as opposed to evenly spacing the individuals present, is a contentious issue in ecology. Never-theless, neighbors presumably all have equal opportunities for surrepti-tious trespass, so none can benefit at a neighbor's expense. The contri-butions of Stripe-backed Wrens do not seem to fit this scenario for mutual benefits of partial defections. Another possibility is variation in helping associated with position in the queue. Individuals near the head of the cue would stand to benefit sooner from recruiting new members than would those farther back. There might also be mutual advantages to having some diversity in the kinds of contributions. Some might per-form more difficult tasks less frequently. Diversity might also result from a mixture of Tit for Tat and Snowdrift cooperation. The defining example for the latter case of cooperation involves two travelers on either side of a snowdrift blocking their way. They could reach their respective desti-nations fastest if they both collaborated in clearing the drift. Yet even if one did not cooperate, the other would still get to its destination faster by clearing the entire drift than by refusing to work in a pique of spite. In this form of cooperation, two noncooperating partners each have a lower payoff than does a cooperator with a noncooperating partner. In Snowdrift situations the "sucker's payoff" is not the lowest payoff, while in Tit-for-Tat situations it is. The Snowdrift conditions permit greater chances for the evolution of cooperation. In queuing, as in a Tit-for-Tat scenario, noncooperating (challenging instead of waiting) makes the sucker's payoff lower than that for two noncooperators. In contrast, by helping to raise new members for a group, noncooperating (resting instead of working) matches Snowdrift situations, with the lowest payoff for two noncooperators. It is clear that we have more to learn about queuing in Stripe-backed Wrens if the chance to resume such a study ever arises.

Noisy communication also affects how these forms of cooperation might evolve. Before considering this possibility, one further observation about Stripe-backed Wrens is needed.

SOCIAL GROUPS AS SUPERORGANISMS

It seems possible that groups of Stripe-backed Wrens are on the cusp of becoming superorganisms. Wilson's recent emphasis of this concept for social insects and for humans stimulated me to reconsider this possibility for cooperatively breeding birds as well. The wrens can illustrate the issues.

The first challenge in identifying superorganisms comes from lack of clear agreement about the defining features. Bert Hölldobler and E. O. Wilson, in their 2008 book *Superorganism*, emphasize "epigenetic specialization," especially specialization of individuals for reproductive and nonreproductive tasks. They also mention complexity of communication. Reproductive specialization is particularly clear in social insects, many of which have worker or soldier "castes." The distinction between reproductives and nonreproducing individuals is not so absolute as popularly believed, however. In recent decades evidence has accumulated that workers in many species lay some eggs. In termites, specialization of nonreproducing individuals is extreme. Nevertheless, if the reproducing pair in a colony disappears or dies, workers can transform into reproductives. The situation is more extreme than that of nonreproductive wrens waiting for vacancies, but the pattern is the same. The eusocial insects have evolved extreme differentiation of reproducing and nonreproducing individuals in a colony, but cooperatively breeding birds also have distinct behavioral and physiological roles. They also have complex communication. In fact the intragroup sex-specific vocalizations of Stripe-backed Wrens, still of uncertain function in communication, are unprecedented among birds.

Others have proposed other criteria for superorganisms: frequent co-operation and infrequent conflict within groups, rejection of foreign individuals of the same species, and relatedness within groups. As discussed above, all of these features apply to Stripe-backed Wrens and indeed to all cooperatively breeding birds and mammals. John Maynard Smith and Eörs Szathmáry, in their pathbreaking book *The Major Transitions in Evolution*, also emphasize new methods of transmitting genes and irreversibility of evolution (although they allow some exceptions). The social Hymenoptera (ants, bees, and wasps) have an unusual method of sex determination (females develop from fertilized eggs, males from unfertilized), but termites apparently do not have unusual transmission of genes. Humans and other vertebrates with cooperative societies have no known idiosyncrasies in genetic transmission.

Irreversibility of evolution implies that superorganisms have crossed a threshold into a condition from which it is difficult to return. In reflecting on the Stripe-backed Wrens, it seemed to me that the crucial condition promoting irreversibility might be extreme interdependence of individuals for reproduction or survival.

I thus propose that superorganisms are assemblages of individuals characterized by (1) coherence and recognition, (2) internal diversification, (3) proliferation of cooperation, (4) reduction of conflict, and (5) interdependence for survival or reproduction. I further propose that these conditions, once evolved, could lead to (6) escalation of all of the above, (7) irreversibility (at least relatively), and (8) superordinate natural selection.

Stripe-backed Wrens appear to be on the threshold, perhaps just over the threshold, by these criteria. The account above has focused on points 1 through 4. These four features are regular in many other cooperatively breeding birds and mammals. There is a bit more to say about recognition.

The crucial feature of the wrens is point 5, their interdependence for reproduction. As the study of the wrens developed, Rabenold and I often noticed, but never sufficiently focused on, the striking result that pairs

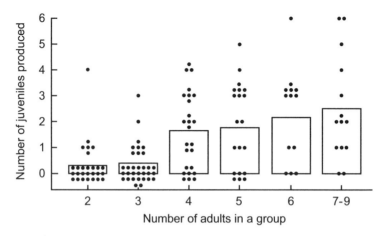

FIGURE 15.3. Small groups of Stripe-backed Wrens have little success in reproduction. Larger groups raise more offspring each year overall. Especially striking are the routine failures of the smallest groups, those consisting of a breeding pair with no helpers or a pair with only one helper. Over the course of their lifetimes, the breeding pairs in these small groups cannot replace themselves in the population. These data come from a single population studied exhaustively, although similar data from pooled studies of several nearby populations also reveal that pairs cannot replace themselves. Adapted from Rabenold (1990).

of wrens cannot replace themselves. Pairs produce on average 0.40 young per breeding season; the annual survival of nonbreeders is 0.61; so the number of one-year-olds produced by a pair per year is $0.40 \times 0.61 = 0.24$. Thus, each member of a pair on average produces 0.12 one-year-olds and thus potential new breeding members of the population. The annual survival of breeders is 0.64 and thus annual mortality is 0.36. A member of a pair is more likely to die each year than to produce a replacement (see Figure 15.3).

In contrast, in the exhaustively studied groups mentioned above, pairs with at least two helpers produce plenty of young. Each such group produces on average 2.40 young in a breeding season; so assuming the same mortality during the subsequent dry season, these groups produce $2.40 \times 0.61 = 1.46$ one-year-olds each year. Each breeding member of a group thus averages 0.73 new adult members of the population each year, much higher than the annual mortality of breeding individuals, 0.36. With a group to help, a breeding wren is much more likely to produce a potential breeder (although most of its progeny join a queue to wait for a reproductive position, as explained above) than it is to die.

If we focus instead on the overall marginal gain in production of young with each additional helper, from pooled observations of less thoroughly studied populations, a pair with three helpers averages 1.5 young each breeding season and thus $1.50 \times 0.61 = 0.90$ one-year-old young each year. Each breeding member of a group averages 0.45 new adult members of population each year, again enough to exceed the annual mortality of breeding individuals. With these pooled observations, two helpers is, however, not enough; such groups average only 1.0 young each breeding season and thus only 0.31 new adults each year. Regardless of which way we analyze the observations, it takes a group with two or three helpers for breeding individuals to replace themselves. The advantage of reproducing with helpers makes it advantageous for a wren to join its natal group and wait for a chance to succeed to a breeding position, even though it might die before this happens.

The persistence of this population of Stripe-backed Wrens thus depends on the formation of cooperative groups. This conclusion does not mean that individual members of groups have not also optimized the spread of their alleles by interactions within groups, as discussed above. Yet the evolution of cooperation in this population of wrens has resulted in a situation in which individual pairs can no longer replace themselves. They are now dependent on membership in a group to reproduce successfully. As explained above, this situation decreases the minimal benefit an individual must receive for membership in a group to 0. Any possibility of succeeding to a breeding position in a group, no matter how small, is enough to justify membership. It seems clear that such situations could favor the further escalation of the forms of cooperation and communication. Queuing, as opposed to scrambling, for breeding positions might evolve as one such form of complex cooperation. Another might be motivation to adopt offspring as new members, whenever possible, and regardless of their relatedness.

These wrens also made us realize that the conditions that maintain the evolution of cooperative breeding might diverge substantially from the conditions that result in the initial evolution of cooperative breeding. As with many complicated evolutionary adaptations, the evolutionary stability of cooperation is a different issue than the evolutionary origin of cooperation. Perhaps initially kin selection had a stronger effect, perhaps in the most direct form of extended parental care for offspring in a population with high adult survival and plenty of resources for reproduction.

Competition with neighbors could also have been important. In our studies, territorial boundaries between groups remained astonishingly stable from year to year, but when changes occurred, even slight ones, it always involved a larger group expanding at the expense of a smaller one. Our study, by good fortune, included a closely related species, the Bicolored Wren, which illustrated an initial transition to minimal cooperative breeding rather than an advanced transition to superorganisms. This species usually reproduces as pairs in areas with sparser trees and many solitary palms. Young usually disperse to establish their own territories at one year of age, but in some situations they remain for a year with their parents and help to raise their parents next brood of young. Such groups have an advantage in protecting their nests from predators. Despite this advantage of groups, breeding pairs of Bicolored Wrens could replace themselves. The advantage of joining a group and helping to raise siblings is marginal in comparison to the advantage of dispersing to nest with a partner.

Stripe-backed Wrens illustrate the interplay between natural selection for an individual's attributes that promote its group as a complex social unit and those that promote its own interests within its group. To the extent that an individual's own interests become dependent on its group, the group becomes a superordinate unit of selection. The process is similar to the transition from single-cell to multicell organisms. Metazoans are in a sense a society of protozoans, in which some cells (only those in the germ line) pass their alleles to the next generation and the others are helpers in this process. In this case, somatic cells outside the germ line have lost all options for reproduction. This extreme "caste" differentiation among cells of metazoans is presumably related to their extreme genealogical relatedness. Unlike societies of ants, humans, or wrens, somatic cells are all nearly identical to germ cells, so kin selection is in full force in assisting the evolution of this cooperation.

Nevertheless, somatic cells occasionally shift to selfish behavior in the form of cancers. Even in this extreme case of cooperation among the cells of an organism, selection within an organism can spread alleles that produce short-term advantages for individual cells despite long-term disadvantages for the organism. Countervailing natural selection also spreads alleles that counteract these tendencies, for instance by repairing damaged DNA or by eliminating cells that do not divide normally. The superordinate selection for an effective organism usually, but not always, prevails over subordinate selection for selfish cells.

A similar interaction between superordinate and subordinate selection must operate in societies of organisms once the reproductive success of groups (organisms, ants, wrens, or humans) requires effective cooperation of units in a group. The critical feature of a superorganism, the feature that generates superordinate selection, is the dependence of the units' reproductive success on cooperative interactions within the group to which they belong. Once individuals become obligately dependent on groups for survival or reproduction, then natural selection should produce adaptations to enhance their efficiency in cooperation. These adaptations might include specializations for communication within groups, such as the remarkable vocalizations for within-group communication in Stripe-backed Wrens. They might also include greater tolerance for group members that do not fully reciprocate or are not close kin. The uncompensated costs of helping other group members cannot, on average, negate the advantages conferred by the group. When individuals cannot survive or reproduce without a group, the acceptable costs approach the ultimate cost of hopelessness. Interactions of individuals within a superorganism, as opposed to interactions in groups that have not reached this criterion, might thus evolve both greater efficiency in cooperation and greater abuses of selfishness.

COOPERATION IN NOISE

The social relationships of Stripe-backed Wrens illustrate the demands on recognition created by complex forms of cooperation. As noted in Chapter 14, cooperation does not require recognition of partners, particularly for sessile organisms. For mobile organisms, random assortment into nearly isolated groups might serve the same purpose. Situations like Snowdrift interactions can promote indiscriminate cooperation. Tit-for-Tat reciprocity, its variations, and indirect reciprocity place cognitive demands on recognition of partners. Indirect reciprocity consists of helping individuals perceived to help others. Unlike the "green beard effect" described in Chapter 14, indirect reciprocity relies on perception of actual cooperative behavior rather than a signal associated with cooperation. Even more than Tit-for-Tat reciprocity it requires recognition of partners.

These forms of cooperation require specificity of recognition. To increase the chances of reciprocation, cooperating individuals must identify individuals likely to reciprocate rather than to cheat. Complex cooperation within a group might also require considerable multiplicity of recognition. Queuing would presumably be stabilized by each individual recognizing all others. Queuing, to emphasize the point again, is not the same as a dominance hierarchy, in which the most successful competitor at any moment dominates all others. Such a hierarchy could result from intrinsic differences in competitive abilities, from learned differences (perhaps as a consequence of preceding wins and losses), or from responses to status signals (signals reliably associated with competitive abilities, such as in many cases mature age). Queuing without overt competition would require recognition of at least sets of individuals ranking higher and lower. If any individual challenges a position in a queue, however, the effect should propagate as challenged individuals lose the advantages of queuing and thus initiate their own challenges. Recognition of all individuals in a queue is thus necessary to stabilize the advantages for any one individual.

Recognition is a form of communication. A signal from one individual (the signaler) is associated with a response by another individual (the receiver). The model of noisy communication in Part II led to the conclusion that no communication can escape noise. The trade-offs for the utility of a receiver's threshold and the exaggeration of signals assures that an equilibrium is reached in which receivers make errors and signalers are not always recognized. Just as noise sets limits to the evolution of communication in general, errors of recognition set inevitable limits to the evolution of cooperation. Models of the evolution of cooperation cannot assume perfect recognition of individuals and thus cannot assume perfect differentiation of cooperating and noncooperating partners. Noise in communication is not just additional variation in responses. It requires an understanding of the trade-offs faced by signalers and receivers in noise, as influenced by all of the parameters identified in Chapters 8 and 9 and listed in Table 10.1.

Noisy communication probably has two important influences on the evolution of cooperative groups. First, it should make the initial stages in the evolution of such groups more difficult. The payoff asymmetries must be greater to compensate for the inevitable errors of recognition. Second, once cooperation has evolved, and particularly if a transition is reached for superorganisms, natural selection should favor elaboration of communication to minimize errors further.

Indirect reciprocity contrasts with the green beard effect in assuming that individuals perceive helping directly rather than a signal contingently associated with helping. If helping is associated with alleles, then helping another individual previously observed helping would assure that helpers focus their behavior on other helpers. On the other hand, indirect reciprocity would resemble a green beard effect if helping one recipient was only contingently associated with helping another. The complex recognition required to keep track of the behavior of multiple potential social partners might allow helpers to direct their help to particular recipients or in particular situations that benefited the actors most. It is even possible that an individual might choose to help only when no costs were incurred or when the helper's apparent cost could mislead viewers about the real cost. Among humans it is easy to imagine situations in which apparently helpful individuals might not reveal the actual costs and benefits of their actions. How much should I admire a philanthropist who spends millions on social welfare when any compromise of his prosperity is vanishingly small and the publicity might even produce a compensatory increase in the philanthropist's income? Even if I were to grant a helper these two points, could I be assured that willingness to help someone else would also apply to me or whether my record of helping others would influence his decision? Also germane is the question of whether I could ever accurately assess any of these possibilities. The distinction between indirect reciprocity and the green beard effect is thus bridged by a continuum of situations in which third parties perceive signals that are more or less contingently associated with real (costly) helping.

These subtly differentiated possibilities mean that the properties of communication have fundamental consequences for the evolution of cooperation. The reliability of communication and perceptions are essential ingredients of the expected utilities for cooperation and selfishness. An understanding of complex cooperation requires an understanding of noisy communication.

16

MOLECULAR SIGNALS

NOISE IN AN ORGANISM'S BODY

THE ULTIMATE CASE of genealogically related signaler and receiver are the individual cells of a multicellular organism. With the exception of occasional mutations, all cells in an organism have the same chromosomes and the same alleles. It is thus not surprising that cells in an organism express high levels of cooperation. They constitute a corporate entity distinct from other such entities. Individual cells in an organism specialize in ways that preclude their own independent existence in order to promote their joint survival. Yet only a few cells in an organism, its germ line, actually pass their genes to the next generation. The other cells just help, the ultimate case of helpers sacrificing their own interests to promote the evolutionary interests of others.

It is also now well established that cells routinely sacrifice their lives for the survival of the organism. Programmed cell death is triggered by errors in duplicating chromosomes, for instance—a form of suicide for the good of the organism. Cells also sacrifice themselves to nourish an organism's offspring. The nutritive fluids produced by the reproductive tracts of some live-bearing fish, the "trophic eggs" (that is, eggs laid among growing offspring solely to provide food for those offspring), the crop "milk" produced by both sexes of all doves and pigeons, and all milk from mammary glands of female mammals—all of these consist of whole cells that have stored nutrients and then ruptured, sacrificing themselves, for the nourishment of the germ cells' offspring.

To the extent that all cells in an organism have the same alleles, their genealogical relatedness is 1.0, and Hamilton's condition for the evolution of cooperation becomes

$$B - c > 0.$$

The benefits for the receiving cells must simply exceed the costs to the acting cells. Just as argued previously for interactions of related individuals, the interactions of cells within any one individual organism still must make the trade-offs of signalers and receivers in noise. Hamilton's rule just shows that the trade-off is more lenient for cells of an organism, not that a trade-off does not exist.

Communication is the way most cells interact most of the time. This is not to say that in some ways the interactions of cells, like the interactions of organisms, are not communication. Their strong physical connections require that together they provide the power (in the form of synthesis of the required molecules) to bind themselves together physically. Again, Hamilton's rule leads us not to be surprised that each organism's cells bind more tightly than the cells in the bacterial films discussed in Chapter 14. Close genealogical relatedness justifies greater costs in cooperating. Some cells engulf others. In essence, they prey on them. Nevertheless, most of the interactions of cells are forms of communication by means of signals that do not provide all of the power to effect the response of the recipient cell. Neurons communicate in most cases by means of chemical synapses. The signaling neuron secretes molecules that then interact with molecules on the surface of the target or receiving neuron. This receiving neuron then provides the power to generate its own action potential. Even in the case of the less frequent electrical synapses between neurons, the impulse in the downstream neuron is generated by its own power. An animal's nervous system is a vast system of impulses communicated from one cell to the next. Hormones are also molecular signals secreted by special signaling cells.

These hormonal signals then alter molecular receptors on receiving cells, often after transport to distant parts of the body.

IMMUNE RESPONSES AS SIGNAL DETECTION

Molecular mechanisms for resistance to disease and parasites depend crucially on signals. All animals and plants have an innate immune system, which challenges pathogens (such as viruses, bacteria, or fungal cells) invading the organism. Cells of the innate immune system respond to signals (proteins or parts of proteins or other molecules) produced uniquely (or at least predominantly) by pathogens. Vertebrates (other than lampreys and hagfish) also have an elaborate system of acquired immunity to pathogens. This acquired immunity, one of the most complicated adaptations of vertebrates, is based on an especially complex system of signals and backup signals that allow specific cells to identify foreign proteins in a highly specific way. Early in the development of the acquired immune system, while the organism itself is still embryonic, each T cell and B cell rearranges specific locations in its genome (requiring risky double-strand breaks and random repairs in a restricted region of DNA) to produce a molecule on its surface that, if its structure is just right, can bind to a particular bit of protein (a particular sequence of amino acids in a particular conformation). The chromosomal rearrangements produce a large variety of these potential receptors, some of which bind to pieces of foreign (nonself) proteins. Others bind to pieces of the organism's own (self) proteins. Also early in development, these variously equipped T cells selectively accept programmed cell death if their receptors happen to bind to pieces of self-proteins. Only those that do *not* respond to self-signals survive. Each B cell likewise initiates its own death if its receptors bind a self-protein, but only if the binding is strong. Many B cells that weakly bind self-proteins escape.

Once past this early stage of development, if all goes well, surviving T cells (and some B cells) bind only foreign proteins. Pathogens are recognized by other cells (macrophages) that ingest any bacteria, virus, or unusual cells they encounter (including parasitic or damaged cells) and digest them into pieces. They present these pieces (called antigens) on their surfaces. When a B cell binds to an antigen and also can bind to a T cell that recognizes the same antigen, the T cell stimulates the B cell to proliferate and to produce copious amounts of its specific receptor (antibody) for the antigen. The antibodies bound to the antigens inactivate them and thus kill the bacteria or virus that are the source of the antigen.

The preceding account does not begin to do justice to the complexity of the interactions involved in acquired immunity. There are multiple subtypes of T and B cells, each with somewhat different receptors and responses, and the receptors themselves are produced by rearrangements at over a dozen possible locations in the genome. Furthermore, the exact nature of the responses of all of these cells and their receptors to each other and to both self-proteins and foreign proteins is currently under vigorous investigation. It is already clear that part of the complexity comes from adaptations to reduce errors in the responses of B cells to self-proteins by requiring a T cell to provide a "second opinion." It is also clear that errors in the plethora of signals and responses can have adverse consequences for the organism, consequences that range from annoying (seasonal allergies) to catastrophic (aggressive autoimmune diseases).

The immune system is thus subject to the two kinds of errors that characterize all signal detection. Allergies are false alarms of the immune system, particular molecules that activate the immune system but are not pathogens. Autoimmune diseases are also false alarms, self-molecules that should not activate the immune system but nevertheless do. Missed detections include deceptive pathogens with strategies for making their signals less easily detected in the molecular noise. In some cases they rapidly change the proteins on their surfaces (for instance, influenza and HIV viruses). In others they quickly hide inside normal cells that lack the capability of recognizing them (malaria parasites and, again, HIV viruses). The missed detections that are the complements of the false alarms of autoimmunity are cancers. Pathologists often can detect these cells by examination under a microscope or by distinctive molecules they produce, but the immune system misses many of them because of their superficial similarity to normal cells. Cancers consist of cheating or selfish cells, which proliferate at the expense of the organism (the group) to which they belong.

Clearly immune systems are signal detectors subject to the trade-offs discussed in this book. The details of many of these errors are under current investigation. Yet the trade-offs of signal detection attract little discussion. Why do we suffer from allergies and (occasionally but disastrously) from autoimmune diseases? Why do we suffer from cancers? Is there an unavoidable trade-off between these false alarms and missed detections? How do our immune systems make such mistakes? How might we shift the threshold for responses in our favor? It is not too rash to suggest that signal detection thinking is needed to answer these questions.

MORE MOLECULAR TRADE-OFFS IN SIGNALING

All communication among cells is likely to be noisy. Evolution is as unlikely to produce perfect communication among cells as it is to produce perfect communication among organisms. Here as elsewhere, diminishing returns for approaching perfection are likely to end in an equilibrium with a compromise between irreconcilable errors of false alarm and missed detection. The genetic identity between cells of an organism alters the conditions for costs and benefits, so cells take on somewhat more cost in approaching perfection in communication than would unrelated individual organisms. Nevertheless, the adjusted costs do not abolish the trade-offs signalers and receivers face in noisy communication. Signals and responses evolve to provide advantages to individuals on average, but not in every instance.

There are indications that investigation of these trade-offs in communication among the cells of our own bodies is beginning to receive some traction among molecular biologists. The molecules that produce the receptors on the surfaces of cells are, it seems, often "sticky" in the sense that proteins other than appropriate signals sometimes adhere to them. They seem incapable of excluding all false alarms. They must also miss some detections when molecular signals, for instance, do not align correctly with receptors. The literature on "sticky" receptors, still in its infancy, has taken the first steps in analyzing the trade-offs in molecular signaling.

The "stickiness" addressed in this literature pertains to the scope of molecules that bind to a receptor and thus the possibilities for false alarms by a receiving molecule. It must be distinguished from another aspect of molecular stickiness—how strongly molecular signals bind to receptors. Some molecular receptors and their signals are known to bind much less strongly (stick together less) than others. For instance, the acetylcholine molecule, which transmits signals across chemical synapses between two neurons, only binds weakly to the receptor on the target neuron. A weak association results in the signal falling off the receptor quickly, which allows the receiving (target) neuron to respond quickly to another molecule of acetylcholine. It allows neurons to react and thus to communicate rapidly.

This example illustrates that the strength of binding is related to its time course. A strong bond corresponds to a long time constant, and a weak bond to a short time constant for response. The time for a response, as described in Chapter 3, determines the temporal resolution of a receptor, which is inversely related to its frequency resolution. In molecular signaling a short time must allow discrimination of only a relatively small number of molecular configurations. Acetylcholine is, after all, a relatively small signaling molecule. A small number of atoms can assume only a small number of configurations and thus permit correspondingly limited scope for discriminations. Rapid responses at a synapse thus come with low selectivity. This inevitable trade-off for any receiver applies to molecular communication just as it does to all other cases. Every receptor evolves a criterion for response that maximizes the utility of its balance between selectivity and sensitivity and thus between slow or rapid response.

Another avenue of research on cellular metabolism also seems destined to lead to issues of signal detection. In this case the focus is on networks of proteins in cells. This work depends on high-speed computing and recent techniques for measuring the production of many proteins in a cell at once. An analysis of these networks provides an alternative to the prevailing approach of investigating how proteins produce one response at a time. This approach sees each protein, whether a signal, a receptor, a switch, or an amplifier, as one step in a chain of responses. Noisy communication leads us to expect instead that proteins in different chains should interact with each other to various degrees. Sometimes particular sets of proteins might form a module incorporated into several different chains of reactions. Probably more often they interfere with each other's activities as signals or responses in ways that produce errors in signaling. The alternative approach of isolating signals and responses is, consequently, unlikely to lead to an understanding of the behavior of cells, any more than isolating signals and responses from noise is likely to explain adaptations in the behavior of organisms.

An essential feature of molecular signaling is the tuning of receptor molecules. This subject is sometimes called noise suppression in the literature of molecular biology. As has been noted in Chapter 7, all receivers make decisions by means of thresholds or other criteria for response. For instance, consider G proteins, which form the backbone of many cellular receptors. They extend through the cell membrane (the "skin" of the cell) in order to bind with signaling molecules outside and also to initiate, with the help of additional proteins, responses inside. Some of these helpers, called regulator of G signaling proteins, or RGS proteins, serve to adjust the scope of external molecules to which the G protein binds. This tuning of a molecular receptor is analogous to the tuning of a frog's or insect's hearing to the sounds of its own species. It is like the tuning of cochlear responses to particular frequencies of sound.

As described in Chapter 7, the selectivity of a receiver is inversely related to its sensitivity. A receptor with a high threshold for response has few missed detections of appropriate signals; it is highly selective (restrictive or fastidious). One with a low threshold has few false alarms but many missed detections; it is highly sensitive (permissive or gullible). The problem created by noise in communication is not whether to tune or not to tune but *how much* to tune. Should it be permissive (and have few missed detections) or should it be restrictive (and have few false alarms)? Noise results in a trade-off for any receptor. If the thesis of this book is correct, noise is universal, even in molecular signaling. So is this trade-off.

THE MOLECULAR BASIS OF ANIMALS' SIGNALS

The unavoidable trade-off between the two kinds of error by molecular receptors creates a situation now familiar in our discussion of noisy

communication: the impossibility of perfection, the impossibility of reducing both kinds of errors simultaneously. To reduce the false alarms
of allergies, presumably immune systems would have to accept more
missed detections of pathogens. To attack more cancerous cells might
require more autoimmune attacks on normal cells. A threshold for response can only optimize, not eliminate, the balance between the two
errors.

I do not sense, however, that an awareness of the inevitability of error
in cellular communication has sunk deep into thinking about human
physiology and disease. Yet if the analysis of noisy communication in
this book is even approximately correct, it does not seem likely that a
full understanding of any of the multitudinous forms of inter- and intracellular signaling in an organism's body can be well understood
without exploring the implications of signal detection in noise. What
elicits any response—in any noisy communication, whether between organisms or between cells—cannot be completely understood without
also understanding what else in the environment can also elicit a response. Responses to signals cannot be understood without understanding responses to noise. Properties of signals cannot be understood
without understanding the properties of noise. Furthermore, because the
components of every organism's receptors (sensors, gates, and amplifiers)
are all composed of cells, all communication by organisms comes down
to communication at the level of cells. At this level all signals and all noise
produce their effects by molecular interactions.

The evolutionary adaptations of signals are thus fundamentally adaptations of molecular signals. Yet we cannot conclude that the study of
molecular signals should supersede the study of animals' signals. As in
every aspect of biology, investigations at different levels of organization
are complementary. Study of societies identifies problems that must be
explained by interactions of individuals, and study of individual organisms identifies problems to be explained by interactions of the cells
and molecules that compose them. Nevertheless, it is remarkable how
difficult it is to proceed in the opposite direction. Study of cells explains
the way individuals behave, but in practice it is much more difficult to
extrapolate the properties of an organism solely from investigating
the properties of its cells—or of a society from investigating the prop-

erties of its constituent individuals. Our knowledge of a higher level of organization is asymptotically underdetermined by our knowledge of any lower level. It is as if knowledge were fractal. Complexity of the patterns at any one level of organization is underlain by ramifying complexities at a lower level. It is also as if knowledge had entropy. It is easy to analyze the patterns at one level into isolated components at a lower level, but difficult to synthesize the isolated components at a lower level into the patterns at a higher level. It is easy to identify the trees in a forest, but difficult to construct a forest by studying its trees. Likewise for the amino acids in a protein. In every case, the analysis is spontaneous, the synthesis laborious. In the case of signal detection, the patterns at the organismal level have just begun to be analyzed at the molecular level. The prospect of synthesizing the patterns of communication of organisms in societies from the properties of communication by molecules is still unimaginable. Perhaps a reason for this asymmetry between analysis and synthesis is noise. The responses of components in isolation lack the constraints that arise in a composite, and these constraints include noise in signaling.

PART FOUR

FAR HORIZONS

PERHAPS THE MOST UNEXPECTED consequences of investigating noisy communication are the extraordinary implications for how I must think about myself—as a participant in communication and also as a perceiver of my surroundings. Two aspects of noisy communication force these consequences on me.

First, there is a receiver's inevitably constrained view of what is happening in communication. A receiver is aware of only two situations each time it pays attention to its input—either a signal appears to have occurred or not. These two situations depend on its criterion for a response. Nevertheless, at each moment, a receiver only knows one of two states of the world. Its criterion classifies the activity of its receptors into two categories, either appropriate for response or not. Each time it checks its receptors a receiver becomes either responsive or not. It knows only two possible outcomes at that moment. As emphasized in Chapter 1, an independent observer with a special vantage or special apparatus might comprehend all four possible outcomes for a receiver, but the receiver itself cannot.

Likewise a signaler knows only two states. In as much as it produces signals in response to its own input, it too only knows two situations: those appropriate for a signal and those that are not. Like the receiver, at each moment a signaler only knows two outcomes. Of course, either a signaler or a receiver might correct its previous awareness as a result of attending to subsequent input. These corrections are, however, contingent on a second-order awareness.

Second, my perception of my surroundings has the essential features of noisy communication. My perceptual thresholds or criteria for classifying input determine how I experience everything around me. Again, at each moment, there are four outcomes each time I classify my sensory input, and two of the possibilities are errors—false perceptions and missed perceptions—analogous to false alarms and missed detections in communication. Nevertheless, only an independent observer with an ideal vantage can discern these four possibilities. I myself only know two of them: either my input justifies a response (acceptance and memory of a perception) or it does not. Each of my perceptions thus seems absolutely clear to me, although in fact there are other possibilities. Whether a receiver or a perceiver, I am in the same predicament. The chapters of Part IV consider some of the implications of noisy communication, on the one hand, and noisy perception, on the other.

Above I said that an awareness of these consequences of pervasive noise was thrust upon me. When I first began to think of animals' adaptations for communication, I never expected to look so far afield. There were plenty of pressing questions at hand about the nature of communication by animals other than humans. Yet, as I always reminded my students, it is well to keep an eye on the horizons around even the most focused question. There are fruits to be gathered and seeds to be sown in distant lands. The challenge, of course, is that these horizons, not so distant ones even, are already well populated by indigenous experts. Intrusion inevitably meets resistance, although mutual invigoration sometimes ensues. It is thus with aspiration, as well as trepidation, that I embark for the realms of philosophy. Hesitation might miss a fair wind.

The predicaments that noise creates for receiving and for perceiving undermine standard ways of thinking about ourselves. They leave us more nearly helpless than we might want to accept. But in the end, I will suggest in these final short chapters, they indicate an ennobling way forward.

17

HUMAN COMMUNICATION

IS HUMAN COMMUNICATION PECULIAR?

NOT ONLY IS NOISE PERVASIVE in a vague way, it is unavoidable. The optimization of receivers' and signalers' behavior does not result in escape from the consequences of noise. The possibility of escape has diminishing returns. As signalers and receivers approach perfection in communication, the disadvantages of greater exaggeration and higher thresholds mount faster than the advantages. Optimization by human design must face the same pattern of diminishing returns for receivers and signalers, so both evolution and human design are expected to fall short of perfection in communication or perception. Communication can approach perfection but not attain it. Noise is not just pervasive, in a contingent way, but is absolutely inescapable. Noise is thus doubly annoying.

So far this book has not explicitly addressed human communication. Before proceeding, some basic questions arise. Is human communication, in particular human language, like other forms of communication? Is thinking a form of communication? And does language evolve by natural selection?

Language is not the only way in which humans communicate with each other. Facial expressions, gestures, and "body language" are obviously important as well. It is a remarkable finding that some human gestures, especially facial expressions, are produced and understood in essentially the same ways by all human populations. I say "essentially" because careful studies have provoked some debate about the details. Nevertheless, photographs of facial expressions by people from diverse parts of the earth are understood in broadly similar ways by people everywhere. In most places on earth these days, the juxtaposition of peoples from diverse origins makes this conclusion a matter of everyday experience. Languages are different. Even the most extravagant polyglot can communicate in only a handful of languages, and people like me in only one or two. Yet everybody can communicate fundamental states of mind with every other human by means of facial expressions. This communication requires no thinking or learning, either for signaling or receiving.

What happens during communication with facial expressions? Each individual produces particular facial expressions associated with particular states of mind—smiles when happy, scowls when angry, and so forth. A state of mind is a physiological state of an individual's brain (often influenced by the physiological states of other organs of the body). A state of the brain can produce responses by facial muscles that together produce a particular expression. Another human seeing this expression, just as unthinkingly, associates the perceived expression with a response appropriate for the signaler's state of mind. A receiver's response, as emphasized repeatedly, might be immediately overt, but it might also be delayed or covert. Often one person's state of mind, as evidenced by facial expressions, appears to potentiate a similar state of mind, with the associated responses, in the receiver. All producers of television programs know that laughter is contagious. Likewise, anger often provokes anger in others. Communication by means of facial expressions is at least partially involuntary and unconscious.

Nevertheless, it might be objected that people can learn to dissemble. Con artists sometimes manipulate gullible targets. Even more interesting, however, are actors and actresses. They manage to produce the body language, facial expressions, and gestures, quite aside from the spoken language, needed to convince an audience of a suite of false mental states. Intensive training can evidently make a person adept at manipulating receivers' responses. This skill is a form of deception, communication that exploits receivers' responses to signals. Convincing acting is, however, not easily accomplished. It also has some limits. An actor playing Macbeth does not, I presume, fully acquire the mentality of a murderer. Furthermore, although practice is often needed to dissemble convincingly, no learning is necessary for every human to produce and to respond to normal facial expressions.

Successful deception might become possible only after some self-deception, as Robert Trivers has recently argued. A person must first be deceived before convincingly deceiving others. Trivers's arguments seem convincing, albeit (as a result of their reflexive nature) perhaps deceptive! They share a claim with my arguments that the responses of any organism, including a human, reflect its state of mind. In this broader perspective, they apply to more than deception. Anyone trying to sell anything must realize the importance of first accepting the arguments in favor of a sale. In general it is difficult to argue convincingly for any position unless convinced of its correctness. In other words, the internal state of a signaler must have some correspondence with the signals emitted. A convincing argument requires a self-convinced signaler. A deceptive argument requires a self-deceived signaler.

Trivers's arguments diverge from mine in failing to emphasize that the association between a signaler's mental state and its signals (between an organism's internal state and its behavior) is statistical if we accept the pervasiveness of noise. Noise creates variation in the relationship between brain and behavior. It precludes any exact correspondence. There are other differences between Trivers's position and mine. My approach does not exclude the possibility that an organism's brain is compartmentalized, at least diffusely, so that self-deception need not reach totality. Persuasion might require self-conviction only in whatever compartment has control for the moment. Doubt could resurface as another compartment resumed control. Individuals might differ in their abilities to compartmentalize their states of mind and associated behavior.

The main lesson here is that facial expressions fit all the conditions for communication: a statistical association of a signal with states of the signaler and with responses by a receiver. It is evidently difficult, although not impossible, for most people to dissemble convincingly with expressions and gestures. In this respect, facial expressions of humans are much like communicatory displays of nonhuman animals, largely unlearned. Such communication is often labeled emotional as opposed to cognitive.

Language, in contrast, is the epitome of cognitive communication. Although facial expressions develop nearly uniformly in all humans without learning, the development of language is strongly influenced by learning. Furthermore, the mental states associated with instances of language are highly specific, whereas those associated with facial expressions are presumably less so. Nevertheless, language fits the same conditions for communication. An internal physiological state of a signaler is associated with a signal and this signal is perceived and associated with a response (or suite of responses) by a receiver. It makes no difference whether language is spoken, gestured, or written.

Every step in the use of language includes noise. When a person says (or writes or signs) something, this behavior results from a particular state of the brain. The brain is vastly complex, so the state of the brain associated with a particular utterance is no doubt minutely specific. People make mistakes in usage of language, sometimes unintentionally and involuntarily, so the association between minutely specified states of the brain and utterances is not perfect. It is instead statistical. This conclusion is not qualitative. It is not a question of whether a particular response is determined by a state of the brain. Instead it is a question of the magnitude of the association between each response and states of the brain. These associations presumably vary for different responses, different utterances, different people, and different situations. Nevertheless, human involuntary mistakes make it unlikely that these associations ever have zero variation. Zero is a difficult number to measure, in as much as the task requires exclusion of all other possible magnitudes. No matter how small, any nonzero variation in the association of states of the brain and responses is noise. Language fits the conditions for noisy communication.

Human language is also predominantly conversational. Two people often alternate utterances, each intending to evoke understanding in the other. Each individual serves alternately as signaler and receiver. The conditions for communication (associations between signals and states of mind) still apply. Because any individual can both produce and receive signals and because responses to signals can include covert changes in a receiver's internal state, it follows that the state of a signaler's brain might have resulted from a covert response to a previously received signal.

The question then is, How do states of a person's brain correspond to states of the world? Philosophy of language in the past century or so has often emphasized the importance of a "referent" for human utterances. A speaker might have in mind a black dog seen previously, for instance, and describe it in words. These words are intended and often in fact do evoke an image of a black dog in the mind of a listener. What is the relationship between the signaler, the receiver, and the black dog? Ferdinand de Saussure early in the last century and C. K. Ogden and I. A. Richards a few decades later proposed a triangular relationship among the speaker, the listener, and the referent for both the speaker's utterance and the listener's comprehension (in this case a black dog).

This model of communication is partly compatible with the one I have adopted here. A black dog perceived by one individual elicits a specific state of that individual's brain (with some noise). That condition is associated (with some noise) with an utterance. The speaker has thus been a perceiver (receiver of sensory input) and then a signaler. The utterance perceived by the listener (with some noise) is associated (with some noise) with a specific state of that individual's brain. Although it seems unlikely that the state of the listener's brain becomes identical to the state of the signaler's brain, the correspondence is often enough to allow them to proceed with their business. Divergence from my approach arises when the referent becomes a specific state of mind corresponding precisely with an external object (a particular black dog). This reification allows the field of semiotics to range widely from the paths I propose here. One can imagine such fabulous abstractions, but physiology does not require them.

Noisy communication is also at variance with Descartes's "Je pense donc je suis" (in a later translation, "Cogito ergo sum," or "I think, therefore I am"). This expostulation follows from his resolution to accept as true only those propositions that are absolutely clear to his mind. In noisy communication, perceptions are absolutely clear. On the other hand, the possibility of an absolutely certain proposition about a perception never arises. A person, in the role of a receiver, knows that sensory input meets a criterion for a response (perhaps covert—a perception or memory) but not that it corresponds exactly to reality or to a signaler's mind. There is no certainty that a proposition about a perception is a correct detection rather than a false alarm. A correct proposition of this sort requires zero variation in associations of brain and behavior (mind and utterance of signalers or sensation and reality of receivers), no errors or noise. When we realize that perception, like communication, ineluctably includes noise, absolutely clear propositions do not occur. Even complex propositions, such as "I think" or "thinking entails some other proposition," must fail the test of complete clarity. Complex propositions, including these, are second- or higher-order associations of perceptions and covert responses in the nervous system, a point addressed in the next section of this chapter.

Descartes's position is also affected by his distinction between mechanism and reason. For him, animals (and human bodies) are mechanisms, not unlike human tools designed to accomplish tasks and inerrant in their operation. To save "free will" for humans, he maintains that reason lacks this mechanistic nature. He recognizes that human reason can alter emotions and emotions can alter reason. Nevertheless, Descartes maintains, human reason (*l'âme,* a word he appears to use interchangeably for "reason" or "soul"), but not animal mechanisms, can decide whether or not a proposition is absolutely clear to mind. His concern to preserve human cognitive freedom becomes less pressing if noise is pervasive. Noisy communication and perception preclude certainty absolutely, but save "free will" (perhaps "capriciousness" is a better term) by introducing variation in associations of mind and behavior. Absolute (mechanical) predictability of behavior is excluded. In a later section it becomes apparent that each individual organism can, nevertheless, acquire a refined awareness of its own individuality, confidence, and decisiveness.

IS THINKING COMMUNICATION?

Like most philosophers, my approach to noisy communication accepts that thinking depends on language. Thinking in this view is concatenation of covert responses. Human brains are complex enough that we can generate specific linguistic responses to perceptions. As emphasized earlier, these responses might be immediately overt in the form of speech, writing, signing, or typing, yet they might also be covert in the form of memory or some other inconspicuous change in physiological state. If these internal states are also compartmentalized—in other words, controlled by different components of nervous systems—it is plausible that some of these covert responses could compose concatenated chains of signals and responses. Just as an organism might emit a signal after receiving a particular sensory input (such as an alarm signal after sighting a predator), one compartment of the nervous system might produce a signal propagating to another compartment after receiving a signal from a third.

The past half century has made it clear that one internal state can lead to another. All complex coordination exemplifies chains of muscular responses, in which one momentary state leads to another. A completely unconscious but remarkably complex coordination of humans (and most animals) is swallowing. In addition, locomotory coordinations in most animals include endogenously generated chains of muscular coordination—endogenously in the sense that the pattern can proceed without sensory input. Although normal locomotion requires modification of these endogenous patterns by continuous sensory input, nevertheless additional input is not required to complete the coordination. Thinking, I suppose, is analogous to locomotion. One mental state and its associated covert responses lead to the next. When language is involved, the result is thinking. One covert sentence (or phrase or word) leads to another. Indeed, all sorts of internal states can evidently lead to all sorts of subsequent covert responses. Images can lead to sentences, sentences to remembered odors, and any of these to emotional states such as anger or happiness. It would be reasonable to suppose that dreams are similar, although perhaps less structured, as a result of a sleeping rather than awake brain. These covert responses sometimes erupt partially into overt movements. I sometimes (but not always) find myself making incipient movements of tongue and lips as I think, the sort of mini-movements that J. B. Watson hypothesized would always accompany thought. Presumably the signs of "dreams" in animals are analogous, partial overt movements in response to covert chains of signals in a partially activated brain. No other animals have language like humans, so the covert signals in their nervous systems presumably result from nonlinguistic signals and responses, either learned or unlearned. Thus thinking presumably consists of covert chains of linguistic signals and responses—of great importance for us humans but not fundamentally different from chains of covert signals and responses of other sorts in all organisms.

A standard objection to any mechanistic proposal for human thinking is that sensory input cannot alone produce meaning. Perceptions are not just sensations. Every sensation, it is argued, has significance for the individual. Individuals construct meanings for all sensory input. Indeed, no sensory input is fundamentally singular. Each individual perceives objects, not unitary sensations. Newton's discovery that white light is an aggregation of all colors initially suggested that a perception ("whiteness") was a summation of unitary sensations (all colors), but Goethe and Helmholtz subsequently conclusively refuted this possibility. Goethe first noticed that colors are altered by adjacent colors. Juxtaposed black and white appear blacker and whiter near the boundary than away from it (now a well-known physiological consequence of lateral inhibition in vision). Goethe's color wheel, later refined by Helmholtz, made his point: although Newton's spectrum was linear, human perception of color is circular. The color magenta results from red combined with blue, hues at the opposite ends of the visible spectrum. Human cognition, we would now say, constructs this shade of purple by aggregating sensations of red and blue.

The fundamental components of a receiver in noisy communication incorporate this phenomenology. Any response, as emphasized throughout this book, results from a sensation and a decision. Language and thought are thus not solely the result of sensations. The decision-making component is also essential. The criterion that associates a particular sensory input with a particular response might consist of a simple threshold, but it could also require complex cognition. The logical relationships in language, universal among human languages, are easily viewed as components of a mechanism for decisions. The essential components, like the analogous transistors in computers, are the relations *and, or,* and *not.* These components of decision-making mechanisms are fundamental to all discrimination, learned or otherwise. They are also sufficient for synthesis of inputs as well as analysis. Objects, patterns, perceptions—all require aggregation of sensory inputs, all of which derive from specific cellular and molecular events. These components of decisions also permit inference and, in combination with concatenation, conclusions of causality. They might lead to Descartes's conclusion above.

It is sometimes noticed that digital computers easily handle dichotomous operations. As a consequence, they naturally (are most easily programmed to) organize information hierarchically (in directories within directories within directories). Human brains, in contrast, naturally (easily) jump from one diffuse association to another. Digital programs are getting better at diffuse association, but at the cost of complex programming. Internet search engines have, however, found commensurate benefits in the form of advertising. Humans, on the other hand, can master complex dichotomies, although the task appears more complex and requires practice. I suspect that this difference results from aggregation of inputs either by definition or by "family resemblance."

Ludwig Wittgenstein successfully excluded definition as a basis for human language (an argument resumed in Chapter 18) and instead suggested family resemblances as an alternative mode for aggregation of objects. To see the distinction, consider the following seven instances of sensation (inputs to a receiver's sensors), each of which has three properties indicated by letters.

1	2	3	4	5	6	7
A	B	C	F	G	H	E
B	C	D	G	H	I	F
C	D	E	H	I	J	G

One might define an object [1 + 2] by a definition, "an object with properties B and C." Another object might be [5 + 6] with a definition, "an object with properties H and I." One could instead recognize objects by family resemblance, for instance, [1 + 2 + 3] and [4 + 5 + 6]. These objects have no definitions short of complete enumeration, yet they are distinctly different. Even if we include sensation 7 in the second object, the object is still recognizable although no longer distinct. When a teacher, for instance, meets someone on the street and recognizes a student from class but cannot at that moment remember anything else about the student, presumably recognition is based on a family resemblance among the students in class.

How our brains might store family resemblances is unknown. One of ethology's most important contributions was the idea of a "releasing stimulus" and a corresponding "releasing mechanism." Some responses of animals are elicited by a few parameters of the total stimulation reaching the animal. The classic example is a newly hatched Herring Gull's response to the red spot on its parent's bill, which results in re-gurgitated food for the chick to eat. The newly hatched gull has, of course, had no opportunity to learn this association between a red spot and food. The response in the first hours of life is thus innate (in normal circumstances). It turns out that the young gull pays no attention to the shape or coloration of the rest of the adult gull and does not even need a spot on a beak. A moving red object about the width of the adult's red spot is sufficient. Although the neurophysiology of vision has never been investigated in Herring Gulls, what we know of other vertebrates makes it easy to imagine that young gulls are born with cells in their brains (or even in their retinas) that respond selectively to moving red spots and stripes of the appropriate size. As noted elsewhere, many striking examples of selectively responding cells are now known in the brains of many organisms. Such cells are neural representations of defi-nitions. They respond to sensations with a defined set of features.

Despite these remarkable cells, it seems likely that many of our mental states do not depend on "definition cells." Neural networks might store information in a way more like their analogs in computers. These net-works, as programmed on computers, consist of layers of interconnected nodes that can inhibit or excite each other, much as neurons do. They can learn to discriminate complex inputs by trial and error. Connections are strengthened or weakened depending on previous success of the network in discriminating inputs. In the end, each complex object is identified by activity in a network of nodes, without any one node responding in a defi-nitional way to the sensory input and without any necessity for the com-plex objects to have defining features. They can, in other words, classify sensations (discriminate sets of inputs) by their family resemblances.

Aristotelian syllogisms are easily derived from definitions but not from family resemblances. To the degree that human nervous systems classify sensations by family resemblances rather than by definitions, the associations between words and sensations are likely, as Wittgenstein noted, to do so as well. If so, it is futile to expect that language would have a strict correspondence with Aristotelian (or any axiomatic definitional) logic.

The difference between definition and family resemblance fades, however, in noisy communication or noisy perception. Objects might have properties (with noise) that result in sensations (with noise) for a perceiver who then associates them (with noise) with a response (perhaps covert). Family resemblance, as Wittgenstein perhaps suspected, is thus the only way to classify (aggregate and discriminate) sensations in noise. We reach this conclusion by the simple step of noting that noisy communication (or perception) requires a three-component receiver, with sensors, switches (decision makers), and amplifiers. By accepting the possibility of covert as well as overt responses and the possibility of compartmentalization and concatenation of covert responses, the basic properties of perception and thinking emerge. Neurophysiology has much left to explain, but enough is known now to see these possibilities. Human thought is complex, but its basic properties are those of noisy communication and perception.

Chapter 3 has presented evidence that discrimination of inputs is a fundamental form of decision by all nervous systems and is clearly documented in many nonhuman animals. Aggregation of inputs is also a fundamental form of decision. Pattern recognition and object constancy are the result. No doubt our brains develop these capabilities because they have allowed us for hundreds of generations to deal with circumstances encountered in the course of our lives. Natural selection, in other words, has favored the development of useful brains, ones that promote our survival and successful reproduction. Or at least they did so in our ancestors' generations.

DOES LANGUAGE EVOLVE BY NATURAL SELECTION?

Natural selection leads to the last of our three preliminary questions. The crux of this question is the relationship of natural selection to learning. Human language is strongly influenced by learning. Does this property preclude natural selection? Chapter 7, which introduced these evolutionary issues, emphasized that learning is a form of individual development and that all development results from an interaction of genetic and environmental influences. The following paragraphs reinforce the conclusion that all development, including learning, results from an interaction of genes and environment. Consequently even capabilities and limitations of learning evolve by natural selection.

Learning is no longer considered a definitive property of human language. It is now clear that some animals learn the signals they produce. Songs of birds in the suborder Passeri (true songbirds) are the prime example. In all species of this group, as far as known, young birds reared in acoustic isolation do not develop normal singing. Other members of their species do not respond normally, or in many cases do not respond at all, to these isolates' songs. Tutoring with recordings of songs or exposure to singing adults (sometimes in particular circumstances) results in normal development of the species' song.

Learning applies to receiving, as well as producing, signals. Receivers learn to associate particular responses with particular signals. All experimental studies of learning in animals show this capability. A rat or a pigeon, for instance, learns to press a key when a red light appears (provided subsequent delivery of food makes this response worthwhile). Demonstrations of learning by receivers in natural communication of animals are infrequent. The ability of territorial birds to discriminate the songs of several neighbors is perhaps the best evidence for receivers' associative learning. Learning does not introduce new properties for human communication.

A claim that human language (and thus all culture) is subject to evolution, just as human anatomy and physiology are, is a claim that the development of individuals is influenced by the evolution of populations. The genes that influence an individual's development are subject to all the processes of evolution in the population that includes that individual. Evolution, a change in allele frequencies in a population, proceeds uniquely for every population. Differences in the survival and reproduction of individuals can change frequencies of alleles in successive generations of each population. Development, an interaction between the alleles inherited by an individual and its particular environment, produces each unique organism. Even when differences between individuals depend strongly on the environments in which they develop, the capacity for these developmental differences depends on genes. In any specified environment, differences in the development of individuals depend on differences in their genes. Any measurement of the influence of genes on development thus applies to a particular environment. Conversely, for individuals with the same genes, differences in development depend on differences in their environments. Any measurement of the influence of environments on development applies to particular genes. These measurements are notoriously difficult to make on humans. Nearly all humans are genetically unique (aside from identical siblings, to a close approximation), and each develops in its own unique physical and social environment. The differences between two humans in any feature thus cannot be attributed to genes or environment alone. Nevertheless, all evidence about development allows us to conclude that each individual develops as a result of an interaction between its particular genes and its particular environment. Everything about every individual is thus influenced by its genes in relation to its environment (to some degree that we can sometimes estimate), and everything is influenced by its environment in relation to its genes (also to a degree that we can sometimes estimate).

Learning is an example of development obviously influenced by specifics of the environment. Yet, as argued in Chapter 7, it is difficult to imagine how learning can proceed in an organized way if there were not innate predispositions to channel the initial course of learning. Innate—in the sense that these predispositions develop regardless of the specifics of the environment. Predispositions for learning are sometimes discussed as constraints on learning. The difference in emphasis depends on whether the focus is on what tends to be learned or on what tends to be ignored.

Humans learn an enormous amount from others they live among and in particular from those they grow up with. Although it is impossible to associate specific accomplishments in learning with survival or reproduction, the capacity to learn from others in an organized way is clearly important for successful relationships with others. Language is a clear example. Communication would be drastically impoverished if humans did not learn the language used by their frequent associates. It is difficult to understand another culture (another human population with different learned traditions) without learning that culture's language.

This ability to learn human languages must result in part from our genes, because other closely related species have much less capability for language. Discoveries in recent decades that great apes, especially Common and Bonobo Chimpanzees, seem capable of rudimentary expression of many basic attributes of language only underscores how wide the gap is between their capabilities and our own. Only by probing the difference did its magnitude become demonstrable. Furthermore, all human languages seem to be comparably complex, so all cultures have comparable capabilities to learn a human language. Indeed, to a large extent individuals with parents from widely separated populations, such as Europeans and Asians, seem approximately equally capable of mastering each other's languages as children. We still do not know whether different human populations have adapted, just slightly, to learning the local language. After all, the anatomy of different populations differs slightly, so it is possible, for instance, that some phonemes or some grammatical constructions are more easily produced by some populations than by others. Nevertheless, it is clear that all normal humans can learn as children any human language with a high degree of skill.

The capabilities for learning languages are not limitless, however. We know not only that some congenital neurological conditions preclude normal mastery of language but also that all humans have some constraints on learning languages. For instance, the evidence from several well-described feral children, those not exposed to language until well into their teens, indicates that they never learn to speak and comprehend language even approximately normally. Sometime in late childhood the ability to master language ends. It does, to some degree, for all of us. After our teenage years, it becomes difficult to master new phonemes. Although later in life people can acquire the vocabulary and grammar of a new language, the phonemes become much more challenging. Perhaps even the grammar and vocabulary become more difficult to master as well, but the difficulty of mastering new phonemes is most apparent.

In many immigrant families, the arriving parents speak their new language with an obvious accent to the end, while their young children in their new environment flawlessly master the new phonemes and lingo as fast as children of indigenous parents.

This aspect of language is one of the striking parallels between humans and birds in learning complex vocalizations of their population. Learning proceeds rapidly without obvious reinforcement or teaching at an early age, but later only with difficulty and in response to laborious teaching and practice, if at all. For human language these constraints are qualitatively clear but little effort has been made to measure them. Measurement always supports insight.

Yet another argument for genetic influences on human language comes from the necessity for some innate focusing for any form of learning. If learning consisted of nothing other than associations of juxtaposed stimulation, it is difficult to imagine how a predictable outcome could unfold. Consider young birds learning their species-specific songs. As a rule, young birds are exposed to many other species' songs as well, not to mention many other sounds in their environments. Some of these sounds could not easily be produced by their vocal tracts, so anatomical constraints on production might limit each species to a certain range of sounds. Yet other species often have songs that do not differ much in their physical properties. Neural limitations seem more likely—either limitations on performance or limitations on perception or acceptance of songs as models for learning. Experiments suggest that physical limitations on performance are not sufficient to guarantee learning of the species-specific song. When young songbirds (suborder Passeri) are raised without hearing other birds' songs, they usually do not produce fully normal songs. Their simplified songs often diverge so far from the species' norm that they are not recognized by other members of their species when played back in the field. Some species, when these young isolates are tutored with recordings, readily learn to sing close matches of other species' songs. In some experiments, young birds have focused their learning on individuals that interact closely with them—often those that attack them and dominate them. Since species usually direct their aggression toward other members of their own species, these experiments suggest that these young birds focus their learning on songs of dominant adults of their own species. They learn the songs of successful members of their own species. In other species, such as parrots and some finches, vocalizations are learned only from close affiliates—normally just mates.

In some species, young birds preferentially learn recorded songs of their own species rather than others, even without social interactions. The most illuminating examples are Swamp Sparrows and Song Sparrows, two close phylogenetic relatives that often defend territories and sing within hearing of each other in the northern United States and Canada. Swamp Sparrows produce songs of uniform syllables, each consisting of a few brief notes (too rapid for humans to hear distinctly), in a series that lasts more than a second. Song Sparrows, in contrast, have songs with a complex pattern, often several short series of notes introduced by a distinctive sequence of about three accelerating notes. Experiments with recordings presented to isolated young birds show that Swamp Sparrows only learn to produce notes that match those in their own species' songs, regardless of the pattern in which the notes are arranged. They pay no attention to the pattern. Instead, they reject all notes that are not Swamp Sparrow notes. They then assemble the notes they have learned into typically uniform Swamp Sparrow series. Song Sparrows, on the other hand, learn either Song Sparrow or Swamp Sparrow notes, provided they are presented in a typical Song Sparrow pattern of short series with an accelerating introduction. They learn any notes in a Song Sparrow pattern. In both species it seems unlikely that physical constraints on performance could explain the preferential learning of their own species' songs. Instead, neural constraints on the perception or performance of songs must limit them to their own species' songs. As a result, a young bird in the field, exposed often to both kinds of songs, learns only those of its own species.

In Noam Chomsky's legendary 1959 critique of behaviorist theories of language, the same point is emphasized about human acquisition of language. Children learning their first language are not explicitly taught or rewarded for correct usage by adults. To some extent they might receive some reward when approximations to correct usage produce useful results. Yet as many observers note, during the period of rapid acquisition of vocabulary and grammar, the addition of new features of language is so fast that some innate perceptual focus on the features of language seems necessary. This observation has suggested that there must be some universal features of human languages that could provide a focus for children learning their first language. Like Song and Swamp Sparrows, some acoustic feature of phonemes (like notes for Swamp Sparrows) or grammar (like patterns of notes for Song Sparrows) must focus learning on just the correct sorts of perceptual input. Of course, close social interaction, even dominance or affiliation, might contribute a focus also. In some way, innate predispositions seem necessary to explain such focused learning.

Noise creates hurdles in the accurate acquisition of language. Presumably all parents notice errors in their children's usage of words or phrases, and some of these persist long past childhood. I confounded the words *temerity* and *timidity* until I was nearly 20, although in normative usage they have opposite meanings. Many people confound *comprise* and *compose*, although they have opposite meanings. The perception of language by children must also be fraught with noise. Adults do not always enunciate clearly nor construct sentences in standardized ways. Yet children, at least for the most part, learn the patterns, not the noise, to which they are exposed. The same is true for young birds learning their songs. When experimentally exposed to songs degraded by noise, young birds learn to produce songs free of noise. Some of this correction of noisy input might depend on constraints on production, but some, such as the regularity of notes, must result from predispositions in learning.

NOISE AND IMPERFECTION

The accumulated evidence thus indicates that human language can evolve adaptations just as much as other features of humans and other organisms. Noisy communication leads to the same trade-offs for receivers (listeners) and signalers (speakers) and the same consequences of optimizing thresholds for response and exaggeration of signals. Optimizing thresholds no doubt involves learning, but it must also require innate predispositions. Such predispositions depend on genes with effects sufficiently canalized (innate) in normal environments to focus learning in appropriate directions. Evolution of human language should result in optimal criteria for comprehension (threshold for responses) and optimal emphasis or clarity of speech (exaggeration of signals). Listeners should sometimes make errors, and speakers should sometimes find frustration. Noise is inescapable for humans, for the same reasons that it is for noisy communication in general, and perfection in communication is unattainable.

18

TRUTH IN LANGUAGE

THE HUMAN IMPERATIVE

FROM ANCIENT TIMES humans have been agitated to determine how we can know the truth. Since the time of Descartes, attention has often focused specifically on how we can know that a statement is true. The issue is the relationship between language and our sense of logic. Bertrand Russell's theory of descriptions brought this issue to the fore. His examples indicated that some statements, although correctly phrased, had no clear relationship to truth or falsity. Subsequently, philosophers have tried to resolve this incongruence between language and logic in different ways. Some accept that statements can be meaningless. They thereby reject Aristotle's premise of the excluded middle, that a statement or its negative must be true. Some have identified two ways language is used, sometimes to make affirmations, which are true or false, and sometimes to perform tasks, which are neither true nor false but instead either accomplished or not. Others have separated the absolute meaning of a sentence, which is true, false, or meaningless, from the speaker's meaning, which depends on the speaker's intention for a particular audience. The speaker's stance also isolates statements with private references not accessible directly to others, such as statements about one's own pain or other feelings or perceptions. Some have advocated recognizing that statements might refer to possible or impossible states of the world, as well as to true, false, necessary or contingent states of the world.

Still others have approached this problem of the incongruence of logic and language by insisting that statements have meaning only if they are verifiable (or falsifiable, a notion widespread among scientists). Otherwise they are either tautologies or meaningless. The crux of verification is a link between language and the state of the world. A statement is verifiable (or falsifiable) if it corresponds (or not) with our perceptions of the world. To overcome the unreliability of perception alone, some propose to temper this link by combining perception with intuition. Another argument suggests that meaning, like any state of mind, is revealed by behavior. If so, the meaning of a statement consists of behavioral responses to it. As Alan Turing argued in 1950, we would have to conclude that a machine had consciousness if its responses did not differ from a conscious person's. It helps the argument to add the qualification "in the long run," inasmuch as a few simple responses might not provide enough information to distinguish mechanism and consciousness. To save free will, while acknowledging that minds are like machines, Daniel Dennett proposed that decisions require two stages, a summoning of alternatives (which reflects a person's circumstances) and then reasoning to select one of the alternatives (the source of freedom in decisions). Such complications are unnecessary. Noise guarantees free will to minds and unpredictability to machines. Living, evolving minds, as argued in Chapter 19, can communicate with each other and, despite the attendant noise, can develop a degree of confidence in their decisions.

Language requires agreements on rules in a community of communicators but, as Ludwig Wittgenstein suggested, rules are not completely shared and can lack logical consistency. Instead, the rules of language, in actual practice as opposed to the books of grammarians, only unite instances of usage by family resemblances (a sharing of some but not all possible features, as described in Chapter 17). Wittgenstein also considered the possibility of private languages. Not last nor least is the issue of translation or, as Willard Van Orman Quine calls it, radical translation. How can a person amid people whose language is not understood come to communicate with them? Indeed, how can anyone come to communicate with anyone else? Crucial in this situation are indexical signals, pointing combined with naming.

TRANSLATION

Translation raises all of these problems at once. The issue is the indeterminate nature of definition, even by a combination of pointing and naming. Perhaps when pointing toward a rabbit while uttering a word, a speaker might have in mind a particular rabbit, an adult rabbit, any lagomorph, or any mammal. Or perhaps a rock behind the rabbit, or food, or a memory from yesterday, or a god, or even something that is not a rabbit. In this pool of possibilities surface all the issues of verification, affirmation, intention, intuition, decision, rules, morality, and private minds. Quine asserts that radical translation of this sort is ultimately impossible—at least not with exact precision. Donald Davidson maintains that repeated interactions between the speaker and the translator can eventually narrow the possibilities to a certainty.

Noisy communication creates a problem here. Furthermore the problem is larger than translation of unknown languages. It is, albeit in grand form, the same as the everyday problem of knowing somebody else's state of mind or communicating your own state of mind to somebody else. In fact, the problem is understanding each other. What precision is possible?

No doubt some comprehension of a previously unknown language is possible. After all, anthropologists in this situation have returned home to write books about people with cultures that differ from their own. On the other hand, subsequent generations of anthropologists sometimes argue that their predecessors misinterpreted their informants, or were misled by them. These allegations suggest that the process of translation is not perfect, especially when it is not clear which of the two alternative interpretations is most nearly correct. These disagreements are never complete negations, however. The arguments tend to focus on the aspects of human minds that are most difficult to confirm by means of communication, such as intentions.

Elsewhere I have argued that there are two fundamentally opposed ways to approach the question of how a person can know another person's state of mind. The same approaches apply to a person's ability to know the mind of a nonhuman animal or one animal's ability to know the mind of another. One way is to accept that another organism's state of mind is revealed by its behavior, either overt or covert, either immediate or eventual. Minds are thus reduced to the physical (and hence physiological) state of their parts (including the neurons of brains). You know another organism's mind to the degree that you can predict its behavior in any possible situation in the future.

Some would argue that a person's mind is not completely revealed by behavior. There is some plausibility to this argument. How could someone know my pain, for instance, or my intentions? It is possible that some time in the future (assuming I live that long) it might be possible to specify the state of every neuron in my brain. New technology is making large strides toward identifying activity in the brain at greater and greater resolution, and computation is making large strides in acquiring and reducing all the information generated by each such snapshot of a brain. The president of the United States has recently launched a program to investigate the workings of human brains to an unprecedented precision. Nevertheless, to confirm that I have a pain or an intention, as opposed to a physiological state of my brain, I would have to respond affirmatively to a question such as "Do you feel pain?" or "Do you intend to tell the truth?" It is possible that I would feel pain but nevertheless respond "No." It is possible that sufficient knowledge of the physiological state of my brain might reveal these higher-order states of mind. Yet this possibility does not seem to be utterly certain. At any rate, it might prove even more difficult to make this sort of information available to communicating humans in real time. In my previous discussion, I was thus left with a necessary uncertainty about whether mind could be entirely reduced to behavior or not.

If it is not, then Quine is correct: radical translation cannot be achieved, and any success in communicating a state of mind must rely on empathy. To the extent that I can perceive similarities between myself and another person, or any other organism, I might infer a similar state of mind. For example, if another person has responses similar to my own in a situation that arouses happiness or fear in me, then I might infer that that person also feels happiness or fear. This process, however, involves extrapolation from incomplete information. Empathy has its limits in inferring states of mind. It also has a dark side. If I can perceive similarities between myself and others I might accept empathy with them. Perceiving similarities, however, implies that I might also perceive dissimilarities and thus diminish empathy with another person.

These approaches—empathy and observation of behavior—are in practice hardly distinguishable. For two interacting people trying to understand the states of each other's minds, both procedures come down to extrapolation from incomplete observations of behavior. Empathy does not necessarily develop from perceptions of similarities in behavior, as opposed to other features, but if a person were interested in another's state of mind it would probably be best to focus on similarities of behavior in similar situations. In any case, more information about behavior in diverse situations should improve the accuracy of predicting states of mind and thus future behavior. It is no wonder that negotiations of all sorts so frequently involve face-to-face meetings, even meals, athletic ventures, and socializing—all interactions that present situations with many dimensions for possible responses. Negotiating by telephone or by mail involves flattening the dimensionality of possible situations and responses.

If noise is added to the interactions of two individuals, then communication of a state of mind from one person to another, with certainty, becomes impossible. As discussed in Part II, receivers and signalers succeed with correct detections and effective signals on some occasions. The proportion of occasions with success depends on the trade-offs faced by both parties. Despite some success, enough to promote the evolution (or design) of thresholds and signals, at equilibrium receivers cannot avoid some error, and signalers cannot avoid some frustration. Communication is on average honest and effective, but not always.

In the case of the anthropologist encountering a population with an unknown language, learning the meanings of words in the new language must involve communication—even at the very first stages. The anthropologist as translator might ask a question to elicit a response. A question might consist of pointing toward an object and uttering a sound with a rising inflection, "Eh?" In so doing the anthropologist would be relying on universal signals for human communication. The gesture of pointing is understood by adults in all human populations. Very young children and most animals fail to understand this gesture and instead pay attention to the pointing finger itself rather than to the direction it indicates. Children acquire the adult response to pointing at an early age without any evident teaching or practice. This innate predisposition is crucial for subsequent mastery of language. Likewise the vocal gesture of a rising inflection indicates a question in many human languages, although in some, such as Asian languages related to Chinese, it can also mark different phonemes. Rising terminal inflection also became within recent decades a cultural fad of teenage girls in some parts of Australasia, England, and North America.

A person pointing with a rising inflection of voice includes all sources of noise previously identified. No doubt the audible background would interfere with accurate perception of the utterance, and the visual background would obscure the exact direction of the anthropologist's finger. There is variability in the production of the signals, the straightness of fingers, the accuracy of pointing, and the distinctness and exaggeration of the inflection. There is variation in the receiver's classification of possible directions of pointing and of terminal inflections. Finally, perhaps most important, there is variation in the attitude of the two communicators toward their interaction—as a negotiation of friendship or rivalry or as a comparison of definitions of utterances. These attitudes might influence the payoffs for possible outcomes of their interaction. In other words, the receiver might not be in a state appropriate for the signal or, even if it were, the receiver might not respond correctly. The states of mind of the signaler and receiver might not make a complementary match, or if they did the receiver might err in response and the signaler fail to elicit a beneficial response. For any of these reasons, any possible combination of gestures can result in failure of the anthropologist and the informer to attain consilience of their minds.

One need only revisit Michelangelo's *Creation of Adam* on the ceiling of the Sistine Chapel to realize that pointing can have both equivocal and ambiguous significance. What the exact gesture is and what it might mean to me, to you, or to Michelangelo is, I dare say, infused with uncertainty.

As a result, the precision of translation would have a limit. Even if behavior revealed a subject's state of mind, communication with that subject could never determine its state of mind without some error. The anthropologist could repeat the question with the same or different objects (as categorized by the signaler), yet successive application of procedures for noise reduction would fail to reach certainty in any finite length of time. Like empathy, behavior would only lead to an approximate understanding of another's state of mind. The behavior of any subject would inevitably retain an element of unpredictability associated with noise—both external noise and noise internal to the signaler and the receiver.

This example points out another limitation to both procedures: the necessity of innate predispositions for any organized learning including translation. Innateness, to reiterate, is similar development of some feature of an organism regardless of a range of environmental circumstances. Unlike learning, an innate trait develops without an organism's exposure to specifically relevant situations. Innate predispositions are necessary to focus an organism on the range of potential situations for learning. Otherwise, associational learning would lead to chaotic associations among the features of impinging stimulation. We are so used to noticing the detailed associations that form the basis of complex skills and complex thoughts that we usually overlook the vast number of possible associations that we do not learn. Innate predispositions are the result of canalized development, influenced by genes more than environment—at least within a range of environments and a range of alleles. Innate dispositions in behavior are thus influenced by the evolution of allele frequencies in populations. They can evolve adaptations to particular environmental circumstances in which populations live. When we attempt to gain understanding of other species' behavior, our predispositions might lead us astray. Our widespread tendency to approach mate choice as a sensory discrimination among possible mates might not, as argued in Chapter 13, apply to other species. Our predisposition to respond to behavior mimicking that of dependent children might blind us to the evolutionary role of many domesticated animals as commensal parasites of humans—especially those domesticated in the etymological sense—incorporated in our homes. Cats and dogs in their own distinct way manage to hijack our predispositions by convincing us they are like human children. Our predisposition to cooperate with those who gain our empathy and to challenge those who do not might condition our conclusions about the naturalness of cooperation, on the one hand, or competition, on the other, in other species. Our predispositions are revealed in our anthropomorphisms.

In addition to evolutionary adaptations of predispositions, which can result in differences between species and populations, the genetic influences on predispositions must often vary among individuals within populations. Any variation in predispositions for responses to signals and exaggeration of signals would make problems for knowing other people's minds analogous to the problems just mentioned for knowing the minds of other species. Our ability to use empathy and observation to understand somebody else's mind is shunted if not derailed by our differences in predispositions.

PROBLEMS OF CONSILIENCE

The previous discussion encompasses more than just translation of unknown languages. The same problems arise in attaining consilience among any group of humans in definitions of words, or in general in use of language. How can I be sure that the word *magenta* is associated with the same sensory input as another person's use of this word? How can I be sure of the meanings of the words *truth* or *falsify* or even an expression in mathematical or logical symbols? In every case, consilience requires communication. In every case, noise in communication limits precision to nonzero variation. In every case, individual differences in predispositions for categorizing sensory input thwart consilience. The strategies described in Chapter 6—contrast, redundancy, and predictability of signals—can reduce the consequences of noise. Highly formalized signals, such as mathematical or logical symbols, help. Repeated signals help. Focus on situations in which the possible range of signals is more predictable, such as a conference on a specified topic, no doubt also helps to promote consilience in communication. None of these tactics can reduce noise to zero, whenever participants adjust the exaggeration of their signals and the criteria for their responses by their utilities. Noise is unavoidable.

This conclusion brings both despair and hope. The despair is a realization that noise cannot be completely circumvented, that complete consilience in communication is unattainable. The hope is a realization that communication does result in a degree of consilience—in many cases, a high degree. As the calculations in preceding chapters have revealed, communication is honest on average, even if not ideal. When communication is about communication itself, the analogous conclusion is that consilience is possible to a degree, even if not to perfection.

Wittgenstein, we might conclude, was right in thinking that language lacked complete consilience and that categorization of meaning had the properties of family resemblance rather than definition. This position might not have troubled him so much if he had fully appreciated the consequences of noisy communication.

19

SUBJECTIVITY

SPLIT AWARENESS

Noisy communication creates a disparity between subjectivity and objectivity. As emphasized throughout this study, the possibility of errors in noisy communication defines the basic properties of any receiver, any mechanism that responds to signals. Because signals do not provide all the power for a receiver's response, the receiver must decide whether to respond or not. Any receiver must adopt a criterion for a response to signals in noise. A receiver's criterion imposes discrimination and aggregation on sensory input—in other words, it requires recognition of patterns in noisy input. By adopting a criterion, a receiver distinguishes two states of the world—activity in its sensor exceeds the criterion or it does not. An external observer, in contrast, with a special vantage or special equipment, can often distinguish four states of the world, the four possible outcomes of any decision by the receiver (correct decision or false alarm when a receiver responds, correct rejection or missed detection when a receiver does not respond). A receiver at any moment knows one of just two possible states of the world, although in fact there are four possible states.

This study of communication has relied primarily on examples in which receivers might make immediate, overt responses to signals, such as mating with an apparently optimal partner, taking cover when apparently warned of a nearby predator, or attacking apparently palatable prey. Yet this study has repeatedly emphasized that responses need not be immediate nor overt. There is no compelling reason to exclude covert responses that consist of changes in the internal state of the receiver, its neural or other physiological states that can influence responses to future signals including future covert responses to signals. Changes in internal state can include memory (both retrievable and latent memory) and other neural changes not usually called memory, including sensory adaptation and central habituation or sensitization. They can include other physiological alterations of the receiver, such as endocrine changes and their consequences. Endocrine changes can alter a receiver's sensors (the proliferation and sensitivity of sensory neurons), its criteria for future responses (its "motivation" to respond to particular stimulation), and its capacity to produce particular responses (especially the mass of muscles or structures that produce the receiver's responses, such as structures that alter the efficacy of sounds or visual displays).

The internal state of a receiver might include higher-order responses. Second-order responses, for instance, might summarize or classify first-order changes in the receiver's internal state or associate categories of first-order changes. For a computer composed of integrated circuits of transistors, a simple second-order response might be nothing more than a report of the total memory remaining available. More complex second-order responses might include calculations of the number of operations performed each second. All searching and indexing operations are also second-order changes in the internal state of a computer. A computer does, of course, meet our specifications for a receiver. It detects input, associates each input with a response, and provides the power to effect the response. The output of its power supply, some of which is lost as heat, provides the power to alter the computer's internal state, its banks of memory, which in turn can alter responses to future input.

An organism's brain could also produce such second- or higher-order responses. The ultimate summary of these responses is presumably the basis for self-awareness. In this sense, self-awareness amounts to a receiver's detection, summarization, and comparison of its own altered internal states. Psychologists have shown that humans are not aware of all changes in their internal states. Latent and "repressed" memories, much proprioception, and even numerous habits seem to lurk below the horizon of awareness. To the extent that we do detect some of them, we have a sense of self-awareness. Our relatively large brains suggest that we might have greater capacity for higher-order responses than other organisms with relatively smaller brains or than other mechanisms with fewer and less complex connections of components—including even the most elaborate among current computers.

Self-awareness is an internal state of a receiver not necessarily associated with overt immediate responses and thus with signals that others might easily detect. This view of self-awareness is mechanistic, in the sense that self-awareness corresponds to a particular physiological state of the brain. Nevertheless, it is not apparent to me that this state can be confirmed by another person, even by one equipped with the latest equipment for evaluating the state of a person's brain, or even by a future person equipped with even more sophisticated equipment for precise evaluation of the state of every neuron in the brain. The problem arises because the only evidence for self-awareness must be a report by the subject. In other words, it must involve another layer of communication. No doubt solutions to these cascading problems can be proposed. Whether or not any are persuasive enough to be durable remains to be seen.

If responses to stimulation produce covert responses in the internal state of a receiver, including higher-order responses, then noisy communication creates the possibility that each receiver, in assessing its own internal state, discovers that it is only partially congruent with its external situation. As suggested in the introduction to Part IV, perception is precisely analogous to the operations of a receiver in communication. Perceiving consists of responses, often covert, that result from a receiver's decisions. The environment provides signals (with noise), energy sufficient to influence a receiver's sensors, but insufficient to produce a response. The output of a receiver's sensors (with noise) is then associated (with noise) with a possible response, and amplifiers produce the power for a response (with noise). A receiver engaged in noisy perception of its external environment knows only one of two possible states of the world, although in fact there are four possible states. Sensory input relevant to a situation either exceeds a threshold (meets a criterion) for a response by the receiver, or it does not.

Consider a particular case for the possibilities for supervention of second-order associations. If a vigilant receiver detects no warning signal, all it knows is that there is no evidence of danger nearby. Even if a predator subsequently appears, the receiver still cannot know whether there had actually been no warning signal or whether it had missed the detection of one. Another perception might change this appraisal though. The receiver (perceiver) might have seen another individual take cover before the predator appeared. It could take this perception as evidence that this individual had detected a warning signal when the perceiver had not. Such a receiver might conclude, in a self-conscious assessment of its own internal states, that its own detections were subject to error and that its own responses differed from other apparently similar receivers. A self-conscious receiver in noisy communication thus might conclude that it has a cumulative set of perceptions and secondary associations not shared by any other individual. It has an internal subjectivity that is apparently distinct from its external objectivity.

Emergent properties of minds are bound to arise in a subject's categorization of its own internal states. Every word is an emergent property of a receiver's subjective experience in noisy communication. Repeated exchange of signals can lead to a conclusion that use of a word by another person indicates a distinct state of that person's mind. Noisy communication and noisy perception make this process an approximation, one that suggests to any self-conscious receiver a subjectivity distinct from its external objectivity. It is noise that creates the possibility of a distinct subjectivity as an emergent property of the mind.

Can animals without obvious analogues of human language also form higher-order associations and thereby become aware of subjective experience and achieve self-awareness? Even without language, higher-order associations might result from covert nonverbal responses to perceptions. Human languages have a quantitative advantage, however. The much greater specificity and complexity of signals permits correspondingly greater specificity and complexity in communication and thus in thinking. For instance, labeling higher-order associations must facilitate exploring their relationships and thereby identifying (and labeling) progressively superordinate relationships. Lesser capabilities for higher-order thinking presumably result in lesser degrees of self-awareness. If so, self-awareness in other animals is likely to admit of many gradations, and we should resist dichotomous assumptions.

Perhaps those few animals with brains most similar to humans' are capable of comparable communication, thinking, and self-awareness. Dolphins, for instance, have brains nearly the same size as humans. Because their overall mass is also approximately the same as humans, their brains are nearly the same size as humans even in relation to their overall size. With its deeply folded cortex, a dolphin's brain even looks astonishingly like a human's. Some of the great apes have brains that are absolutely larger than humans', but all are relatively somewhat smaller. It is known that organisms that routinely move in three dimensions, in the sea, in the air, in the canopies of forests, have larger brains for their body sizes than those confined mostly to two dimensions on the ground. Some of dolphins' larger brains might thus serve to coordinate three-dimensional locomotion in the sea. Nevertheless, they are astonishingly quick learners and can master some language-like tasks. As for chimpanzees and other great apes, it remains to be discovered how they use these capabilities in their natural lives. Their large brains might or might not prove to be capable of complex higher-order associations of covert mental states. We no doubt have more to learn about how the brains of animals work—including our own brains. Nevertheless, even if the cognitive differences between humans and other animals are incremental, they add up to a large difference in mental capabilities, especially in the use of language for communication, thinking, and perhaps self-awareness.

OTHER MINDS

Is this subjective assessment of one's own state also private? Can a signaler reveal its own internal state to another? Or know the internal state of another signaler? These questions raise the problem of radical translation again. To what extent is a receiver's subjective sense of its internal state private? In Chapter 18, I suggested that the answer to such questions depends on whether or not an organism's behavior completely reveals its state of mind. If it does, observation of its behavior can determine its state of mind. If we suppose that a state of mind might be complicated, an observer would presumably have to record sufficient nuances of behavior to determine another's state of mind. An investigation of an individual's state of mind might go even further. It does not change the problem conceptually if we expand our investigation to include the neurological state of an individual's brain (or other physiological states), as well as its overt behavior.

Communication, as well as perception, might help in assessing the state of another's mind. An observer, in the role of a signaler, might probe another individual's state of mind by producing signals to which the subject, in the role of a receiver, could respond. To the extent that behavior—or the physiological state of the nervous system—is mind, then states of mind are revealed by behavior (or physiology). This is the assumption behind Alan Turing's test for whether or not a machine can think like a human being—if it behaves like a human, it thinks like a human. The point that needs emphasis here is that any distinct conclusion about the state of another individual's mind is subject to the limitations of noisy communication and perception. Through repeated observations or communication our observer might approximate correct conclusions about another's state of mind, as discussed in Chapter 6, yet such conclusions are unlikely ever to reach perfection. Noisy communication sets limits on the accuracy of conclusions about other minds.

If, on the other hand, behavior and brain do not completely reveal mind, then observations and communication cannot completely determine another individual's state of mind. An individual attempting to know another's state of mind would have to resort to some extrapolation. If behavior and brain only partially indicate the internal state of a signaler, an observer might use empathy to draw conclusions. Empathy is a sense of similarity with another object. To the extent that a receiver sensed similarity with a signaler, in whatever way seemed salient to the receiver, this receiver might be disposed to conclude that the signaler's state of mind resembled its own state of mind when producing similar responses to similar situations. Empathy presumably has no definite rules, so it presumably depends on the particular observer's own state of mind. When mind is not completely revealed by behavior, empathy provides an uncertain extrapolation from observations of a signaler's behavior to its mind, on the basis of an individual's subjective assessment of the relationships between its own mind and behavior.

In summary, this section argues that noisy communication and noisy perception produce a sense of self-awareness, a distinct subjectivity, not completely congruent with external objectivity. They also set limits to the accuracy of conclusions about others' minds. We can expect access to others' minds, as well as their access to ours, but without perfection in either direction.

NEURAL MECHANISMS, NOISE, AND FREEDOM

Identifying the basic processes of all receivers clarifies the basic operations of the nervous system of organisms: sensation, association, and amplification. These operations correspond to the three components of any receiver—receptors or sensors, decision makers or switches, and amplifiers. They recur at every level of the nervous system. Signaling is the prevailing operation of the nervous system from bottom to top.

The properties of sense organs determine the features of external stimulation that evoke responses. Many of these properties of sensory analysis have been carefully studied by now. It is clear that, for any species, sense organs respond to a particular scope of external stimulation, often limited to a portion of the variation present externally. The cochlea of our inner ears, discussed in Chapter 3, provide an example of receptors that respond only to a limited range of external stimulation—in the case of humans, sound frequencies from 50 to 20,000 Hz at best. Other organisms have receptors that respond to lower or higher frequencies of sound. In many insects and amphibians sensory cells for hearing respond only to relatively narrow ranges of frequencies tuned to the frequencies in the species' vocalizations or to sounds produced by the movements of predators. The sensory cells in our retinas respond only to wavelengths of light from red to violet, although many other organisms can also see ultraviolet. Sensory cells, as has been noted throughout this book, are the first receivers of any nervous system. Their responses become the signals for neurons that associate progressively more specific features of the sensory input.

Association of sensory input begins as soon as the responses of sensory cells converge. Even at this early stage in the nervous system, sensory cells behave as signalers sending signals to second-order neurons. Even within the retinas of all vertebrates this process begins. Second- and third-order retinal neurons are excited (or inhibited) by input from many sensory cells located in one tiny area of the retina. These neurons respond to light at particular spots on the animal's retina and furthermore have antagonistic responses to light (either inhibitory or excitatory, respectively) in the immediately surrounding area. In some vertebrates (though not primates) neural connections in the retina produce neurons that respond to movement of a stimulus across the retina. In those areas of the brain that first receive input from the retina, neural connections produce cells that respond to strips of light, with particular orientations in small areas of the retina. Neurons in immediately adjacent areas mutually inhibit each other. This lateral inhibition of adjacent neurons sharpens visual images in almost all organisms. Further associations of sensory input in other areas of the human brain are now known to respond selectively to human faces or hands. Sensory association thus results in neurons that respond to progressively more specific features of sensory input. The increasing specificity of neurons forms the basis for discriminations between features of sensory input. At each step in the process of association, lower-level neurons (signalers) send action potentials (signals) to higher-level neurons (receivers). Progressive selectivity of neurons results in responses to specific combinations of features in impinging stimulation, which can produce complex criteria for responses.

Responses to signals from these associational neurons can lead to even higher-order associations including potentially those the result in self-awareness in humans, as discussed in the preceding section. They can also lead to overt responses, either innate or learned. Learned associations between sensations and overt responses are another example of signaling in the nervous system that requires discrimination and association of features of sensory stimulation. Furthermore, as discussed in Chapter 17, complex learning can initially require some guidance from predispositions. These predispositions develop when the organism requires an initial focus not provided by specific features of the environment. They require sensory associations sufficient to provide a criterion for an initial response and also association of this criterion and the initial response. These predispositions of the nervous system are innate in the sense that their development is relatively canalized (stabilized over a relatively broad range of normal environmental conditions). The development of stripe detectors in the visual cortex of kittens before their eyes open is a clear example. The pathways leading from the retinas to these neurons organize themselves, in the absence of any patterned visual input, in a chain of signals and responses to produce neurons in the brain that respond specifically to the criteria for a tiny stripe of light at a particular orientation and in a particular place on the retina. The information for this chain of associations must come from properties of the neurons specified by genes activated within the scope of normal environmental conditions for a developing kitten. Perhaps these predispositions, or their eventual modifications by experience, become for a self-aware receiver the "natural" way to respond to sensory input.

As an organism develops throughout its lifetime, all physiological responses result from an interaction of its current internal state and its current sensory input. Whether an overt behavioral response or a covert physiological response, all result from an interaction between a receiver's sensory input and its current internal state. From moment to moment its internal state is updated by its covert responses to preceding stimulation, including memory, learned associations, sensory adaptation, and habituation. Its external situation is updated by its overt responses that alter its environment. Chapter 7 discussed the interaction between nature and nurture, genes and environment, in the development of organisms. The process of perception is also an interaction of internal state with external stimulation. The moment-to-moment updating that controls responses to external stimulation is just the instantaneous updating of the developmental process that begins at conception.

Applying this view of behavior to receivers in noisy communication changes our view of mechanistic determinism. If organisms are viewed as machines entirely explained by Newtonian mechanics, even if we allow quantum mechanics and relativity at very small and very large scales, it appears that no freedom of response is left. Every organism's fate is determined at least on a macromolecular scale. This sort of conclusion is especially disturbing for moral decisions.

Noisy communication throughout the nervous system makes this sort of interpretation more difficult. The four outcomes a receiver faces with each decision are the basis for a self-conscious receiver's sense of a distinct subjectivity. The limitations of noisy communication make knowing the mind of another person a process of approximation. For each self-aware receiver, its decisions lack determination and instead more or less often fail to conform to subsequent experience. Each time a decision is made, a receiver knows only two possibilities. Its sensory input either justifies a response or it does not. Subsequent experience might reveal to a self-aware receiver that the previous situation had been more complex. Four outcomes, not two, were possible, and subsequent evidence might suggest that the receiver had made an error of commission or omission (a false alarm or a missed detection). At the time of each decision, however, all that a receiver can know is the sufficiency or insufficiency of its input.

The sense of distinct subjectivity is thus also a sense of the importance of decisions, the importance of one's own criterion for situations in which a response is appropriate and those in which it is not. Observation of other individuals and inferences about their minds, fraught as they are with uncertainty, can suggest a similar sense of the importance of their decisions as well. Noisy communication by self-aware receivers thus creates an emergent sense of freedom and responsibility. It can also create a sense of intuition, a sense that certain decisions are more "natural" than others.

SCIENCE AND ART FOR EVERYONE

Scientists and artists have often seemed to have different mental processes, at least to some degree. Final thoughts in this chapter on subjectivity and objectivity should reconsider these two mental poles. The analogies developed between perception and communication suggest that scientists and artists, although unlikely to differ fundamentally, might diverge in how much they focus on communicating internal subjectivity or external objectivity. Both perspectives result from the effects of noise on signals from external situations during perception and communication. The effects of noise apply to everybody. Perhaps it makes more sense, therefore, to distinguish scientific and intuitional modes of thought by any person, corresponding to perception and communication, rather than to separate scientists and artists as distinct categories of people. All humans in their scientific mode rely on repeated observations, careful measurements, probing interactions, and stylized communication with others in the same mode to identify their modal perceptions in noise. They seek consilience in perception. All humans in their intuitive mode focus on the unique, the fleeting, the categorical, and intuitive communication with others to appreciate the scope of perceptions in noise. They seek freedom in perception.

In their professional modes, scientists and artists are both limited in their objectives by noise in perception and communication. Both might do well to study the consequences of noise for their efforts to express, respectively, their sense of objective reality or their sense of subjective reality. At other times, in our leisure modes, we might all enjoy the degrees of both subjectivity and objectivity that noise provides.

20

VERIFICATION

CONSILIENCE

NOISE CREATES A DISTINCTION for every individual, in its role as a receiver or perceiver, between subjective and objective experience. Communication, on the other hand, saves individuals from solipsism. Chapter 19 showed that noisy communication permits a degree of consilience between signalers and receivers. Chapter 10 showed that natural selection leads to a joint optimum for signaler and receiver at which signals are often honest and both parties benefit on average, even though perfect performance cannot be achieved by either party. Two individuals communicating reciprocally can thus achieve some consilience, even if not perfect.

Consilience in communication is necessary to assure that signals are mutually understood. In other words, situations associated with a particular signal by one party are likewise associated with that signal by the other. This process requires two steps. A signal as described in Chapter 1 is a pattern of sensory input. So the first step for communicating parties is to classify their inputs in analogous ways, by recognizing and discriminating features of signals and other sensations similarly. In other words, they must adopt similar criteria for correct detections. The second step is to associate these patterns of input with similar (perhaps covert) responses. Both steps must deal with noise.

One way such consilience in communication might arise is by natural selection. Populations can evolve strong predispositions to respond to particular situations in standardized ways so that receiving individuals classify and respond to signals from signaling individuals in ways that benefit both parties on average. Signalers, as discussed in the Part I of this book, evolve signals that optimize their performance, and receivers respond to signals in ways that optimize theirs. In many instances of animal communication no learning occurs in the development of signaling or receiving. Natural selection has produced an optimal level of coordination.

This sort of coordination of signaling and receiving is perhaps not what we might want to recognize as consilience. Examples of imitation come a step closer. Young songbirds learn to sing normally by listening to adults. This learning is communication. A young bird's response is memorizing an adult's song. At least in some species, only weeks later does the young bird use this memorized template to develop its own songs with features matching those of the adult it previously heard. In this case, communication has resulted in one individual transferring its state of mind to another individual, often with a high degree of accuracy. So striking is the consilience in many cases that nobody has paid much attention to the noise, the errors (however slight) in this communication. Other cases of imitation, none so well studied, probably have the same property of consilience in communication. To what extent, for instance, do young animals learn complex hunting techniques from adults? There are many cases among birds and mammals in which young accompany hunting adults and progressively develop their skills. In some of these cases, rough experiments indicate that young animals master similar skills even when separated from adults. Either innate maturation of the nervous system or individual learning by trial and error could produce this result. Common Chimpanzees provide more suggestive possibilities. Population-specific techniques for cracking hard nuts and for "fishing" for termites are plausible examples of imitation as a result of consilience in communication. As argued previously, learning by imitation or otherwise depends on innate predispositions to respond in particular ways to particular sensations. Nevertheless, the details of each individual's specific environment have a strong influence on its development. Young birds might have an innate tendency to attend to particular kinds of sounds or to particular interacting individuals, but the details of the songs they eventually sing result from learning.

Human language opens further possibilities. It permits complex descriptions of a person's perceptions and states of mind—not only features of immediate sensory input but also their relationships in space and time. To communicate effectively with language requires some consilience in producing and responding (even covertly) to linguistic signals. Learning has a strong influence, in many ways like birdsong. Predispositions are also necessary to explain the development of language, as discussed in Chapter 17. Trial-and-error learning no doubt occurs as individuals learn which words elicit particular responses from listeners. Yet the expansion of a child's linguistic competence is so fast that there must be a predisposition simply to imitate what others say in particular situations. Every child acquires phonemes, phrases, and uses of language that match those of the adults and peers with which it interacts.

Learning this complex skill for describing perceptions and states of mind by imitation might not be enough to produce satisfactory consilience. Imitating others' verbal responses in particular situations is a case of perception, which as emphasized already is fraught with noise. The problem of translation resurfaces here. How can perception of another individual and its signals be sufficiently precise? Judging another individual's state of mind by perceiving its behavior, although not entirely fruitless, must have limits. Such a judgment, we could say, is underdetermined by perceptions.

An ability to communicate about one's perceptions by describing them in space and time should help in this predicament. Two individuals engaging in mutual communication might revise and refine their uses of verbal signals—both their production of signals and their (often covert) responses. In a particular situation they might compare signals and become aware of possible discrepancies. This communication is of course still noisy. As in translation from an unknown language, striving for consilience in communication even with the most intimate partner has limits. Yet as we noticed in Chapter 18 for radical translation and for noisy communication in general, redundancy mitigates the consequences of noise.

SOLIPSISM RESCUED

Consilience in use of language has another important consequence, in addition to facilitating communication with others about perceptions and states of mind. It provides the only way to assure stability in thinking about ourselves. If there were no consilience in communication, our use of linguistic signals for thinking would have no constraints beyond innate predispositions. The discussion of radical translation in Chapter 19 indicated that noisy communication limits the precision with which two individuals can develop agreement in use of language. Yet communication (combined with some shared predispositions for language) permits some consilience. In contrast, internal communication in isolation has only predispositions to shape usage. As a result many possible associations among perceptions or between perceptions and responses have no opportunity for comparison with an external standard. In isolation, subjective comparisons of thoughts have no boundary conditions. By comparing usage of language internally, without some stable model, there is no way to determine agreement. Apparent agreement or disagreement could always result from noise in perception. As a result of noise, an isolated perceiver can never determine that any single perception is correct. Any choice between alternatives would be a guess without possibility of correction. It is hard to imagine that the ensuing chaos could allow the development of a stable complex language. This situation would even extend to labeling our own internal states. Noise would generate irreducibly erratic patterns of signals. Without recourse to communication with others sharing the same language, even with some unavoidable noise, there is no way to confirm that complex usage of language has any consistent associations with either our external or our own internal situations. Stability would be reduced to innate predispositions.

Human thinking is so complex it obviously does not result solely from innate predispositions. Human brains have nearly 100 billion neurons (approximately an order of magnitude more than the Common Chimpanzee, despite the comparatively slight difference in overall size), and these neurons average about 10,000 to 100,000 synapses each. In comparison, our genome has a mere 3 billion base pairs, the four constituents of DNA, organized into about 20,000 protein-encoding genes. These numbers do not permit any precise comparisons between genes and brains because everything depends on how the constituents, neurons or base pairs, are organized. There is evidently plenty of DNA to allow our genes to predispose, to nudge, the development of our brains in particular ways in particular environments, but interactions with our environments take over from there. Consequently, stability in our thinking can only come from consilience in language, which in turn can only come from communication, noisy though it is.

This argument, that stability in thinking about ourselves presupposes consilience in communication, is related to the possibility of a private language. The issue is not the invention of a novel language. People invent idiosyncratic ways to encode observations and experiences all the time. In the process they are still thinking in the language they had previously acquired. The issue instead is the development of language from scratch, *ab initio*, without concurrent participation in communication. The previous paragraphs argue that human language cannot develop in this way. Our innate predispositions for language are not sufficient to generate a complex yet stable human language. Consilience in communication provides the additional information for this process.

Nevertheless, a case has been made for the possibility that language can develop *ab initio* in private. The evidence comes from children without hearing and without exposure to people proficient in a sign language. In these circumstances, a child develops gestures collectively called home sign that mediate interactions with his or her family. Susan Goldin-Meadow and her colleagues report that such children produce combinations of gestures with distinctions among nouns, pronouns, verbs, and adjectives and that they can designate objects or events not immediately present. Because home sign retains a degree of individual idiosyncrasy, it might qualify as a private language. The complexity of this language (or these languages) is nevertheless rudimentary compared to those used by most children. There are also two difficulties in interpreting these reports. First, adults often respond to a deaf child's gestures with a series of tentative actions until the child indicates it is satisfied. Indeed investigators use this procedure to plumb the child's intentions. Although these interactions are nonverbal, the child has an opportunity to develop some consilience with its interlocutors in the use of gestures. Second, the children in these studies, although not exposed to adults employing a sign language, attend preschools with other deaf children. Yet it is known that interactions among deaf children promote the development of language-like signs. Both of these demurrals suggest that the children under study were not completely deprived of communication. Although limited, this communication could allow some consilience and thus facilitate development of at least some rudiments of language.

When deaf children interact with each other, they soon develop a gestural language with the complexity of a spoken language. The emergence of Nicaraguan Sign Language is the exemplary case. When a new school for deaf children opened in Managua in 1977, those attending had had no previous exposure to a Deaf community with a sign language. Apparently each child had only its idiosyncratic home sign. Within a decade the children had produced a new sign language with the complexities of other natural languages, including morphological, grammatical, and semantic conventions. There remain disagreements about how rapidly this process transpired, about the nature of home signs from which it began, and about what it all means for the development of language in other children. Other examples of people who failed to develop human language in the absence of communication, such as Helen Keller in her first seven years or children severely abused by abandonment or confinement, are even less well documented and more problematic. Nevertheless, the evidence so far is not inconsistent with a conclusion that development of a complex language requires communication.

Children are capable of language, even predisposed to language, to judge by how swiftly they acquire it in the usual circumstances. It seems appropriate to conclude that they are innately predisposed to develop language by a process that is remarkably canalized but not predetermined. "Canalized" because children ignore some input, even explicit parental corrections, in intermediate stages of language development. Not predetermined, "resilient," as Goldin-Meadow says because children can after all learn any of a large number of possible languages, in any of several sensory modalities, from any of several sets of relatives, peers, or rivals. It also seems appropriate to conclude that children acquire language only in communication with other people. If so, a person cannot think like a human without communication with humans. An isolated child's perceptions of objects and events are insufficient to produce a language, at least not beyond the rudiments.

The situation is different in songbirds. These birds, as described in Chapter 17, must hear other members of their species in order to develop normal songs. Social interactions usually facilitate this learning. In these respects the parallel with human language is compelling. Yet birds only require a model for imitation, although the degree of imitation, as opposed to improvisation, varies among species. Humans in contrast do not just improvise on imitations. Instead they abstract the recombining components of an open-ended language. At least in some birds there is a further rudimentary parallel with human language. Each song pattern in an individual's repertoire is sung preferentially in a particular situation, such as proximity to a territorial boundary or nest or presence of a rival or mate. The association of songs with contexts has similarities to the adjustment of human language to different situations. In birds we do not yet know whether or not these associations are learned, nor how complex they might be. Nevertheless, there is no evidence to suggest that birds use their songs to think in open-ended ways about their perceptions. For birds, the issue of consilience in communication arises, as far as we know, only in the most rudimentary way at best.

For humans, on the other hand, consilience is essential for language. Agreement between signalers and receivers about usage of signals reinforces consistency in usage. In privacy, there is no imperative for consistency. An isolated individual could learn or remember that some responses to a situation were beneficial and others were not. These responses might generalize to similar situations. A private language might label sets of response or situations. Because situations and responses have complex constituents and gradations, the labeling would be idiosyncratic. Even more important, noisy perception of situations and noisy production of responses would compound the idiosyncrasy of labels. Alternative labels would often have equal utility for the individual.

If the complexity of an idiosyncratic private language permitted introspection, it could provide no assurance of either accuracy or utility. The crucial opportunity to surmount these problems comes with communication. Communication, even if noisy, can generate enough consilience of labeling (reference) to mitigate individuals' idiosyncrasies. It also allows individuals to counteract noisy perception. By comparing notes, individuals in communication with each other can reduce the noise in their own perceptions.

There is another way that agreement in communication could arise. Innate predispositions to respond in particular ways in particular situations could result in a "language" for communication. Canalized development in normal environments might provide sufficient agreement in use of signals. Placing this sort of "language" in quotation marks indicates that such communication, for the reasons described above, is constrained to less complexity than human language. Nonhuman animals can form associations and abstract generalizations. Many produce combinations of signals and vary the composition of the mix. A few species with especially large brains accomplish such tasks as adeptly as young children. In these species, establishing communication with humans seems to promote the development of these rudimentary language-like capabilities. Yet no nonhuman animal naturally develops a language of recombining components for open-ended communication about its perceptions and introspections. If this line of reasoning is correct, even approximately, it requires an adjustment in the way we normally think of language in relation to communication in humans. It is natural to assume that language is a prerequisite for communication. Yet the converse is also true: communication is a prerequisite for language. It specifically enables the development of consilience for a stable complex language. Without communication, there is no such language, and without such a language, there is no complex thought, neither circumspection nor introspection.

This train of thought leads to a realization that communication rescues each person's sense of individuality. The consilience gained in communication assures, to a degree, our confidence in our thoughts about ourselves. One might conclude, to paraphrase Descartes, "Je communique, donc je suis"—with emphasis.

We humans are thus unique individuals, each with a unique sense of ourselves, and our subjectivity, but saved from the chaos of solipsism by a degree of consilience in communication. Our sensations (with noise) are perceived (with noise) and confirmed by communication (with noise). We are stuck between a Locke (who would uncritically accept sensations) and a Berkeley (who could imagine each of us encapsulated within our private brains).

SCIENCE

This intermediate position has relevance for science. A standard view of science is the practice of a community of experts who confirm each other's observations of patterns in the external world. Often the emphasis is on verification or falsification of hypotheses. If we accept the logic of the excluded middle (that every proposition or its negative must be true and the other false, with no middle ground), as Aristotle first suggested, then an emphasis on falsification makes sense. After all induction from observation can never absolutely verify a statement, but a single observation might falsify it.

This emphasis on the logical relationships of hypotheses relies on Aristotle's syllogisms. Chapter 17 has already made the point that a distinction between verification and falsification is clouded by noisy communication and the resulting classifications of terms by family resemblance rather than definition. The application of syllogisms to deduce logical consistency in perceptions is likewise obfuscated by noise. These cases, as others before, are not hopeless. Noisy communication is often honest, even if never perfectly so. And, likewise, noisy perception is often correct, even if not perfectly so. Falsification and syllogism work, just not as inerrantly as Aristotle and Karl Popper have assumed.

Confirmation of others' observations is also affected by noisy communication. As just discussed, noisy communication about each others' perceptions and states of mind (hypotheses and conclusions, for instance) is expected to achieve some, but not perfect, consilience. Furthermore, scientific observations and hypotheses have always required special expertise not successfully attained by everybody. The work of physicists exploring the nature of subatomic particles or the background radiation of the universe cannot possibly be confirmed by just anybody—nor can the work of molecular geneticists. In fact, despite assumptions of some to the contrary, even the sort of work I do on behavior of animals in the field requires a degree of expertise in anyone I would accept for confirmation of my results. Michael Polanyi has emphasized the importance of expertise in confirmation of scientific results, but it is a slippery slope. Is science essentially a congregation of confirmed believers?

Notice that partial consilience as a result of noisy communication terminates this downward slide. Because consilience is achieved by mutual acceptance of signals during noisy (and thus imperfect) communication, receptivity to consilience must balance resistance to consilience—susceptibility must balance skepticism. The boundary between susceptibility and skepticism is analogous to a threshold in noisy communication. Too much skepticism precludes much consilience about states of mind (too many missed detections). Too much susceptibility promotes unwarranted consilience (too many false alarms) and thus invites individual idiosyncrasy and collective fads. A community of scientists, I suggest, consists of people with a high threshold for consilience.

Expertise is a subordinate issue. Scientists are those with a high level of skepticism. Acquiring expertise is something a person does as a result of skepticism about others' observations. In particular, scientists have high skepticism about observations and hypotheses in their particular field of interest. Or they should. Just below, I argue that everybody is a scientist at times and that there is no real division between the process of science and other human ventures. Scientists do not always maintain high levels of skepticism even in their own fields. The benefits of consilience can tempt anyone. Promotions, grants, and administrative positions can depend to a degree on a scientist's consilience with others in the field. The history of science has its share of fads and bandwagons. Furthermore, increased competition for positions and funds is likely to increase the temptations of consilience among colleagues.

It has become commonplace to scrutinize the human foibles of scientists. It is just as likely, however, that at times we are all scientists, with high thresholds for consilience, and at other times groupies, with low thresholds. It seems likely that most of us compartmentalize our internal states to some degree, so that on weekends we might succumb to excessive consilience for communal reassurance, but on Monday awake to strong skepticism for solving problems. Some problems probably allow greater compartmentalization than others. For instance, a mechanic must think like a scientist to repair machinery. A high degree of compartmentalization is possible here because machinery comes with repair manuals. As noisy communication, these manuals require some skepticism. Less compartmentalization is possible for mechanisms that do not have manuals, such as the human body. Physicians and nurses must presumably maintain a wider sphere of skepticism to test their hypotheses. Like repair, construction also requires solving problems. Construction of tangible items, such as machinery, toys, buildings, and bridges, can require great expertise, essentially writing the manual or drawing the plans, but the sphere of necessary skepticism is not great. Scientists, when doing their job correctly, construct explanations for the patterns of the world and thus require a high threshold for consilience inasmuch as the unexplored possibilities have indefinite limits. Even mathematics and logic can change to fit the observations. Nevertheless, scientists do not always—although in my experience some do—work every day of the week.

A low threshold for consilience risks acceptance of perceptions and hypotheses that in the end cannot be verified by others. Even different groups, each with members in agreement among themselves, might fail to confirm each others' observations. Too much consilience can lead to such fads but also, as a result of relaxed skepticism, to acceptance of thoughts that others would consider supernatural or miracles. On the other hand, too little consilience can lead to mental chaos, as suggested above, to unverified subjectivity.

Not only do all, or at least most, people employ scientific thinking at times, they also put conciliatory thinking to use at other times. Consilience is no doubt the primary objective of much human activity. Group solidarity is important for humans and has been through our evolution. Gossipers, salespeople, politicians, commanders, and artists (not to mention the Stripe-backed Wrens in Chapter 15) presumably rely strongly on abilities to achieve consilience with their audiences. No doubt we all operate mentally at some place along the spectrum from low to high thresholds for consilience, from receptivity to skepticism—all because our perceptions are noisy and our attempts at confirming them require noisy communication. We must set our thresholds to optimize the utility of our mental states in each situation. The required agility presumably has some limits. Some people are no doubt more agile than others.

Noisy communication and noisy perception create our sense of subjectivity and individuality. Noise creates the problems, but communication, despite the noise, saves us from the mental chaos of solipsism. We are left balancing the optimal levels of consilience in communication, balancing susceptibility and skepticism, as best as we can to suit the situations we face.

NATURAL SELECTION REDUX

Having followed the implications of noisy communication to an examination of the process of science, we have come full circle. Applications of scientific procedures—both empirical and theoretical—to understand noisy communication in Parts I, II, and III of this book have now, in Part IV, circled back to science itself as noisy communication. The argument is based on evolution by natural selection, a discovery of scientists. Where does this circularity leave the argument?

Natural selection has only been recognized as a natural process for some 150 years. It is one of four mechanisms of evolution. Evolution—any change in allele frequencies in a population from generation to generation—can result from mutation, migration of individuals into and out of a population, and random influences on survival or reproduction (called genetic drift), as well as from natural selection and its variants, such as sexual selection, group selection, or kin selection. Natural selection is a mechanism of evolution because all organisms have special features. First, they reproduce themselves (presumably because like any mechanism they gradually wear out until it becomes easier to replace them than to repair them); furthermore, they reproduce with limited mutation. If no mutation occurred, no selection would occur. If mutation were too great, selection would likewise not occur because progeny would never have much in common with parents. Second, there is selective retention of progeny based on their performance. Those that survive and reproduce more often than others transmit their parents' features more often to the following generation. So those features associated with retention accumulate in the population.

These principles, reproduction with limited variation and selective retention of progeny, apply in other contexts too. For instance, trial-and-error learning shares these features of natural selection. Evolutionary computing in recent decades has adapted natural selection to the solution of otherwise impossibly complex problems. A program is allowed to "spawn" a number of variants with limited mutations of parameters. These progeny are then compared to some goal the programmer has in mind. Those that come closest are allowed to reproduce again (they survive and reproduce, the others do not). The population of variant programs gradually accumulates those features successful in meeting the goal. In natural selection there is no programmer making the comparisons. Survival and reproduction in a particular environment is, as Charles Darwin realized, a sufficiently stringent test to produce natural selection. Natural selection as clearly described by Darwin does not depend on any particular genetic mechanisms. In fact, Darwin—who lived before Gregor Mendel's rediscovery and who failed to reach any major insights in his own extensive experiments with genetics—had no concept of modern genetics.

The preceding is not to say that the particular properties of natural selection in the evolution of living organisms has no need of further investigation. All the details are important in any particular application. Notice that even the rate of mutation should evolve adaptation, inasmuch as neither too much nor too little seems optimal. The evolution of evolution is up for investigation. The basic principle of natural selection, however, seems undeniable. Although first proposed for biological organisms, it is a principle as fundamental as any other logical principle and as widely applicable.

INTERACTIVE DEVELOPMENT REDUX

The other major premise for the arguments in this book is the interaction of genes and environment in development. This finding is confirmed by all relevant experiments. It is routine, for instance, to show that genes influence learning, and also that the influence of genes on learning depends on the situation (environment). Furthermore, learning is a form of plastic phenotypic development, one that is influenced particularly intricately by the individual's experience (environment).

The interaction of genes and environment has, like natural selection, analogues in human experience. How can anything be constructed except by means of plans in a particular context? The expression of the plans depends on the context, the materials available, the lay of the land, the climate, or even the weather. On the other hand, the result in any one context depends on the plans. It is not possible in principle to construct anything by plans alone. Furthermore, a context alone has too many options. The same is true of the development of an organism, including learning.

These two principles, natural selection and interactional development, have generality far beyond the scope of biology or even science. Like our sense of number or our sense of time, they seem to be undeniable principles. They do not seem to require confirmation for the reason that it is impossible to think of alternatives. They seem as basic as the concepts of number and entropy. Indeed, those two concepts together might provide at least part of an explanation for both natural selection and interactional development.

A LESSON

What follows from these premises about noisy communication does invite confirmation. My deductions of the fundamental properties of signals and receivers rest on the nature of noise. Subsequent calculations suggest that noise is inescapable. I have not presented an analytical proof, however. All the details are up for investigation. Nobody has so far measured most of the parameters identified in Chapter 9 for a minimal analysis of communication in noise.

The arguments appear tentative for an even more fundamental reason. The present study is itself an attempt at communication. By its own arguments, I must accept that it is inevitably noisy. I can hardly conclude that communication is imperfect while claiming an exception for this book. As a signal it no doubt incorporates its own dose of noise, from obfuscation to error. Any adaptations on my part to mitigate noise through adjustments of contrast, redundancy, and predictability are unlikely to eliminate all error. In addition, the physical transmission might contribute more noise. And you, dear readers, you *receivers,* no doubt add a share to the noise. Consilience about noisy communication has yet to develop. Perhaps this book can start the conversation.

This study also pleads for a position between extremes. Communication, it argues, should arrive at an optimum, by evolution or design, that assures honesty on average but precludes escape from occasional error. It predicts satisfactory but imperfect consilience. For some situations, however, this intermediate conclusion does not appear at all satisfactory. In some circumstances we must hope for perfection. Can communication prevent internecine warfare? Can injustice be eradicated? Can public policies prevent the looming environmental upheaval? Like all social interactions, warfare, justice, and public policy depend primarily on communication—in these cases in diplomacy, in courts, and in government. We should hope, but, if there is a lesson in this book, we should never rest.

If, as I have argued, perfection in communication is illusory because of the pervasive effects of noise, we cannot expect to know each other's minds without error. The inevitability of misunderstanding is the price of communication. Our rescue from solipsism is its reward. There are ways to reduce the consequences of noise in communication and in perception. In critical situations, it behooves us to heed them. The calculations presented in this book indicate that we can do well by communicating with each other. Yet we cannot expect perfection. We indeed "see through a glass, darkly" and (contradicting Paul) *not* "face to face." We "know in part" and (although Paul anticipates otherwise) "even as also we are known." What stance should we then take toward each other?

I suggest that humility and tolerance are in order. But the greater of these is humility. Perhaps love needs nothing more.

CONCLUSION

THE ARGUMENT IN THIS BOOK proceeds from natural selection and an observation that communication is often noisy. It shows that noisy communication requires decisions by a receiver as well as a signaler. The operation of natural selection on these decisions produces a trend toward a joint optimum for signaler and receiver that does not escape noise. Communication, if the argument is correct, is inescapably noisy. Human language is no exception. Furthermore, perception is a form of communication, and it too is inescapably noisy. Noisy perception produces our sense of subjectivity and thus individuality. Communication, albeit noisy, allows a degree of confirmation of our use of language and thus of our thinking about ourselves. It thus rescues us from the mental chaos that would otherwise result from solipsism. It also allows a degree of confirmation for our observations of patterns in the external world. In so doing, we adjust our threshold for consilience between susceptibility and skepticism. In the end, this argument shows that we are the result of both the noise and the signals. Because of noise, as already emphasized, each of us can paraphrase Descartes, "Je communique, donc je suis"—"I communicate, therefore I am."

BIBLIOGRAPHIC AND OTHER NOTES

1. NOISE AND SIGNALS INTRODUCED

Claude Shannon's theory of information was developed at the Bell Research Laboratory, the sort of corporate institution no longer favored in these days of short-term profit maximization . . .

Shannon, C. E. 1948. The mathematical theory of communication, I and II. *Bell System Technical Journal* 27: 379–423, 623–656.

Wide dissemination of the theory followed its presentation in a delightfully readable version . . .

Shannon, C. E., and W. Weaver. 1963. *The mathematical theory of communication.* Urbana: University of Illinois Press.

My own take on information in communication follows Shannon's lead . . .

Wiley, R. H. 2013. Communication as a transfer of information: Measurement, mechanism and meaning. In U. Stegmann, ed., *Animal signals and communication: Information and influence,* 113–129. Cambridge: Cambridge University Press.

Wiley, R. H. 2013. Signal detection, noise, and the evolution of communication. In H. Brumm, ed., *Animal communication in noise,* 7–30. Berlin: Springer-Verlag.

Hooded Warblers sing in part to communicate with territorial neighbors, a possibility my students and I have investigated . . .

Wiley, R. H., R. Godard, and A. D. Thompson Jr. 1994. Use of two singing modes by Hooded Warblers as adaptations for signalling. *Behaviour* 129: 243–278.

Godard, R., and R. H. Wiley. 1995. Individual recognition of song repertoires in two wood warblers. *Behavioral Ecology and Sociobiology* 37: 119–123.

Godard, R. 1991. Long-term memory of individual neighbours in a migratory songbird. *Nature* 350: 228–229.

Godard, R. 1993. Tit for tat among neighboring Hooded Warblers. *Behavioral Ecology and Sociobiology* 33: 45–50.

Nevertheless, they do not always stay on their own territories and sometimes create problems for their neighbors. Neighboring males are thus rivals (although evidence in Chapter 14 indicates that they also cooperate) . . .

Pitcher, T. E., and B. M. Stutchbury. 2000. Extraterritorial forays and male parental care in Hooded Warblers. *Animal Behaviour* 59: 1261–1269.

Stutchbury, B. J. M., et al. 1997. Correlates of extra-pair fertilization success in Hooded Warblers. *Behavioral Ecology and Sociobiology* 40: 119–126.

For birdsong in general, there are now several excellent introductions to its role in communication . . .

Kroodsma, D. 2005. *The singing life of birds: The art and science of listening to birdsong.* Boston: Houghton Mifflin.

Catchpole, C., and P. J. B. Slater. 2008. *Bird song: Biological themes and variations,* 2nd ed. Cambridge: Cambridge University Press.

Marler, P., and H. W. Slabbekoorn, eds. 2004. *Nature's music: The science of birdsong.* San Diego: Elsevier Academic. This and the next two volumes have chapters by many experts on vocal communication by birds.

Kroodsma, D., and E. H. Miller, eds. 1982. *Acoustic communication in birds,* vols. 1 and 2. New York: Academic Press.

Kroodsma, D., and E. H. Miller, eds. 1996. *Ecology and evolution of acoustic communication in birds.* Ithaca, NY: Cornell University Press.

For general perspectives on the sources of noise for animals . . .

Narins, P. M. et al., eds. 2006. *Hearing and sound communication in amphibians.* New York: Springer.

Luther, D., and K. Gentry. 2013. Sources of background noise and their influence on vertebrate acoustic communication. *Behaviour* 150: 1045–1068.

Naguib, M. 2013. Living in a noisy world: Indirect effects of noise on animal communication. *Behaviour* 150: 1069–1084.

Brumm, H., ed. 2013. *Animal communication in noise.* Berlin: Springer-Verlag.

For comparisons of diverse ways to encode information . . .

Hailman, J. P. 2008. *Coding and redundancy: Man-made and animal-evolved signals.* Cambridge, MA: Harvard University Press.

For communication in choruses of frogs and toads (anurans) . . .

Ryan, M. J. 2001. *Anuran communication.* Washington, DC: Smithsonian Institution Press.

Gerhardt, H. C., and F. Huber. 2002. *Acoustic communication in insects and anurans.* Chicago: University of Chicago Press.

And in crowded colonies of birds . . .

Bretagnolle, V. 1996. Acoustic communication in a group of nonpasserine birds, the petrels. In D. E. Kroodsma and E. H. Miller, eds., *Ecology and evolution of acoustic communication in birds,* 160–178. Ithaca, NY: Cornell University Press.

Aubin, T., and P. Jouventin. 1998. Cocktail-party effect in King Penguin colonies. *Proceedings of the Royal Society B: Biological Sciences* 265: 1665–1673.

Mackin, W. A. 2005. Neighbor-stranger discrimination in Audubon's Shearwater *(Puffinus l. lherminieri)* explained by a "real enemy" effect. *Behavioral Ecology and Sociobiology* 59: 326–332.

There is also noise generated by human activity, such as traffic on highways . . .

Patricelli, G. L., and J. L. Blickley. 2006. Avian communication in urban noise: Causes and consequences of vocal adjustment. *Auk* 123: 639–649.

Warren, P. S., et al. 2006. Urban bioacoustics: It's not just noise. *Animal Behaviour* 71: 491–502.

Brumm, H., and H. Slabbekoorn. 2005. Acoustic communication in noise. *Advances in the Study of Behavior* 35: 151–209.

In some cases, noise takes the form of deceptive signals (for references for deception by nonhuman primates see a later chapter) . . .

Munn, C. A. 1986. Birds that cry "wolf." *Nature* 319: 143–145.

Møller, A. P. 1988. False alarm calls as a means of resource usurpation in the Great Tit *Parus major. Ethology* 79: 25–30.

Møller, A. P. 1990. Deceptive use of alarm calls by male swallows. *Behavioral Ecology* 1: 1–6.

Gyger, M., and P. Marler. 1988. Food calling in the domestic fowl, *Gallus gallus:* The role of external referents and deception. *Animal Behaviour* 36: 358–365.

Noise can take the form of spontaneous, irregular activity of neurons. A classic study described this noise basal activity of the nervous system in sensory receptors in the retina of mammals . . .

Barlow, H. B. 1956. Retinal noise and absolute threshold. *Journal of the Optical Society of America* 46: 634–639.

In birds this spontaneous activity even has a tendency to match the periodicity of the frequency that best excites the sensory neuron . . .

Manley, G. A., ed. 1990. *Peripheral hearing mechanisms in reptiles and birds.* Berlin: Springer-Verlag.

Gleich, O., and G. A. Manley. 2000. The hearing organ of birds and Crocodilia. In R. J. Dooling, R. R. Fay, and A. N. Popper, eds., *Comparative hearing: Birds and reptiles,* 70–138. New York: Springer.

The importance of signals that allow individuals to recognize members of their own species was emphasized decades ago and is still an important focus of research on animal communication . . .

Lorenz, K. 1970. *Studies of human and animal behaviour,* vols. 1 and 2. Cambridge, MA: Harvard University Press.

Marler, P. 1957. Specific distinctiveness in the communication signals of birds. *Behaviour* 11: 13–39.

Seddon, N. 2005. Ecological adaptation and species recognition drives vocal evolution in neotropical suboscine birds. *Evolution* 59: 200–215.

Pfennig, K. S., and D. W. Pfennig. 2009. Character displacement: Ecological and reproductive responses to a common evolutionary problem. *Quarterly Review of Biology* 84: 253–276.

2. PRODUCING ACOUSTIC SIGNALS IN NOISE

The physics of sound and the mathematics of waves are presented in many textbooks. Two especially clear treatments are . . .

Kinsler, L., and A. Frey. 1962. *Fundamentals of acoustics,* 2nd ed. New York: Wiley. A classic introduction to the physics and mathematics of sound. More recent editions expand the topics presented but not the exposition of basics.

Everest, F. A., and K. C. Pohlmann. 2009. *Master handbook of acoustics,* 5th ed. New York: McGraw-Hill. Equally authoritative but with more illustrations and practical applications and less math.

The physics of musical instruments (as well as basic concepts of acoustics) are admirably explained by . . .

Fletcher, N. H., and T. D. Rossing. 1998. *The physics of musical instruments,* 2nd ed. New York: Springer.

The source-filter model of the human voice is now well established. The classic description is . . .

Fant, G. 1960. *Acoustic theory of speech production.* The Hague: Mouton.

Recent work focuses on variants of the basic voice, especially singing. The fundamental sources are . . .

Titze, I. R. 2008. The human instrument. *Scientific American* 298(1): 94–101.

Titze, I. R. 2000. *Principles of voice production.* Salt Lake City: National Center for Voice and Speech.

Sundberg, J. 1987. *The science of the singing voice.* DeKalb: Northern Illinois University Press.

Here is a selection of recent articles describing interactions of the larynx and vocal cavities in trained singers . . .

Titze, I. R., and B. H. Story. 1997. Acoustic interactions of the voice source with the lower vocal tract. *Journal of the Acoustical Society of America* 101: 2234–2243.

Titze, I. R. 2004. Theory of glottal airflow and source-filter interaction in speaking and singing. *Acta Acustica united with Acustica* 90: 641–648.

Joliveau, E., J. Smith, and J. Wolfe. 2004. Tuning of vocal tract resonances by sopranos. *Nature* 427: 116. A clear introduction to the issues.

Joliveau, E., J. Smith, and J. Wolfe. 2004. Vocal tract resonances in singing: The soprano voice. *Journal of the Acoustical Society of America* 116: 2434–2439.

Garnier, M., N. Henrich, J. Smith, et al. 2010. Vocal tract adjustments in the high soprano range. *Journal of the Acoustical Society of America* 127: 3771–3780.

Garnier, M., N. Henrich, L. Crevier-Buchman, et al. 2012. Glottal behavior in the high soprano range and the transition to the whistle registers. *Journal of the Acoustical Society of America* 131: 951–962.

Doellinger, M., et al. 2012. Effects of the epilarynx area on vocal fold dynamics and the primary voice signal. *Journal of Voice* 26: 285–292.

Sundberg, J., F. M. B. La, and B. P. Gill. 2013. Formant tuning strategies in professional male opera singers. *Journal of Voice* 27: 278–288.

Some jazz clarinetists and trumpeters can make their instruments "talk" in a way. It appears that Richard Wagner realized the constraints on enunciation by coloratura sopranos and set his lyrics to music in a way allows them to "talk" much as a jazz musician might make a clarinet talk (by using melodies that match source frequencies to the expected formant frequencies of normal speech) . . .

Smith, J., and J. Wolfe. 2009. Vowel-pitch matching in Wagner's operas: Implications for intelligibility and ease of singing. *Journal of the Acoustical Society of America* 125: EL196-201; doi: 10.1121/1.3104622.

Joe Wolfe and colleagues in the Acoustics Group at the University of New South Wales maintain well-illustrated web pages on all aspects of vocal and instrumental acoustics. Particularly relevant links for the human voice and for sopranos in particular are . . .

http://newt.phys.unsw.edu.au/jw/voice.html
http://newt.phys.unsw.edu.au/jw/soprane.html

No one has yet definitively determined whether or not the fundamental frequency of the syrinx in songbirds (order Passeriformes) is affected by resonances in the vocal tract. If so, the resonances of the vocal tract (perhaps the entire system of air sacs throughout a bird's body) would entrain the frequencies of the syrinx rather than filter them. A songbird would resemble in this way an operatic singer but not a human speaking normally. It is important to emphasize, on the other hand, that birds, including the true oscine songbirds, have extremely diverse vocalizations and considerable variation in the anatomy of their syringes. Roderick Suthers's lab has focused on species such as the Northern Cardinal and the Brown Thrasher that produce loud, nearly tonal songs for long-range advertisement of territories and interactions with mates. Another songbird often used for research on sound production, the Zebra Finch, does not defend territories nor interact with mates at a distance. It has no loud tonal vocalizations and instead produces harmonically rich sounds for communication at close range. It is thus possible that it does not have the syringeal structures needed for loud tonal sounds. The mechanisms of avian vocalization might differ as much (or more) among species or orders of birds as the mechanisms of human singing, speaking, and whispering. Sources for my discussion of vocal production by songbirds include . . .

Brackenbury, J. H. 1982. The structural basis of voice production and its relationship to sound characteristics. In D. E. Kroodsma and E. H. Miller, eds., *Acoustic communication in birds,* vol. 1, *Production, perception, and design features of sounds,* 53–73. New York: Academic Press.

Nowicki, S. 1986. Vocal tract resonances in oscine bird sound production: Evidence from birdsongs in a helium atmosphere. *Nature* 325: 53–55.

Nowicki, S., and P. Marler. 1988. How do birds sing? *Music Perception* 5: 391–426. Further discussion of the experiments with helium and also an early suggestion that birdsong might share some properties of human singing.

Suthers, R. A., F. Goller, and C. Pytte. 1999. The neuromuscular control of birdsong. *Philosophical Transactions of the Royal Society of London. Series B: Biological Sciences* 354: 927–939.

Larsen, O. N., and F. Goller. 1999. Role of syringeal vibrations in bird vocalizations. *Proceedings of the Royal Society of London. Series B: Biological Sciences* 266: 1609–1615.

Nelson, B. S., G. J. L. Beckers, and R. A. Suthers. 2005. Vocal tract filtering and sound radiation in a songbird. *Journal of Experimental Biology* 208: 297–308.

Riede, T., et al. 2006. Songbirds tune their vocal tract to the fundamental frequency of their song. *Proceedings of the National Academy of Sciences of the United States of America* 103: 5543–5548.

Fletcher, N. H., T. Riede, and R. A. Suthers. 2006. Model for vocalization by a bird with distensible vocal cavity and open beak. *Journal of the Acoustical Society of America* 119: 1005–1011.

Sitt, J. D., et al. 2008. Dynamical origin of spectrally rich vocalizations in birdsong [of Zebra Finches]. *Physical Review E* 78; 011905. The focus is on short-range vocalizations.

Suthers, R. A., and S. A. Zollinger. 2008. From brain to song: the vocal organ and the vocal tract. In H. P. Zeigler and P. Marler, eds., *Neuroscience of birdsong,* 78–98. Cambridge: Cambridge University Press.

Riede, T., and F. Goller. 2010. Functional morphology of the sound-generating labia in the syrinx of two songbird species. *Journal of Anatomy* 216: 23–36.

Suthers, R. A., E. Vallet, and M. Kreutzer. 2012. Bilateral coordination and the motor basis of female preference for sexual signals in canary song. *Journal of Experimental Biology* 215: 2950–2959.

Riede, T., N. Schilling, and F. Goller. 2013. The acoustic effect of vocal tract adjustments in zebra finches. *Journal of Comparative Physiology A* 199: 57–69.

Goller, F., and T. Riede. 2013. Integrative physiology of fundamental frequency control in birds. *Journal of Physiology—Paris* 107: 230–242.

The syrinx of doves (order Columbiformes) produces a wide-spectrum of harmonics, which is passively filtered by the vocal tract to produce a nearly pure tone . . .

Fletcher, N. H., et al. 2004. Vocal tract filtering and the "coo" of doves. *Journal of the Acoustical Society of America* 116: 3750–3756.

Beckers, G. J. L., R. A. Suthers, and C. Ten Cate. 2003. Pure-tone birdsong by resonance filtering of harmonic overtones. *Proceedings of the National Academy of Sciences of the United States of America* 100: 7372–7376.

The importance of opening the bill on production of high frequencies by songbirds was first explored by Stephen Nowicki and his associates . . .

Westneat, M. W., et al. 1993. Kinematics of birdsong—functional correlation of cranial movements and acoustic features in sparrows. *Journal of Experimental Biology* 182: 147–171.

Podos, J. 1997. A performance constraint on the evolution of trilled vocalizations in a songbird family (Passeriformes: Emberizidae). *Evolution* 51: 537–551.

Hoese, W. J., et al. 2000. Vocal tract function in birdsong production: Experimental manipulation of beak movements. *Journal of Experimental Biology* 203: 1845–1855.

Podos, J., and S. Nowicki. 2004. Performance limits on birdsong. In P. Marler and H. Slabbekoorn, eds., *Nature's music: The science of birdsong*, 318–342. San Diego: Academic.

Nelson, B. S., G. J. L. Beckers, and R. A. Suthers. 2005. Vocal tract filtering and sound radiation in a songbird. *Journal of Experimental Biology* 208: 297–308. Evidence that opening the bill does not change vocal tract resonances. Instead, it attenuates frequencies above about 4 kHz.

The influence of helium mixtures on the resonant frequencies of the oral cavity (the formants) but not on the fundamental frequency of the vibrating larynx is clear evidence for the source-filter theory of the normal human voice. This observation confirms, in a striking way, that the resonant frequencies of the oral cavity passively filter, but do not alter, the frequencies from the source. In view of the importance of this result, it is surprising that it has not been thoroughly studied. The influence of breathing helium mixtures on the formant frequencies of the human voice is easily noticed by anybody, and the constancy of laryngeal vibrations during normal speech is easily confirmed with a spectrograph, but most of the scientific study of the human voice in helium mixtures, conducted decades ago, focused on deep-sea divers who breathed this gas to avoid developing the "bends." In these conditions the voice often becomes almost unintelligible as a result of both the lower density of helium in comparison to nitrogen and also the fourfold (or more) increase in pressure. A number of investigators mention the nearly constant fundamental frequency of the laryngeal vibrations in helium mixtures, despite the increase in atmospheric pressure and the significant changes in resonant frequencies of the oral cavity. The best summary of this work is in a defunct journal . . .

Giordano, T. A., H. B. Rothman, and H. Hollien. 1973. Helium speech unscramblers—a critical review of the state of the art. *IEEE Transactions in Audio and Electroacoustics Technology* AU-21: 436–444.

Ways to increase the efficiency of radiating sound into the atmosphere include . . .

Bennet-Clark, H. 1987. The tuned singing burrow of mole crickets. *Journal of Experimental Biology* 128: 383–409.
Fletcher, N. H., et al. 2004. Vocal tract filtering and the "coo" of doves. *Journal of the Acoustical Society of America* 116: 3750–3756.

Air spaces adjoining the mammalian larynx can also result in impressive intensity and concentration of sound . . .

Koda, H., et al. 2012. Soprano singing in gibbons. *American Journal of Physical Anthropology* 149: 347–355.
Riede, T., et al. 2008. Mammalian laryngeal air sacs add variability to the vocal tract impedance: Physical and computational modeling. *Journal of the Acoustical Society of America* 124: 634–647.

Fitch, W. T., and M. D. Hauser. 1995. Vocal production in nonhuman primates: Acoustics, physiology, and functional constraints on "honest" advertisement. *American Journal of Primatology* 37: 191–219.

Whitehead, J. M. 1995. Vox alouattinae: A preliminary survey of the acoustic characteristics of long-distance calls of howling monkeys. *International Journal of Primatology* 16: 121–144.

Esophageal sacs of the Greater Sage-Grouse have also been investigated . . .

Wiley, R. H. 1973. The strut display of male sage grouse: A "fixed" action pattern. *Behaviour* 47: 129–152.

Dantzker, M. S., G. B. Deane, and J. W. Bradbury. 1999. Directional acoustic radiation in the strut display of male Sage Grouse *Centrocercus urophasianus*. *Journal of Experimental Biology* 202: 2893–2909.

Dantzker, M. S., and J. W. Bradbury. 2006. Vocal sacs and their role in avian acoustic display. *Acta Zoologica Sinica* 52: 486–488.

Krakauer, A. H., et al. 2009. Vocal and anatomical evidence for two-voiced sound production in the Greater Sage-Grouse *Centrocercus urophasianus*. *Journal of Experimental Biology* 212: 3719–3727.

Patricelli, G. L., and A. H. Krakauer. 2009. Tactical allocation of effort among multiple signals in sage grouse: An experiment with a robotic female. *Behavioral Ecology* 221: 97–106.

Measurements of the intensity of birds' vocalizations include . . .

Brackenbury, J. H. 1979. Power capabilities of the avian sound-producing system. *Journal of Experimental Biology* 78: 163–166.

Nemeth, E. 2004. Measuring the sound pressure level of the song of the Screaming Piha *Lipaugus vociferans:* One of the loudest birds in the world? *Bioacoustics* 14: 225–228.

Brumm, H. 2009. Song amplitude and body size in birds. *Behavioral Ecology and Sociobiology* 63: 1157–1165.

Brumm, H., and M. Ritschard. 2011. Song amplitude affects territorial aggression of male receivers in chaffinches. *Behavioral Ecology* 22: 310–316.

3. RECEIVING ACOUSTIC SIGNALS IN NOISE

Joseph Fourier's theorems on the equivalence of a waveform and its spectrum, Harry Nyquist's theorem on the minimal sampling rate, and the resulting issues in the computation of frequency spectra are discussed in many texts. For particularly clear discussions of the reciprocity between temporal resolution and frequency resolution in spectral analysis (with all the relevant calculus) see . . .

Briggs, W. L., and V. E. Henson. 1995. *The DFT: An owner's manual for the Discrete Fourier Transform*. Philadelphia: SIAM.

Schwartz, M. 1990. *Information transmission, modulation, and noise*, 4th ed. New York: McGraw-Hill. See especially Chapter 2 of this volume.

There are a number of programs available for producing spectrograms and spectra such as those in this book. PRAAT is a free program widely used by linguists and students of human voice. The illustrations in the present volume were produced by WildSpectra, a free program designed especially for analysis of animals' vocalizations . . .

Boersma, P, and D. Weenink. PRAAT: Doing phonetics by computer; http://www.fon.hum.uva.nl/praat/.

Wiley, K., and R. H. Wiley. Wildspectra; http://www.unc.edu/~rhwiley/wildspectra.

For reciprocity of location and velocity in subatomic particles . . .

Krauss, L. M. 2012. *A universe from nothing: Why there is something rather than nothing*. New York: Atria.

For this reciprocity and an introduction to wavelet analysis . . .

Hubbard, B. B. 1998. *The world according to wavelets*, 2nd ed. Natick, MA: A. K. Peters.

Studies of avian hearing have been thoroughly reviewed . . .

Manley, G. A. 1990. *Peripheral hearing mechanisms in reptiles and birds*. Berlin: Springer-Verlag.

Dooling, R. J., B. Lohr, and M. L. Dent. 2000. Hearing in birds and reptiles. In R. J. Dooling, R. R. Fay, and A. N. Popper, eds., *Comparative hearing: Birds and reptiles*, 308–359. New York: Springer.

Dooling, R. 2004. Audition: Can birds hear everything they sing? In P. Marler and H. Slabbekoorn, eds., *Nature's music: The science of birdsong*, 206–225. San Diego: Elsevier.

Pohl, N. U., et al. 2009. Effects of signal features and environmental noise on signal detection in the Great Tit, *Parus major*. *Animal Behaviour* 78: 1293–1300.

Dooling, R. J., and S. H. Blumenrath. 2013. Avian sound perception in noise. In H. Brumm, ed., *Animal communication and noise*, 229–250. Berlin: Springer-Verlag.

For recent work on the temporal resolution of avian hearing . . .

Dooling, R. J., et al. 2002. Auditory temporal resolution in birds: Discrimination of harmonic complexes. *Journal of the Acoustical Society of America* 112: 748–759.

Lohr, B., R. J. Dooling, and S. Bartone. 2006. The discrimination of temporal fine structure in call-like harmonic sounds by birds. *Journal of Comparative Psychology* 120: 239–251.

Henry, K. S., and J. R. Lucas. 2008. Coevolution of auditory sensitivity and temporal resolution with acoustic signal space in three songbirds. *Animal Behaviour* 76: 1659–1671.

Henry, K. S., et al. 2011. Songbirds tradeoff auditory frequency resolution and temporal resolution. *Journal of Comparative Physiology A* 197: 351–359.

Gall, M. D., L. E. Brierley, and J. R. Lucas. 2012. The sender-receiver matching hypothesis: Support from the peripheral coding of acoustic features in songbirds. *Journal of Experimental Biology* 215: 3742–3751.

Sound localization with pressure-difference (pressure-gradient) ears is described in . . .

Kühne, R., and B. Lewis. 1985. External and middle ears. In A. S. King and J. McLelland, eds., *Form and function in birds*, vol. 3, 227–221. London: Academic Press.

Klump, G. M. 2000. Sound localization in birds. In R. J. Dooling, R. R. Fay, and A. N. Popper, eds., *Comparative hearing: Birds and reptiles*, 249–307. New York: Springer.

Larsen, O. N., R. J. Dooling, and A. Michelsen. 2006. The role of pressure difference reception in the directional hearing of Budgerigars *(Melopsittacus undulatus)*. *Journal of Comparative Physiology A* 192: 1063–1072.

Christensen-Dalsgaard, J. 2011. Vertebrate pressure-gradient receivers. *Hearing Research* 273: 37–45.

For the power (intensity) of avian vocalizations, see the notes for Chapter 2.

4. TRANSMISSION OF ACOUSTIC SIGNALS

The pioneering paper on the implications of sound transmission for animal communication is . . .

Morton, E. S. 1975. Ecological sources of selection on avian sounds. *American Naturalist* 109: 17–34. This paper put the lid on the coffin of any idea that animals' signals evolved in completely arbitrary ways. If there were such features of organisms, evolving entirely by random mutation and genetic drift, without adaptation to the environment, they would exactly reveal the phylogeny of populations. In the early days of ethology, between 1930 and 1950, this assumption sparked unquestioning use of similarities in animals' communicatory signals to deduce close phylogenetic relationships. Correction of this assumption began in 1955, when Peter Marler reported convergent evolution of birds' alarm calls. This report initiated a search for adaptations of signals to the environment or situation in which communication occurred, a trend that culminated in Eugene Morton's paper. As with all other features of organisms, randomness influences the evolution of communicatory signals, but care is needed to separate similarities that result from ancestry and those that result from convergence in similar environments.

Marler, P. 1955. The characteristics of certain animal calls. *Nature* 176: 6–7.

The discussion in this chapter is based on my previous reviews . . .

Wiley, R. H., and D. G. Richards. 1978. Physical constraints on acoustic communication in the atmosphere: Implications for the evolution of animal vocalizations. *Behavioral Ecology and Sociobiology* 3: 69–94.

Wiley, R. H., and D. G. Richards. 1982. Adaptations for acoustic communication in birds: Sound transmission and signal detection. In D. H. Kroodsma and E. H. Miller, eds., *Acoustic communication in birds,* vol. 1, *Production, perception, and design features of sounds,* 131–181. New York: Academic Press.

The physics of sound in natural environments, including the influence of boundaries, is presented in . . .

Piercy, J. E., T. F. W. Embelton, and L. C. Sutherland. 1977. Review of noise propagation in the atmosphere. *Journal of the Acoustical Society of America* 61: 1403–1418.

Frequency-dependent attenuation of sound in forests has been measured many times. Particularly clear examples include . . .

Marten, K., and P. Marler. 1977. Sound transmission and its significance for animal vocalization. I. Temperate habitats. *Behavioral Ecology and Sociobiology* 2: 271–290.

Marten, K., D. Quine, and P. Marler. 1977. Sound transmission and its significance for animal vocalization. II. Tropical forest habitats. *Behavioral Ecology and Sociobiology* 2: 291–302.

Waser, P. M., and C. H. Brown. 1984. Is there a "sound window" for primate communication? *Behavioral Ecology and Sociobiology* 15: 73–76.

Frequency dependence of reverberation in forests has also been measured . . .

Richards, D. G., and R. H. Wiley. 1980. Reverberations and amplitude fluctuations in the propagation of sound in a forest: Implications for animal communication. *American Naturalist* 115: 381–399.

Waser, P. M., and C. H. Brown. 1986. Habitat acoustics and primate communication. *American Journal of Primatology* 10: 135–154.

Reverberation in forests might augment the intensity of sound . . .

Slabbekoorn, H., J. Ellers, and T. B. Smith. 2002. Birdsong and sound transmission: The benefits of reverberations. *Condor* 104: 564–573.

Naguib, M. 2003. Reverberation of rapid and slow trills: Implications for signal adaptations to long-range communication. *Journal of the Acoustical Society of America* 113: 1749–1756.

Nemeth, E., et al. 2006. Rainforests as concert halls for birds: Are reverberations improving sound transmission of long song elements? *Journal of the Acoustical Society of America* 119: 620–626.

Alternatively, forests might have some properties of waveguides . . .

Waser, P. M., and M. S. Waser. 1977. Experimental studies of primate vocalization: Specializations for long-distance propagation. *Zeitschrift für Tierpsychologie* 43: 239–263.

For the propagation of sound in water . . .

Urich, R. J. 1983. *Principles of underwater sound,* 3rd ed. New York: McGraw-Hill. Propagation of sound in shallow water is described on pp. 172–182.

Au, W. W. L., and M. C. Hastins. 2008. *Principles of marine bioacoustics.* New York: Springer. A briefer treatment of sound in shallow water is on pp. 112–114.

The classic description of light in natural environments is . . .

Endler, J. A. 1993. The color of light in forests and its implications. *Ecological Monographs* 63: 1–27.

The transmission of olfactory (and other) signals is clearly described by . . .

Dusenbery, D. B. 1992. *Sensory ecology: How organisms acquire and respond to information*. New York: Freeman.

5. ADAPTATIONS TO NOISE IN DIFFERENT ENVIRONMENTS

Estimates of the active space of signals in natural environments include these pioneering studies . . .

Waser, P. M., and M. S. Waser. 1977. Experimental studies of primate vocalization: Specializations for long-distance propagation. *Zeitschrift für Tierpsychologie* 43: 239–263.

Brenowitz, E. A. 1982. The active space of Red-winged Blackbird song. *Journal of Comparative Physiology* 147: 511–522.

Brown, C. H. 1989. The active space of Blue Monkey and Grey-cheeked Mangabey vocalizations. *Animal Behaviour* 37: 1023–1034.

Naguib, M., et al. 2008. The ecology of vocal signaling: Male spacing and communication distance of different song traits in nightingales. *Behavioral Ecology* 19: 1034–1040.

The active space of a male frog in a large chorus has been measured by . . .

Gerhardt, H. C., and G. Klump. 1988. Masking of acoustic signals by the chorus background noise in the Green Tree Frog: A limitation on mate choice. *Animal Behaviour* 36: 1247–1249.

Wollerman, L. 1999. Acoustic interference limits call detection in a neotropical frog *Hyla ebraccata*. *Animal Behaviour* 57: 529–536.

Wollerman, L., and R. H. Wiley. 2002. Background noise from a natural chorus alters female discrimination of male calls in a neotropical frog. *Animal Behaviour* 63: 15–22. This study revealed that, in addition to requiring a signal 3 dB above background noise to detect a single male, females did not discriminate between more preferred males and less preferred males of their species unless the males' calls were at least 9 dB above the background.

The ability of animals to estimate the location of a source of sound has attracted the attention of several members of my research group . . .

Richards, D. G. 1981. Estimation of distance of singing conspecifics by the Carolina Wren. *Auk* 98: 127–133.

Whitehead, J. 1987. Vocally mediated reciprocity between neighbouring groups of Mantled Howling Monkeys, *Alouatta palliata palliata*. *Animal Behaviour* 35: 1615–1627.

Whitehead, J. M. 1989. The effect of the location of a simulated intruder on responses to long-distance vocalizations of Mantled Howling Monkeys, *Alouatta palliata palliata*. *Behaviour* 108: 73–103.

Whitehead, J. M. 1990. Gaining access to the black box. *Behavioral and Brain Sciences* 13: 413–414. Summarizes experiments that failed to confirm deceptive calls in the repertoire of Mantled Howling Monkeys.

Wiley, R. H., and R. Godard. 1996. Ranging of conspecific songs by Kentucky Warblers and its implications for interactions of territorial males. *Behaviour* 133: 81–102.

Naguib, M. 1995. Auditory distance assessment of singing conspecifies in Carolina Wrens: The role of reverberation and frequency-dependent attenuation. *Animal Behaviour* 50: 1297–1307.

Naguib, M. 1996. Ranging by song in Carolina Wrens *Thryothorus ludovicianus:* Effects of environmental acoustics and strength of song degradation. *Behaviour* 133: 541–559.

Naguib, M. 1997. Ranging of songs with the song type on use of different cues in Carolina Wrens: Effects of familiarity. *Behavioral Ecology and Sociobiology* 40: 385–393.

Naguib, M. 1997. Use of song amplitude for ranging in Carolina Wrens, *Thryothorus ludovicianus*. *Ethology* 103: 723–731.

Naguib, M. 1999. Mechanisms of auditory distance perception in passerine birds. *Journal of the Acoustical Society of America* 105: 1202–1203.

Naguib, M., and R. H. Wiley. 2001. Estimating the distance to a source of sound: Mechanisms and adaptations for long-range communication. *Animal Behaviour* 62: 825–837.

Our experiments on birds' abilities to judge the distance of neighbor's songs stirred up a minor disagreement with the pioneer of avian acoustical adaptations, Eugene Morton. After Douglas Richards's study had been published (1981), Morton (1986) argued, on theoretical grounds, that a bird could not judge the distance to a singing neighbor unless it had acquired the ability to sing exact copies of the neighbor's songs. Only in that way would the listener have a reference for the degraded songs from a distant singer. This situation, Morton argued, explained some puzzling aspects of bird song. It explained why, in many songbirds, neighbors learn identical song patterns (in order to be able to keep track of where their neighbors were). With the additional assumption that it was never advantageous for a singing bird to reveal its location to neighbors, he could also explain why, in many species, individuals have repertoires of so many different songs. Each individual, he proposed, was in a sort of arms race with its neighbors. Adding new songs to its repertoire in order to avoid singing those its neighbors had already learned.

Morton, E. S. 1986. Predictions from the ranging hypothesis for the evolution of long distance signals in birds. *Behaviour* 99: 65–86.

Our results (see references above) failed to confirm the basic prediction that birds must be able to sing a pattern in order to be able to range it. This disagreement came to a crux after Renee Godard and I experimented with a species, the Kentucky Warbler, in which each individual sings but one pattern and neighboring individuals never sing the same pattern (although, as with other species, they all have songs falling within definite limits). So, by Morton's prediction, they should not be able to judge the distances of singing neighbors. Nevertheless, they did. Their responses to artificially degraded songs were about the same as the responses of Hooded Warblers and Carolina Wrens, denizens of the same forests, with complex repertoires of songs, always including some that matched their neighbors'. We concluded that Kentucky Warblers can range their neighbors' songs without recourse to learning to sing the songs first and can do so about as well as species that do learn to match neighbors' songs.

As suggested in this section, my students and I have felt that territorial individuals do not always benefit by concealing their locations from neighbors. Chapter 14 on communication during cooperation describes the case for territorial birds in more detail. In addition, why should we expect birds to be incapable of learning to recognize the features of neighbors' songs without learning to sing them as well? In fact, numerous experiments on birds' abilities to recognize other individuals show that they can discern the detailed features of songs and can do so without producing these songs themselves. Birds might have some innate awareness of the general properties of their species' songs (a probability suggested by predispositions to learn their own species' songs), an awareness that could serve them well for comparisons with degraded songs. They might also learn to recognize the alterations produced by reverberation and frequency-dependent attenuation in their species' songs. The syllables of Kentucky Warblers' songs in fact include (in each individual's own way) brief rapid trills, just what is needed for assessing reverberation and thus distance to the source of a song. Furthermore, playbacks to Carolina Wrens did not reveal any effect of familiarity with the song type on distance estimation. These results do not exclude the possibility that familiarity with a neighbor's song might improve the *accuracy* of ranging (for more on this possibility, see Naguib and Wiley 2001). They do, however, exclude the possibility that producing a sound pattern is necessary for effectively ranging it. These issues were all vetted by Morton, Marc Naguib, and myself in 1998. One point we all accept is that many birds can judge the approximate distance to a singing rival of the same species.

The cocktail party effect can raise levels of noise to extremes. The Lombard effect, auditory object recognition and, especially, spatial localization of sound contribute to reducing this noise . . .

Cherry, C., and R. Wiley. 1967. Speech communication in very noisy environments. *Nature* 214: 1164.

Yost, W. A. 1997. The cocktail party problem: Forty years later. In R. Gilkey and T. R. Anderson, eds., *Binaural and spatial hearing in real and virtual environments,* 329–347. Mahwah, NJ: Erlbaum.

Hulse, S. H. 2002. Auditory scene analysis in animal communication. *Advances in the Study of Behavior* 31: 163–200.

Bee, M. A., and C. Micheyl. 2008. The cocktail party problem: What is it? How can it be solved? And why should animal behaviorists study it? *Journal of Comparative Psychology* 122: 235–251.

Aubin, T., and P. Jouventin. 1998. Cocktail-party effect in King Penguin colonies. *Proceedings of the Royal Society B: Biological Sciences* 265: 1665–1673.

Lengagne, T., P. Jouventin, and T. Aubin, 1999. Finding one's mate in a King Penguin colony: Efficiency of acoustic communication. *Behaviour* 136: 833–846.

Farris, H. E., A. S. Rand, and M. J. Ryan. 2002. The effects of spatially separated call components on phonotaxis in Túngara Frogs: Evidence for auditory grouping. *Brain, Behavior and Evolution* 60: 181–188.

Bee, M. A., and K. K. Riemersma. 2008. Does common spatial origin promote the auditory grouping of temporally separated signal elements in Grey Treefrogs? *Animal Behaviour* 76: 831–843. The answer is, yes, but only weakly.

Aubin, T., and P. Jouventin. 2002. Localisation of an acoustic signal in a noisy environment: The display call of the King Penguin *Aptenodytes patagonicus*. *Journal of Experimental Biology* 205: 3793–3798.

Jouventin, P. 2002. Acoustic systems are adapted to breeding ecologies: Individual recognition in nesting penguins. *Animal Behaviour* 64: 747–757.

Brumm, H., and S. A. Zollinger. 2011. The evolution of the Lombard effect: 100 years of psychoacoustic research. *Behaviour* 148: 11–13.

Auditory scene analysis by humans is elucidated by experiments with subjects listening to sounds with one or both ears . . .

Xiang, J., J. Simon, and M. Elhilali. 2010. Competing streams at the cocktail party: Exploring the mechanisms of attention and temporal integration. *Journal of Neuroscience* 30: 12084–12093.

Collins, G. P. 2011. Solving the cocktail party problem. *Scientific American* 304(4): 66–67.

The Acoustic Adaptation Hypothesis has been proposed in various forms and evaluated in many studies, beginning with Morton's classic paper. Those mentioned in this section and a selection of others include . . .

Morton, E. S. 1975. Ecological sources of selection on avian sounds. *American Naturalist* 109: 17–34.

Waser, P. M., and C. H. Brown. 1986. Habitat acoustics and primate communication. *American Journal of Primatology* 10: 135–154.

Wiley, R. H. 1991. Associations of song properties with habitats for territorial oscine birds of eastern North America. *American Naturalist* 138: 973–993.

Boncoraglio, G., and N. Saino. 2007. Habitat structure and the evolution of bird song: A meta-analysis of the evidence for the acoustic adaptation hypothesis. *Functional Ecology* 21: 134–142.

Ey, E., and J. Fischer. 2009. The "acoustic adaptation hypothesis"—a review of the evidence from birds, anurans and mammals. *Bioacoustics* 19: 21–48.

Brumm, H., and M. Naguib. 2009. Environmental acoustics and the evolution of bird song. *Advances in the Study of Behavior* 40: 1–33.

Two studies suggest some novel approaches . . .

Seddon, N. 2005. Ecological adaptation and species recognition drives vocal evolution in neotropical suboscine birds. *Evolution* 59: 200–215.

Henry, K. S., and J. R. Lucas. 2010. Habitat-related differences in the frequency selectivity of auditory filters in songbirds. *Functional Ecology* 24: 614–624.

The dimensions of a source of sound affect its efficiency in producing sounds with different frequencies. The principles are covered by any discussion of the design of loudspeakers (see the references on acoustics in general at the beginning of Chapter 2). Some biological studies have confirmed that these principles apply to animals as well—smaller animals have more difficulty producing sounds with low frequencies . . .

Greenewalt, C. H. 1968. *Bird song: Acoustics and physiology.* Washington, DC: Smithsonian Institution Press.

Bowman, R. I. 1979. Adaptive morphology of song dialects in Darwin's finches. *Journal für Ornithologie* 120: 353–389.

Roberts, J., et al. 1983. Physical considerations in the frequency limits of birdsong. *Acoustics Letters* 6: 100–105.

Wallschläger, D. 1980. Correlation of song frequency and body weight in passerine birds. *Experientia* 36: 412.

Martin, J. P., et al. 2011. Body size correlates negatively with the frequency of distress calls and songs of neotropical birds. *Journal of Field Ornithology* 82: 259–268.

Michael Green's dissertation describes the effects of dropouts and reverberation on recognition of songs by a grassland sparrow is described in his dissertation . . .

Green, M. T. 1992. Adaptations of Baird's Sparrows *(Ammodramus bairdii)* to grasslands: Acoustic communication and nomadism. Ph.D. diss., University of North Carolina–Chapel Hill. *Dissertation Abstracts International B: The Sciences* 53(09): 4540.

Reverberation makes it difficult to distinguish rapidly repeated notes at any one frequency. Consequently, as a rule birds' songs in reverberant environments such as forests tend to avoid rapid repetitions of this sort . . .

Wiley, R. H. 1991. Associations of song properties with habitats for territorial oscine birds of eastern North America. *American Naturalist* 138: 973–993.

Naguib, M. 2003. Reverberation of rapid and slow trills: Implications for signal adaptations to long-range communication. *Journal of the Acoustical Society of America* 113: 1749–1756.

The responses of wildlife, especially singing birds, to urban and other anthropogenic noise is attracting rapidly increasing attention. Reviews include . . .

Warren, P. S., et al. 2006. Urban bioacoustics: It's not just noise. *Animal Behaviour* 71: 491–502.

Luther, D., and K. Gentry. 2013. Sources of background noise and their influence on vertebrate acoustic communication. *Behaviour* 150: 1045–1068.

McGregor, P. K., et al. 2013. Anthropogenic noise and conservation. In H. Brumm, ed., *Animal communication and noise,* 409–444. Berlin: Springer-Verlag.

Studies of anthropogenic noise discussed in this chapter include . . .

Slabbekoorn, H., and M. Peet. 2003. Birds sing at a higher pitch in urban noise. *Nature* 424: 267.

Brumm, H. 2004. The impact of environmental noise on song amplitude in a territorial bird. *Journal of Animal Ecology* 73: 434–440.

Brumm, H., and S. A. Zollinger. 2013. Avian vocal production in noise. In H. Brumm, ed., *Animal communication and noise,* 187–227. Berlin: Springer-Verlag.

Slabbekoorn, H. 2013. Songs of the city: Noise-dependent spectral plasticity in the acoustic phenotype of urban birds. *Animal Behaviour* 85: 1089–1099.

Montague, M. J., M. Danek-Gontard, and H. P. Kunc. 2013. Phenotypic plasticity affects the response of a sexually selected trait to anthropogenic noise. *Behavioral Ecology* 24: 343–348.

Luther, D., and L. Baptista. 2010. Urban noise and the cultural evolution of bird songs. *Proceedings of the Royal Society B: Biological Sciences* 277: 469–473.

Nemeth, E., et al. 2013. Bird song and anthropogenic noise: Vocal constraints may explain why birds sing higher-frequency songs in cities. *Proceedings of the Royal Society B: Biological Sciences* 280: 20122798.

Cardoso, G. C., and J. W. Atwell. 2011. On the relation between loudness and the increased song frequency of urban birds. *Animal Behaviour* 82: 831–836.

Blickley, J. L., and G. L. Patricelli. 2012. Potential acoustic masking of Greater Sage-Grouse *(Centrocercus urophasianus)* display components by chronic industrial noise. *Ornithological Monographs* 74: 23–35.

Naguib, M., et al. 2013. Noise annoys: Effects of noise on breeding Great Tits depend on personality but not on noise characteristics. *Animal Behaviour* 85: 949–956.

Pohl, N. U., et al. 2012. Great Tits in urban noise benefit from high frequencies in song detection and discrimination. *Animal Behaviour* 83: 711–721.

Pohl, N. U., et al. 2013. Why longer song elements are easier to detect: Threshold level–duration functions in the Great Tit and comparison with human data. *Journal of Comparative Physiology A* 199: 239–252.

More on sound in the sea . . .

Finneran, J. J., and B. K. Branstetter. 2013. Effects of noise on sound perception in marine mammals. In H. Brumm, ed., *Animal communication and noise,* 273–308. Berlin: Springer-Verlag.

Tyack, P. L., and V. M. Janik. 2013. Effects of noise on acoustic signal production in marine mammals. In H. Brumm, ed., *Animal communication and noise,* 251–271. Berlin: Springer-Verlag.

Ladich, F. 2013. Effects of noise on sound detection and acoustic communication in fishes. In H. Brumm, ed., *Animal communication and noise,* 65–90. Berlin: Springer-Verlag.

6. REDUCING NOISE, ENHANCING PERFORMANCE

Weber's Law and its approximations are discussed in many books on sensation and perception. For measurements of deviations from this "law" and discussions of the underlying mechanisms . . .

Green, D. M., and J. A. Swets. 1966. *Signal detection theory and psychophysics.* New York: Wiley.

Wier, C. C., W. Jesteadt, and D. M. Green. 1977. Frequency discrimination as a function of frequency and sensation level. *Journal of the Acoustical Society of America* 61: 178–183.

Jesteadt, W., C. C. Wier, and D. M. Green. 1977. Intensity discrimination as a function of frequency and sensation level. *Journal of the Acoustical Society of America* 61: 169–177.

Yost, W. A., and D. W. Nielsen. 1985. *Fundamentals of hearing: An introduction,* 2nd ed. New York: Holt.

Akre, K. L., and S. Johnsen. 2014: Psychophysics and the evolution of behavior. *Trends in Ecology and Evolution* 29: 291–300.

Avoidance of temporal overlap by singing birds and frogs has been widely reported, but not universally. On the other hand, currently disputed evidence indicates that overlap might itself be a threatening signal to a rival . . .

Wasserman, F. E. 1977. Intraspecific acoustical interference in the White-throated Sparrow *(Zonotrichia albicollis). Animal Behaviour* 25: 949–952.

Popp, J. W., R. W. Ficken, and J. A. Reinartz. 1985. Short-term temporal avoidance of interspecific acoustic interference among forest birds. *Auk* 102: 744–748.

Planqué, R., and H. Slabbekoorn. 2008. Spectral overlap in songs and temporal avoidance in a Peruvian bird assemblage. *Ethology* 114: 262–271.

Brumm, H. 2006. Signalling through acoustic windows: Nightingales avoid
    interspecific competition by short-term adjustment of song timing.
    Journal of Comparative Physiology A 192: 1279–1285.
Egnor, S. E. R., J. G. Wickelgren, and M. D. Hauser. 2007. Tracking silence:
    Adjusting vocal production to avoid acoustic interference. *Journal of
    Comparative Physiology A* 193: 477–483.
Vélez, A., and M. A. Bee. 2013. Signal recognition by Green Treefrogs *(Hyla
    cinerea)* and Cope's Gray Treefrogs *(Hyla chrysoscelis)* in naturally
    fluctuating noise. *Journal of Comparative Psychology* 127: 166–178.
Naguib, M., and D. J. Mennill. 2010. The signal value of birdsong: Empirical
    evidence suggests song overlapping is a signal. *Animal Behaviour* 80:
    e11–e15.
Searcy, W. A., and M. D. Beecher. 2011. Continued scepticism that song
    overlapping is a signal. *Animal Behaviour* 81: e1–e4.

Dawn has advantages for communication. It might be a poor time for insectivorous birds to forage and thus a good time to defend their territories by singing. It might also provide better atmospheric conditions for long-range communication, a possibility that has had some support but also encounters some problems . . .

Henwood, K., and A. Fabrick. 1979. A quantitative analysis of the dawn
    chorus: Temporal selection for communicatory optimization. *American
    Naturalist* 114: 260–274.
Waser, P. M., and M. S. Waser. 1977. Experimental studies of primate vocaliza-
    tion: Specializations for long-distance propagation. *Zeitschrift für
    Tierpsychologie* 43: 239–263.
Dabelsteen, T., and N. Mathevon. 2002. Why do songbirds sing intensively at
    dawn? *Acta Ethologica* 4: 65–72. In deciduous forests of Denmark, dawn
    is not the best time for long-range communication.
Pohl, N. U., et al. 2009. Effects of signal features and environmental noise on
    signal detection in the Great Tit, *Parus major*. Animal *Behaviour* 78:
    1293–1300. Dawn choruses mask communication. Signals with energy in
    a narrow band are detected more easily that those with energy spread
    over a wide band.

Wind and rushing water also pose problems for acoustic communication . . .

Lengagne, T., et al. 1999. How do King Penguins *(Aptenodytes patagonicus)*
    apply the mathematical theory of information to communicate in windy
    conditions? *Proceedings of the Royal Society of London. Series B: Biological
    Sciences* 266: 1623–1628.

Narins, P. M., et al. 2004. Old World frog and bird vocalizations contain prominent ultrasonic harmonics. *Journal of the Acoustical Society of America* 115: 910–913.

Feng, A. S., et al. 2006. Ultrasonic communication in frogs. *Nature* 440: 333–336.

Brumm, H., and P. J. B. Slater. 2006. Ambient noise, motor fatigue, and serial redundancy in chaffinch song. *Behavioral Ecology and Sociobiology* 60: 475–481.

Arch, V. A., and P. M. Narins. 2008. "Silent" signals: selective forces acting on ultrasonic communication systems in terrestrial vertebrates. *Animal Behaviour* 76: 1423–1428.

In tropical forests, insects preempt a band of frequencies avoided by both birds and primates . . .

Morton, E. S. 1975. Ecological sources of selection on avian sounds. *American Naturalist* 109: 17–34.

Ryan, M. J., and E. A. Brenowitz. 1985. The role of body size, phylogeny, and ambient noise in the evolution of bird song. *American Naturalist* 126: 87–100.

Waser, P. M., and C. H. Brown. 1986. Habitat acoustics and primate communication. *American Journal of Primatology* 10: 135–154.

Still more sources of extraneous environmental energy can create noise for communication . . .

Hopkins, C. D. 1973. Lightning as background noise for communication among electric fish. *Nature* 242: 268–270.

McNett, G. D., L. H. Luan, and R. B. Cocroft. 2010. Wind-induced noise alters signaler and receiver behavior in vibrational communication. *Behavioral Ecology and Sociobiology* 64: 2043–2051.

Rain imposes major constraints on long-range acoustic communication . . .

Lengagne, T., and P. J. B. Slater. 2002. The effects of rain on acoustic communication: Tawny owls have good reason for calling less in wet weather. *Proceedings of the Royal Society of London. Series B: Biological Sciences* 269: 2121–2125.

Coexisting species in Amazonian forests use sounds for long-range communication that differ from each other more than expected by chance;

in other words, their sounds are overdispersed in signal space. Furthermore, from the receiver's point of view all of the available signal space is occupied. Species also use different hours of the morning for communication (both signalers and receivers) . . .

Luther, D. 2008. Signaller:receiver coordination and the timing of communication in Amazonian birds. *Biology Letters* 4: 651–654.

Luther, D. A., and R. H. Wiley. 2009. Production and perception of communicatory signals in a noisy environment. *Biology Letters* 5: 183–187.

Luther, D. 2009. The influence of the acoustic community on songs of birds in a neotropical rain forest. *Behavioral Ecology* 20: 864–871. Overdispersion of the songs of bird species that share the same stratum of rainforest.

Cardoso, G. C., and T. D. Price. 2010. Community convergence in bird song. *Evolutionary Ecology* 24: 447–461. Minimal overdispersion of calls by coexisting species of Amazonian frogs.

Recognition of individuals has been reported for many species, but the conceptual refinements of recognizing other individuals are not often considered. These refinements can require subtle distinctions between signals and thus create challenges in noisy situations . . .

Wiley, R. H. 2005. Individuality in songs of Acadian Flycatchers and recognition of neighbours. *Animal Behaviour* 70: 237–247.

Wiley, R. H. 2013. Specificity and multiplicity in the recognition of individuals: Implications for the evolution of social behaviour. *Biological Reviews* 88: 179–195.

Moseley, D. L., and R. H. Wiley. 2013. Individual differences in the vocalizations of the Buff-throated Woodcreeper *(Xiphorhynchus guttatus),* a suboscine bird of neotropical forests. *Behaviour* 150: 1107–1128.

Contrast of visual signals follows the same principles as contrast of acoustic signals. The physical principles are well explained in John Endler's classic papers.

Endler, J. A. 1990. On the measurement and classification of colour in studies of animal colour patterns. *Biological Journal of the Linnean Society* 41: 315–352.

Endler, J. A. 1992. Signals, signal conditions, and the direction of evolution. *American Naturalist* 139: S125–S153.

Endler, J. A. 1993. The color of light in forests and its implications. *Ecological Monographs* 63: 1–27.

Specific cases illustrate these principles . . .

Marchetti, K. 1993. Dark habitats and bright birds illustrate the role of the environment in species divergence. *Nature* 362: 149–152.

Endler, J. A., and M. Théry. 1996. Interacting effects of lek placement, display behavior, ambient light, and color patterns in three neotropical forest-dwelling birds. *American Naturalist* 421–452.

Uy, J. A. C., and J. A. Endler. 2004. Modification of the visual background increases the conspicuousness of Golden-collared Manakin displays. *Behavioral Ecology* 15: 1003–1010.

Gomez, D., and M. Théry. 2004. Influence of ambient light on the evolution of colour signals: Comparative analysis of a neotropical rainforest bird community. *Ecology Letters* 7: 279–284.

Peters, R. A. 2013. Noise in visual communication: Motion from wind-blown plants. In H. Brumm, ed., *Animal communication and noise,* 311–330. Berlin: Springer-Verlag.

Fish have provided a number of examples of the importance of visual contrast for communication . . .

Seehausen, O., J. J. M. Van Alphen, and F. Witte. 1997. Cichlid fish diversity threatened by eutrophication that curbs sexual selection. *Science* 277: 1808–1811.

Endler, J. A. 1991. Variation in the appearance of guppy color patterns to guppies and their predators under different visual conditions. *Vision Research* 31: 587–608.

Endler, J. A. 1987. Predation, light intensity and courtship behaviour in *Poecilia reticulata* (Pisces: Poeciliidae). *Animal Behaviour* 35: 1376–1385.

Rick, I. P., M. Mehlis, and T. C. M. Bakker. 2011. Male red ornamentation is associated with female red sensitivity in sticklebacks. *PloS One* 6: e25554; doi: 10.1371/journal.pone.0025554.

Reimchen, T. E. 1989. Loss of nuptial color in Threespine Sticklebacks *(Gasterosteus aculeatus). Evolution* 43 450–460.

Baube, C. L. 1997. Manipulations of signalling environment affect male competitive success in Three-spined Sticklebacks. *Animal Behaviour* 53: 819–833.

Fuller, R. 2002. Lighting environment predicts the relative abundance of male colour morphs in Bluefin Killifish *(Lucania goodei)* populations. *Proceedings of the Royal Society of London. Series B: Biological Sciences* 269: 1457–1465.

Fuller, R. C., and L. A. Noa. 2010. Female mating preferences, lighting environment, and a test of the sensory bias hypothesis in the Bluefin Killifish. *Animal Behaviour* 80: 23–35.

William Dilger's classic comparison of visual and acoustic signals of North American thrushes . . .

Dilger, W. C. 1956. Hostile behavior and reproductive isolating mechanisms in the avian genera *Catharus* and *Hylocichla*. *Auk* 73: 313–353.

Charles Darwin's principle of antithesis in communicatory signals is another consequence of the advantages of contrast in noisy (error-prone) communication . . .

Darwin, C. 1873. *The Expression of the emotions in man and animals*. London: John Murray.

Hurd, P. L., C. A. Wachtmeister, and M. Enquist. 1995. Darwin's principle of antithesis revisited: A role for perceptual biases in the evolution of intraspecific signals. *Proceedings: Biological Sciences:* 201–205.

Repetition of signals, so prominent in much communication, is a simple form of redundancy that reduces errors in noise, although at a price of time and energy. In noisy conditions many animals increase the rate of calling, the number of calls in a bout, or the length of calls . . .

Potash, L. M. 1972. A signal detection problem and possible solution in Japanese quail *(Coturnix coturnix japonica)*. *Animal Behaviour* 20: 192–195.

Brumm, H., et al. 2004. Acoustic communication in noise: Regulation of call characteristics in a new world monkey. *Journal of Experimental Biology* 207: 443–448. A case in which no increase in redundancy occurred in noisy conditions.

Brumm, H., and P. J. B. Slater. 2006. Ambient noise, motor fatigue, and serial redundancy in Chaffinch song. *Behavioral Ecology and Sociobiology* 60: 475–481.

Slabbekoorn, H., and A. Den Boer-Visser. 2006. Cities change the songs of birds. *Current Biology* 16: 2326–2331.

Ord, T. J., et al. 2007. Lizards speed up visual displays in noisy motion habitats. *Proceedings of the Royal Society B: Biological Sciences* 274: 1057–1062.

Jordan Price's comparison of New World blackbirds and relatives shows that repetition evolves to promote communication in error-prone conditions . . .

Price, J. J. 2013. Why is birdsong so repetitive? Signal detection and the evolution of avian singing modes. *Behaviour* 150: 995–1014.

Stereotyped signals, those with little variation, promote contrast and re-dundancy. Their advantages for communication in noise (in other words, with error-prone receivers) have been addressed in . . .

Johnstone, R. A., and A. Grafen. 1992. Error-prone signalling. *Proceedings of the Royal Society B: Biological Sciences* 248: 229–233.

Johnstone, R. A. 1994. Honest signalling, perceptual error and the evolution of "all-or-nothing" displays. *Proceedings of the Royal Society B: Biological Sciences* 256: 169–175.

Multiple signals can serve two purposes: increased information (mul-tiple messages) or redundancy to counteract noise (backup signals) . . .

Johnstone, R. A. 1996. Multiple displays in animal communication: "Backup signals" and "multiple messages." *Philosophical Transactions of the Royal Society of London. Series B:* Biological *Sciences* 351: 329–338.

Multiple signals occur in the context of mate choice, which is of course an example of communication, although the consequences of noise are usually not included in the discussion . . .

Pomiankowski, A., and Y. Iwasa. 1993. Evolution of multiple sexual preferences by Fisher's runaway process of sexual selection. *Proceedings of the Royal Society of London. Series B: Biological Sciences* 253: 173–181.

Iwasa, Y., and A. Pomiankowski. 1994. The evolution of mate preferences for multiple sexual ornaments. *Evolution* 48: 853–867. Iwasa and Pomiankowski focus on relative increase in costs of preferences for multiple traits. Increased costs of joint preferences favor evolution of single preferences.

Loyau, A., et al. 2005. Multiple sexual advertisements honestly reflect health status in peacocks *(Pavo cristatus). Behavioral Ecology and Sociobiology* 58: 552–557.

Doucet, S. M., and R. Montgomerie. 2003. Multiple sexual ornaments in Satin Bowerbirds: Ultraviolet plumage and bowers signal different aspects of male quality. *Behavioral Ecology* 14: 503–509.

Variation in animals' signals for communication and associations be-tween the components of signals have not often been measured. In ex-treme cases the variation is no greater than in individuals' anatomical features. The components of signals sometimes vary independently of each other. Both of these possibilities show up in a couple of my own measurements of animals' displays for communication . . .

Wiley, R. H. 1973. The strut display of male Sage Grouse: A "fixed" action pattern. *Behaviour* 47: 129–152.

Wiley, R. H. 1975. Multidimensional variation in an avian display: Implications for social communication. *Science* 190: 482–483.

I have reviewed alerting signals in . . .

Wiley, R. H., and D. G. Richards. 1982. Adaptations for acoustic communication in birds: Sound transmission and signal detection. In D. H. Kroodsma and E. H. Miller, eds., *Acoustic communication in birds,* vol. 1, *Production, perception, and design features of sounds,* 131–181. New York: Academic Press.

Wiley, R. H. 2006. Signal detection and animal communication. *Advances in the Study of Behavior* 36: 217–247.

Douglas Richards's original experiment showed how adding noise to a complete signal or its isolated components can reveal an alerting component. Subsequently, this procedure has be used to confirm alerting components of other signals . . .

Richards, D. G. 1981. Alerting and message components in songs of Rufoussided Towhees. *Behaviour* 76: 223–249.

Peters, R. A., and T. J. Ord. 2003. Display response of the Jacky Dragon, *Amphibolurus muricatus* (Lacertilia: Agamidae), to intruders: A semi-Markovian process. *Austral Ecology* 28: 499–506.

Ord, T. J., and J. A. Stamps. 2008. Alert signals enhance animal communication in "noisy" environments. *Proceedings of the National Academy of Sciences of the United States of America* 105: 18830–18835.

Alerting signals can include any easily detected signals, even from another species . . .

Waser, P. M., and M. S. Waser. 1977. Experimental studies of primate vocalization: Specializations for long-distance propagation. *Zeitschrift für Tierpsychologie* 43: 239–263.

Greig, E. I., and S. Pruett-Jones. 2010. Danger may enhance communication: Predator calls alert females to male displays. *Behavioral Ecology* 21: 1360–1366. In fairy-wrens of Australia.

Receiver psychology was first described in the influential paper below and its successor . . .

Guilford, T., and M. S. Dawkins. 1991. Receiver psychology and the evolution of animal signals. *Animal Behaviour* 42: 1–14.

Guilford, T., and M. S. Dawkins. 1993. Receiver psychology and the design of animal signals. *Trends in Neurosciences* 16: 430–436.

A peak shift occurs when learning to discriminate between two signals, one of which is disadvantageous (punished). This striking result is explained by Signal Detection Theory . . .

Lynn, S. K., J. Cnaani, and D. R. Papaj. 2005. Peak shift discrimination learning as a mechanism of signal evolution. *Evolution* 59: 1300–1305.

The debate about ritualization, in the context of manipulative or cooperative communication, began with articles such as . . .

Dawkins, R., and J. R. Krebs. 1978. Animal signals: Information or manipulation. In J. R. Krebs and N. B. Davies, eds., *Behavioural ecology: An evolutionary approach,* 282–309. Oxford: Blackwell.

Hinde, R. A. 1981. Animal signals: Ethological and games-theory approaches are not incompatible. *Animal Behaviour* 29: 535–542.

A clear-headed description of ritualization as an increase in the efficiency of communication was eclipsed by this debate . . .

Cullen, J. M. 1966. Reduction of ambiguity through ritualization. *Philosophical Transactions of the Royal Society of London. Series B: Biological Sciences* 251: 363–374.

Niko Tinbergen, a pioneering ethologist and a master at combining observation and experiment, provided an early introduction to ritualization . . .

Tinbergen, N. 1952. "Derived" activities: Their causation, biological significance, origin, and emancipation during evolution. *Quarterly Review of Biology* 27: 1–32. See especially 23–24.

7. SIGNALS, RECEIVERS, AND EVOLUTION

A distinction between cues and signals, on the grounds that only the latter evolve, is emphasized in . . .

Maynard Smith, J., and D. G. C. Harper. 2003. *Animal signals.* Oxford: Oxford University Press.

Bradbury, J. W., and S. L. Vehrencamp. 2011. *Principles of animal communication,* 2nd ed. Sunderland, MA: Sinauer.

Many authors have drawn distinctions among different sorts of signals. Two representative accounts are . . .

Hasson, O. 2000. Knowledge, information, biases and signal assemblages. In Y. Espmark, T. Amundsen, and G. Rosenqvist, eds., *Animal signals: Signalling and signal design in animal communication*, 445–463. Trondheim: Royal Norwegian Society of Sciences and Letters.

Hurd, P., and M. Enquist. 2005. A strategic taxonomy of biological communication. *Animal Behaviour* 70: 1155–1170.

The position presented in this chapter comes from my reviews . . .

Wiley, R. H. 1994. Errors, exaggeration, and deception in animal communication. In L. Real, ed., *Behavioral mechanisms in evolutionary ecology*, 157–189. Chicago: University of Chicago Press.

Wiley, R. H. 2006. Animal communication: Signal detection. In K. Brown, ed., *Encyclopedia of language and linguistics*, 2nd ed., vol. 1, 288–290. Oxford: Elsevier.

Wiley, R. H. 2013. Signal detection, noise, and the evolution of communication. In H. Brumm, ed., *Animal communication and noise*, 7–30. Berlin: Springer-Verlag.

The genetics of behavior is reviewed in a number of recent textbooks, all of which include discussions of the interaction of genes and environment. This is an active field of research, and the number of well-documented cases of gene-environment interaction has increased rapidly. The possibilities include genetic influences on individuals' tendencies to expose themselves to risky environments and epigenetic methylation of genes as a result of environmental influences, as well as genetic influences on susceptibility to environment risks. Two recent reviews present fascinating cases of these interactions . . .

Caspi, A., and T. E. Moffitt. 2006. Gene-environment interactions in psychiatry: Joining forces with neuroscience. *Nature Reviews Neuroscience* 7: 583–590.

Rutter, M. 2007. Gene-environment interdependence. *Developmental Science* 10: 12–18.

The mechanisms of evolution are also presented in recent textbooks; for instance . . .

Zimmer, C. 2014. *Tangled bank: An introduction to evolution*, 2nd ed. Boulder, CO: Roberts.

Zimmer, C., and D. Emlen. 2012. *Evolution.* Boulder, CO: Roberts.

Futuyma, D. J. 2013. *Evolution.* Sunderland, MA: Sinauer.

The definitions of evolution and natural selection presented in this chapter are similar to those elsewhere. Intended to state what is needed and nothing more in a single sentence, these definitions have served me well in teaching undergraduates for the past twenty years or so.

Most theorists who construct mathematical models of evolution have used one of two approaches, either a focus on changes in genotypes in a population from generation to generation or a focus on phenotypes at an optimum for natural selection. The first uses analytical models (nowadays often complemented by simulations of populations by large-scale computation). The second uses game-theoretic models. One text that presents both approaches with great clarity is . . .

Maynard Smith, J. 1989. *Evolutionary genetics.* Oxford: Oxford University Press.

The pivotal introduction of game theory in evolutionary thinking was . . .

Maynard Smith, J. 1982. *Evolution and the theory of games.* Cambridge: Cambridge University Press.

8. OPTIMAL RECEIVERS AND SIGNALERS

Utility as a quantitative basis for rational decision was introduced in . . .

Neumann, J. v., and O. Morgenstern. 1953. *Theory of games and economic behavior.* Princeton, NJ: Princeton University Press.

Excellent introductions to Signal Detection Theory, among many, are . . .

Macmillan, N. A. 1991. *Detection theory: A user's guide.* Cambridge: Cambridge University Press. Reprint, Mahwah, NJ: Erlbaum, 2004.

Macmillan, N. A. 2002. Signal Detection Theory. In H. E. Pashler, ed., *Stevens' handbook of experimental psychology,* 3rd ed., vol. 4, 43–90. New York: Wiley.

The properties of any receiver and the formulation of a receiver's performance in noise follow my earlier suggestions . . .

Wiley, R. H. 1994. Errors, exaggeration, and deception in animal communication. In L. Real, ed., *Behavioral mechanisms in evolutionary ecology,* 157–189. Chicago: University of Chicago Press.

Wiley, R. H. 2006. Signal detection and animal communication. *Advances in the Study of Behavior* 36: 217–247.

Other authors have also proposed combining Decision Theory and Signal Detection Theory . . .

Sperling, G. 1983. A unified theory of attention and signal detection. In R. Parasuraman and D. R. Davies, eds., *Varieties of attention,* 103–181. Orlando, FL: Academic Press.

Lynn, S. K., and L. F. Barrett. 2014. "Utilizing" signal detection theory. *Psychological Science* 25: 1663–1673.

The costs of producing signals have been the subject of much investigation for the past four decades. A review of many examples is provided by William Searcy and Stephen Nowicki (2005), as cited in Chapter 12. For another review of the risks of predation and parasitism as a result of signaling and for more information about the examples mentioned in this chapter, see . . .

Magnhagen, C. 1991. Predation risk as a cost of reproduction. *Trends in Ecology and Evolution* 6: 183–186.

Ryan, M. J. 1985. *The Túngara Frog: A study in sexual selection and communication.* Chicago: University of Chicago Press.

Bernal, X. E., A. S. Rand, and M. J. Ryan. 2006. Acoustic preferences and localization performance of blood-sucking flies *(Corethrella coquillett)* to Túngara Frog calls. *Behavioral Ecology* 17: 709–715.

Cade, W. H., and E. S. Cade. 1992. Male mating success, calling and searching behaviour at high and low densities in the field cricket, *Gryllus integer. Animal Behaviour* 43: 49–56.

Zukl, M., L. W. Simmons, and L. Cupp. 1993. Calling characteristics of parasitized and unparasitized populations of the field cricket *Teleogryllus oceanicus. Behavioral Ecology and Sociobiology* 33: 339–343.

Hartzler, J. E. 1974. Predation and the daily timing of Sage Grouse leks. *Auk* 91: 532–536.

Boyko, A. R., R. M. Gibson, and J. R. Lucas. 2004. How predation risk affects the temporal dynamics of avian leks: Greater Sage Grouse versus Golden Eagles. *American Naturalist* 163: 154–165.

Trail, P. W. 1987. Predation and antipredator behavior at Guianan Cock-of-the-rock leks. *Auk* 104: 496–507.

Balmford, A., and M. Turyaho. 1992. Predation risk and lek-breeding in Uganda Kob. *Animal Behaviour* 44: 117–127.

For matches (and some mismatches) between acoustic signals and hearing of birds . . .

Konishi, M. 1970. Comparative neurophysiological studies of hearing and vocalizations in songbirds. *Journal of Comparative Physiology A* 66: 257–272.

Lohr, B., and R. Dooling. 1998. Detection of changes in timbre and harmonicity in complex sounds by Zebra Finches *(Taeniopygia guttata)* and Budgerigars *(Melopsittacus undulatus). Journal of Comparative Psychology* 112: 36–47.

Henry, K. S., and J. R. Lucas. 2008. Coevolution of auditory sensitivity and temporal resolution with acoustic signal space in three songbirds. *Animal Behaviour* 76: 1659–1671.

Gall, M. D., L. E. Brierley, and J. R. Lucas. 2012. The sender-receiver matching hypothesis: Support from the peripheral coding of acoustic features in songbirds. *Journal of Experimental Biology* 215: 3742–3751.

For frogs . . .

Gerhardt, H. C., and K. M. Mudry. 1980. Temperature effects on frequency preferences and mating call frequencies in the Green Treefrog, *Hyla cinerea* (Anura: Hylidae). *Journal of Comparative Physiology* 137: 1–6.

Gerhardt, H. C. and J. J. Schwartz. 2001. Auditory tuning and frequency preferences in anurans. In M. J. Ryan, ed., *Anuran communication,* 73–85. Washington, DC: Smithsonian Institution Press.

Wilczynski, W., A. S. Rand, and M. J. Ryan. 2001. Evolution of calls and auditory tuning in the *Physalaemus pustulosus* species group. *Brain, Behavior and Evolution* 58: 137–151.

More is known about the genetics of coordinated production and reception of acoustic signals in insects . . .

Gerhardt, H. C., and F. Huber. 2002. *Acoustic communication in insects and anurans.* Chicago: University of Chicago Press.

Boake, C. R. B. 1991. Coevolution of senders and receivers of sexual signals: Genetic coupling and genetic correlations. *Trends in Ecology and Evolution* 3: 173–182.

Butlin, R. K., and M. G. Ritchie. 1989. Genetic coupling in mate recognition systems: What is the evidence? *Biological Journal of the Linnean Society* 37: 237–246.

The prevalence of eavesdropping in animal communication was initially emphasized by Peter McGregor but has by now been documented for a variety of animals and contexts . . .

McGregor, P. K., ed. 2005. *Animal communication networks*. Cambridge: Cambridge University Press.

McGregor, P. K. 1993. Signalling in territorial systems: A context for individual identification, ranging and eavesdropping. *Philosophical Transactions of the Royal Society B: Biological Sciences* 340: 237–244.

Naguib, M., C. Fichtel, and D. Todt. 1999. Nightingales respond more strongly to vocal leaders of simulated dyadic interactions. *Proceedings of the Royal Society of London. Series B: Biological Sciences* 266: 537–542.

Otter, K., et al. 1999. Do female Great Tits *(Parus major)* assess males by eavesdropping? A field study using interactive song playback. *Proceedings of the Royal Society of London. Series B: Biological Sciences* 266: 1305–1309.

Peake, T. M., et al. 2001. Male Great Tits eavesdrop on simulated male-to-male vocal interactions. *Proceedings of the Royal Society of London. Series B: Biological Sciences* 268: 1183–1187.

Peake, T. M., et al. 2002. Do Great Tits assess rivals by combining direct experience with information gathered by eavesdropping? *Proceedings of the Royal Society of London. Series B: Biological Sciences* 269: 1925–1929.

Mennill, D. J., L. M. Ratcliffe, and P. T. Boag. 2002. Female eavesdropping on male song contests in songbirds. *Science* 296: 873–873.

Mennill, D. J., P. T. Boag, and L. M. Ratcliffe. 2003. The reproductive choices of eavesdropping female Black-capped Chickadees, *Poecile atricapillus*. *Naturwissenschaften* 90: 577–582.

Mennill, D. J., and L. M. Ratcliffe. 2004. Do male Black-capped Chickadees eavesdrop on song contests? A multi-speaker playback experiment. *Behaviour* 141: 125.

For deceptive signals by birds, see the references for Chapter 1. Deception by nonhuman primates has also attracted a lot of attention, although reported cases often consist of *withholding* signals rather than producing deceptive ones . . .

Whiten, A., and R. W. Byrne. 1988. Tactical deception in primates. *Behavioral and Brain Sciences* 11: 233–273.

Byrne, R. W., and A. Whitten. 1989. *Machiavellian intelligence: Social expertise and the evolution of intellect in monkeys, apes, and humans*. Oxford: Oxford University Press.

Cheney, D. L., and R. M. Seyfarth. 1990. *How monkeys see the world*. Chicago: University of Chicago Press.

Hauser, M. D. 1992. Costs of deception: Cheaters are punished in rhesus monkeys *(Macaca mulatta)*. *Proceedings of the National Academy of Sciences of the United States of America* 89: 12137–12139.

Mitchell, R. W., and J. R. Anderson. 1997. Pointing, withholding information, and deception in capuchin monkeys *(Cebus apella). Journal of Comparative Psychology* 111: 351–361.

Cheney, D. L., and R. M. Seyfarth. 2007. *Baboon metaphysics: The evolution of a social mind.* Chicago: University of Chicago Press.

9. PAYOFFS FOR PARTICIPANTS

This chapter presents only a small selection of the available measures of costs and benefits of signaling and responding to signals. For "good genes" as a benefit for female choice of mates . . .

Welch, A. M. 2003. Genetic benefits of a female mating preference in Gray Tree Frogs are context-dependent. *Evolution* 57: 883–893.

Doty, G. V., and A. M. Welch. 2001. Advertisement call duration indicates good genes for offspring feeding rate in Gray Tree Frogs *(Hyla versicolor). Behavioral Ecology and Sociobiology* 49: 150–156.

Møller, A. P., and R. V. Alatalo. 1999. Good-genes effects in sexual selection. *Proceedings of the Royal Society of London. Series B: Biological Sciences* 266: 85–91.

Neff, B. D., and T. E. Pitcher. 2005. Genetic quality and sexual selection: An integrated framework for good genes and compatible genes. *Molecular Ecology* 14: 19–38.

Byers, J. A., and L. Waits. 2006. Good genes sexual selection in nature. *Proceedings of the National Academy of Sciences of the United States of America* 103: 16343–16345. Maternal effects not excluded.

For "nuptial gifts," sacs of nutrients and other substances presented by males to females just before copulation in some insects . . .

Gwynne, D. T. 2008. Sexual conflict over nuptial gifts in insects. *Annual Revue of Entomology* 53: 83–101.

Vahed, K., and F. S. Gilbert. 1996. Differences across taxa in nuptial gift size correlate with differences in sperm number and ejaculate volume in bushcrickets (Orthoptera: Tettigoniidae). *Proceedings of the Royal Society B: Biological Sciences* 263: 1257–1265.

For recent discussions of developmental stability, including phenotypic asymmetry . . .

Nowicki, S., S. Peters, and J. Podos. 1998. Song learning, early nutrition and sexual selection in songbirds. *American Zoologist* 38: 179–190.

Spencer, K. A., S. A. Macdougall-Shackleton, and I. R. Titze. 2011. Indicators of development as sexually selected traits: The developmental stress hypothesis in context. *Behavioral Ecology* 22: 1–9.

Little, A. C., B. C. Jones, and L. M. Debruine. 2011. Facial attractiveness: Evolutionary based research. *Philosophical Transactions of the Royal Society B: Biological Sciences* 366: 1638–1659. This paper exemplifies the proliferation of such studies of human preferences in recent years.

Recent evaluations of the interactions of carotenoids, testosterone, and immune responses . . .

Mcgraw, K. J., and D. R. Ardia. 2007. Do carotenoids buffer testosterone-induced immunosuppression? An experimental test in a colourful songbird. *Biology Letters* 3: 375–378.

Peters, A., et al. 2004. Trade-offs between immune investment and sexual signaling in male Mallards. *American Naturalist* 164: 51–59.

Alonso-Alvarez, C., et al. 2008. The oxidation handicap hypothesis and the carotenoid allocation trade-off. *Journal of Evolutionary Biology* 21: 1789–1797.

Svensson, P. A., and B. B. M. Wong. 2011. Carotenoid-based signals in behavioural ecology: A review. *Behaviour* 148: 131–189.

Measuring the costs to females of searching for optimal males has proven more difficult. The available information confirms the relatively high costs of signaling for ectothermic vertebrates and invertebrates in comparison to the relatively low costs for endothermic vertebrates such as birds and mammals.

Gibson, R. M., and G. C. Bachman. 1992. The costs of female choice in a lekking bird. *Behavioral Ecology* 3: 300–309.

Byers, J. A., et al. 2005. A large cost of female mate sampling in Pronghorn. *American Naturalist* 166: 661–668.

Wells, K. D. 2001. The energetics of calling in frogs. In M. J. Ryan, ed., *Anuran communication,* 45–60. Washington, DC: Smithsonian Institution Press.

Costs of producing exaggerated signals might often include increased exposure to predators or parasites attracted to the signals. Another plausible possibility, rarely measured, is a trade-off between the exaggeration of any one signal and the rate of signaling . . .

Endler, J. A. 1980. Natural selection on color patterns in *Poecilia reticulata. Evolution* 34: 76–91.

Godin, J. G. J., and H. E. McDonough. 2003. Predator preference for brightly
colored males in the Guppy: A viability cost for a sexually selected trait.
Behavioral Ecology 14: 193–200.

Vahed, K. 2007. Comparative evidence for a cost to males of manipulating
females in bushcrickets. *Behavioral Ecology* 18: 499–506.

Almost all of the information about the costs and benefits of signaling
and preferences for signals by Barn Swallows comes from the numerous
studies by Anders Pape Møller and his colleagues in several localities in
Europe. Møller's book summarizes the essence of the story . . .

Møller, A. P. 1994. *The Barn Swallow and sexual selection.* Oxford: Oxford
University Press.

Møller, A. P., F. Lope, and J. M. López Caballero. 1995. Foraging costs of a tail
ornament: Experimental evidence from two populations of Barn
Swallows *Hirundo rustica* with different degrees of sexual size dimorphism. *Behavioral Ecology and Sociobiology* 37: 289–295.

Sexual dimorphism in tail length is less pronounced in Spain than in
Denmark. In Spain experimental manipulations of tail length have a
greater influence on foraging for aerial insects. These results indicate that
increased costs of signaling tend to reduce the exaggeration of signals,
as predicted by the approach in this chapter.

10. JOINT OPTIMA FOR SIGNALERS AND RECEIVERS

The calculations described in this chapter come from my recent paper . . .

Wiley, R. H. 2013. A receiver-signaler equilibrium in the evolution of communication in noise. *Behaviour* 150: 957–993.

A few errors in labeling of figures are corrected at http://www.unc
.edu/~rhwiley/refsa/rsequilibrium.html. See also . . .

Wiley, R. H. 2014. Corrigendum. *Behaviour* 151: 1363–1365; doi: 10.1163
/1568539X-00003063.

11. EVALUATION AND EXTENSION OF THE MODEL

More discussion of assumptions underlying the treatment of signals and
receivers in . . .

Wiley, R. H. 1994. Errors, exaggeration, and deception in animal communication. In L. Real, ed., *Behavioral mechanisms in evolutionary ecology,*
157–189. Chicago: University of Chicago Press.

Wiley, R. H. 2006. Signal detection and animal communication. *Advances in
the Study of Behavior* 36: 217–247.

Mimetic neotropical butterflies and their predators . . .

Cook, L. M., L. P. Brower, and J. Alcock. 1969. An attempt to verify mimetic
advantage in a neotropical environment. *Evolution* 23: 339–345.

Chai, P. 1986. Field observations and feeding experiments on the responses of
Rufous-tailed Jacamars *(Galbula ruficauda)* to free-flying butterflies in a
tropical rainforest. *Biological Journal of the Linnaean Society London* 29:
161–189.

Chai, P. 1996. Butterfly visual characteristics and ontogeny of responses to
butterflies by a specialized tropical bird. *Biological Journal of the
Linnaean Society London* 59: 37–67.

Pinheiro, C. E. G. 1996. Palatability and escaping ability in neotropical
butterflies: Tests with wild kingbirds *(Tyrannus melancholicus,*
Tyrannidae). *Biological Journal of the Linnaean Society London* 59:
351–365.

Pinheiro, C. E. G. 2011. On the evolution of warning coloration, Batesian and
Müllerian mimicry in neotropical butterflies: The role of jacamars
(Galbulidae) and tyrant-flycatchers (Tyrannidae). *Journal of Avian
Biology* 42: 277–281.

Theoretical analyses of the evolution of mimicry . . .

Getty, T. 1987. Crypsis, mimicry and switching: The basic similarity of
superficially different analyses. *American Naturalist* 130, 793–797.

Getty, T., A. C. Kamil, and P. Real. 1987. Signal detection theory and foraging
for cryptic or mimetic prey. In A. C. Kamil, J. R. Krebs, and H. R.
Pulliam, eds., *Foraging Behavior,* 525–548. New York: Plenum.

Staddon, J. E. R., and R. P. Gendron. 1983. Optimal detection of cryptic prey
may lead to predator switching. *American Naturalist* 122: 843–848.

Servedio, M. R., and R. Lande. 2003. Coevolution of an avian host and its
parasitic cuckoo. *Evolution* 57: 1164–1175.

A sample of additional theoretical studies of mimicry and crypsis . . .

Edmunds, M. 2000. Why are there good and poor mimics? *Biological Journal
of the Linnean Society* 70: 459–466.

Holen, Ø. H., and R. A. Johnstone. 2004. The evolution of mimicry under
constraints. *American Naturalist* 164: 598–613. This analysis is

distinguished by including all four possible outcomes of a receiver's decision to respond.

Iserbyt, A., et al. 2011. Frequency-dependent variation in mimetic fidelity in an intraspecific mimicry system. *Proceedings of the Royal Society B: Biological Sciences* 278: 3116–3122.

Johnstone, R. A. 2002. The evolution of inaccurate mimics. *Nature* 418: 524–526.

Lotem, A., and H. Nakamura. 1998. Evolutionary equilibria in avian brood parasitism: An alternative to the "arms race evolutionary lag." In S. I. Rothstein and S. K. Robinson, eds., *Parasitic birds and their hosts: Studies in coevolution,* 223–235. New York: Oxford University Press.

Sherratt, T. N. 2002. The evolution of imperfect mimicry. *Behavioral Ecology* 13: 821–826.

Stephens, D. W., and J. R. Krebs. 1986. *Foraging theory.* Princeton, NJ: Princeton University Press.

Additional theoretical analyses that apply the principles of signal detection . . .

Getty, T. 1995. Search, discrimination, and selection: Mate choice by Pied Flycatchers. *American Naturalist* 145: 146–154.

Reeve, H. K. 1989. The evolution of conspecific acceptance thresholds. *American Naturalist* 133: 407–435.

Importance of costs and trade-offs in the evolution of imperfect mimicry has been thoroughly reviewed recently . . .

Kikuchi, D. W., and D. W. Pfennig. 2013. Imperfect mimicry and the limits of natural selection. *Quarterly Review of Biology* 88: 297–315.

Mimicry by the Scarlet Kingsnake in relation to the presence of coral snakes . . .

Pfennig, D. W., W. R. Harcombe, and K. S. Pfennig. 2001. Frequency-dependent Batesian mimicry. *Nature* 410: 323–323.

Harper G. R., Jr., and D. W. Pfennig. 2007. Population differences in predation on Batesian mimics in allopatry with their model: Selection against mimics is strongest when they are common. *Behavioral Ecology and Sociobiology* 61: 505–511.

Harper, G. R., Jr., and D. W. Pfennig. 2007. Mimicry on the edge: Why do mimics vary in resemblance to their model in different parts of their geographical range? *Proceedings of the Royal Society B: Biological Sciences* 274: 1955–1961.

Harper G. R., Jr., and D. W. Pfennig. 2008. Selection overrides gene flow to break down maladaptive mimicry. *Nature* 451: 1103–1106.

Pfennig, D. W., et al. 2007. Population differences in predation on Batesian mimics in allopatry with their model: Selection against mimics is strongest when they are common. *Behavioral Ecology and Sociobiology* 61: 505–511.

Pfennig, D. W., and S. P. Mullen. 2010. Mimics without models: Causes and consequences of allopatry in Batesian mimicry complexes. *Proceedings of the Royal Society B: Biological Sciences* 277: 2577–2585.

Kikuchi, D. W., and D. W. Pfennig. 2010. High model abundance may permit the gradual evolution of Batesian mimicry: An experimental test. *Proceedings of the Royal Society B: Biological Sciences* 277: 1041–1048.

Kikuchi, D. W., and D. W. Pfennig. 2010. Predator cognition permits imperfect coral snake mimicry. *American Naturalist* 176: 830–834.

Akcali, C. K., and D. W. Pfennig. 2014. Rapid evolution of mimicry following local model extinction. *Biology Letters* 10: 20140304.

Coevolution of brood-parasitic mimicry by European Cuckoos and defenses by host species . . .

Davies, N. B., and M. L. Brooke. 1998. Cuckoos versus hosts: Experimental evidence for coevolution. In S. I. Rothstein and S. K. Robinson, eds., *Parasitic birds and their hosts: Studies in coevolution*, 59–79. Oxford: Oxford University Press.

Davies, N. B. 2000. *Cuckoos, cowbirds and other cheats*. London: T. and A. D. Poyser.

Further evidence that the degree of mimicry is related to the frequency of brood parasitism by cuckoos . . .

Stoddard, M. C., and M. Stevens. 2010. Pattern mimicry of host eggs by the common cuckoo, as seen through a bird's eye. *Proceedings of the Royal Society B: Biological Sciences* 277: 1387–1393.

Stoddard, M. C., and M. Stevens. 2011. Avian vision and the evolution of egg color mimicry in the common cuckoo. *Evolution* 65: 2004–2013.

Thorogood, R., and N. B. Davies. 2013. Reed warbler hosts fine-tune their defenses to track three decades of cuckoo decline. *Evolution* 67: 3545–3555.

Additional coevolutionary interactions between cuckoos and hosts . . .

Madden, J. R., and N. B. Davies. 2006. A host-race difference in begging calls of nestling cuckoos *Cuculus canorus* develops through experience and increases host provisioning. *Proceedings of the Royal Society B: Biological Sciences* 273: 2343–2351. Nestling cuckoos learn to match the begging calls of a nestful of the host's young.

Welbergen, J. A., and N. B. Davies. 2011. A parasite in wolf's clothing: Hawk mimicry reduces mobbing of cuckoos by hosts. *Behavioral Ecology* 22: 574–579.

Thorogood, R., and N. B. Davies. 2012. Cuckoos combat socially transmitted defenses of reed warbler hosts with a plumage polymorphism. *Science* 337: 578–580.

Brown-headed Cowbirds (brood parasites of the Western Hemisphere) unlike European Cuckoos, have no known mimicry of their hosts; their young just grow faster than the hosts' young and overpower them.

The importance of peak shift for the evolution of communication has been discussed in . . .

Arak, A., and M. Enquist. 1993. Hidden preferences and the evolution of signals. *Philosophical Transactions of the Royal Society B: Biological Sciences* 340: 207–213.

Lynn, S. K., J. Cnaani, and D. R. Papaj. 2005. Peak shift discrimination learning as a mechanism of signal evolution. *Evolution* 59: 1300–1305.

In particular, peak shift can explain supernormal stimuli, artificial signals that are more effective in eliciting responses than natural signals . . .

Hogan, J. A., J. P. Kruijt, and J. H. Frijlink. 1975. "Supernormality" in a learning situation. *Zeitschrift für Tierpsychologie* 38: 212–218.

Staddon, J. E. R. 1975. A note on the evolutionary significance of "supernormal" stimuli. *American Naturalist* 109: 541–545. Staddon emphasizes that peak shift can result either from directional reinforcement in learning or from directional selection.

One of the few attempts to relate hearing to communication in natural noise is the study . . .

Langemann, U., B. Gauger, and G. M. Klump. 1998. Auditory sensitivity in the Great Tit: Perception of signals in the presence and absence of noise. *Animal Behaviour* 56: 763–769.

12. HONESTY IN COMMUNICATION

Amotz Zahavi summarizes his theory of signaling handicaps in his 1997 book, but the first proposals were two decades earlier.

Zahavi, A. 1975. Mate selection—a selection for a handicap. *Journal of Theoretical Biology* 53: 205–214.

Zahavi, A. 1977. The cost of honesty. *Journal of Theoretical Biology* 67: 603–605.

Zahavi, A. 1993. The fallacy of conventional signalling. *Philosophical Transactions of the Royal Society of London. Series B: Biological Sciences* 340: 227–230.

Zahavi, A., and A. Zahavi. 1997. *The handicap principle.* Oxford: Oxford University Press.

Reviews of the importance of costs (usually called handicaps) for the evolution of signaling include . . .

Johnstone, R. A. 1995. Sexual selection, honest advertisement and the handicap principle: Reviewing the evidence. *Biological Reviews* 70: 1–65.

Johnstone, R. A. 1997. The evolution of animal signals. In J. R. Krebs and N. B. Davies, eds., *Behavioral ecology,* 4th ed., 157–178. Oxford: Oxford University Press.

Searcy, W. A., and S. Nowicki. 2005. *The evolution of animal communication: Reliability and deception in signaling systems.* Princeton, NJ: Princeton University Press. An excellent review of the prevailing issues.

For costs of signaling that exceed those necessary to avoid ambiguity . . .

Maynard Smith, J., and D. G. C. Harper. 2003. *Animal signals.* Oxford: Oxford University Press.

For the prevailing view of costs of signaling in relation to different kinds of signals . . .

Hasson, O. 1994. Cheating signals. *Journal of Theoretical Biology* 167: 223–238.

Hurd, P., and M. Enquist. 2005. A strategic taxonomy of biological communication. *Animal Behaviour* 70: 1155–1170.

For mathematical proofs that honesty requires costs of signaling (and claims that these justify Zahavi's proposals) . . .

Grafen, A. 1990a. Biological signals as handicaps. *Journal of Theoretical Biology* 144: 517–546.

Grafen, A. 1990b. Sexual selection unhandicapped by the Fisher process. *Journal of Theoretical Biology* 144: 473–516.

Maynard Smith, J. 1991. Honest signalling: The Philip Sidney game. *Animal Behaviour* 42: 1034–1035.

Iwasa, Y., A. Pomiankowski, and S. Nee. 1991. The evolution of costly mate preferences II. The "handicap" principle. *Evolution* 45: 1431–1442.

Johnstone, R. A. 1998. Conspiratorial whispers and conspicuous displays: Games of signal detection. *Evolution* 52: 1554–1563.

For use of the term *handicap* by Zahavi and others . . .

Getty, T. 1998. Reliable signalling need not be a handicap. *Animal Behaviour* 56: 253–255.

Getty, T. 2006. Sexually selected signals are not similar to sports handicaps. *Trends in Ecology and Evolution* 21: 83–88.

For graphical interpretations of how signaling costs can produce honest signals, see the preceding two papers, Johnstone (1997), and . . .

Wiley, R. H. 2000. Sexual selection and mate choice: Trade-offs for males and females. In M. Apollonio, M. Festa-Bianchet, and D. Mainardi, eds., *Vertebrate mating systems,* 8–46. Singapore: World Scientific Publishing.

Wiley, R. H. 2013. A receiver-signaler equilibrium in the evolution of communication in noise. *Behaviour* 150: 957–993.

Errors around mean responses to signals (as opposed to the two mutually exclusive forms of error in Signal Detection Theory) have little effect on the evolutionary equilibria for communication. Nevertheless, even with these errors, conditions for perfect honesty are almost never met . . .

Johnstone, R. A., and A. Grafen. 1992. Error-prone signalling. *Proceedings of the Royal Society B: Biological Sciences* 248: 229–233.

Johnstone, R. A., and A. Grafen. 1993. Dishonesty and the handicap principle. *Animal Behaviour* 46: 759–764.

13. SEXUAL SELECTION AS COMMUNICATION

R. A. Fisher's descriptions of sexual selection were introduced in 1915 but primarily described in his 1930 book . . .

Fisher, R. A. 1915. The evolution of sexual preference. *Eugenics Review* 7: 184–192.

Fisher, R. A. 1930. *The genetical theory of natural selection.* Oxford: Clarendon Press. Reprinted with slight modifications, 2nd ed., New York: Dover, 1958. Reprinted with introduction and notes by H. Bennett as *The genetical theory of natural selection: A complete variorum edition.* Oxford: Oxford University Press, 1999. Reprinted with corrections, 2003.

Fisher's verbal description of sexual selection in 1930 is found in three paragraphs on pages 136–137 and is repeated without substantive changes on pages 151–152 in the 1958 edition. The paragraphs preceding this description argue that females, especially in birds, have preferences for mates, that these preferences exert selection on the preferred ornaments of males, and that these preferences by females should themselves evolve. Fisher recognizes that females usually have preferences and males ornaments, although in a minority of species the roles of the sexes are reversed.

In the three central paragraphs, Fisher argues that males with preferred ornaments are more likely to leave progeny than other males. Furthermore, females with these preferences have more sons with these preferred ornaments and thus with their own advantages in mating. As stated in one of his letters to Leonard Darwin shortly afterward, females with preferences leave more grand-offspring than other females. As long as there is a "net advantage" for plumage elaboration, there is also a "net advantage" for enhanced preference. Fisher states that increased ornaments and increased preferences evolve in tandem in an accelerating way. The greater the preference, the greater the advantage for the ornament, and vice versa. The more the preference spreads, the more the ornament spreads, and vice versa. The rate of evolution is proportional to the degree "already attained" so that evolution of ornament and preference "increase with time exponentially" in a "runaway" process. Counteracting selection against the ornament, as a result of greater mortality of males with the ornament, can eventually eliminate any net advantage for the ornament and bring sexual selection to a "standstill." Nevertheless, an ornament with a disadvantage for survival can increase until this disadvantage completely counterbalances the advantage of additional matings.

The problem I see in this description is that it applies equally well to the spread of any signal and response by frequency-dependent selection. A rare signal or a rare response provides little advantage for an individual, because signals only have advantages when appropriate responses occur, and responses have advantages only when appropriate signals occur. As a signal spreads in a population, the advantage of responses increases; conversely, as responses spread, the advantage of signals increases. At least initially, signals and responses of any sort should spread exponentially in proportion to the current frequencies in the population. This process does not require that signalers and receivers mate with each other. Although Fisher focuses on males attracting females for mating, it is not clear from the descriptions in any of his publications that he was thinking, as we now do, of sexual selection as a result of a genetic correlation.

In an exchange of letters during the two years following the publication of Fisher's book in 1930, Leonard Darwin repeatedly questioned Fisher's verbal description of the mechanism of sexual selection. First published by Henry Bennett in 1983, the relevant passages in these letters were assembled in Bennett's 1999 variorum edition of Fisher's book. Leonard Darwin, Charles's grandson and a longtime friend and supporter of Fisher, had repeated difficulties with Fisher's attempts to provide a mathematical formulation of sexual selection. I admit that I do too. In the end Fisher reduces his argument to "the kernel of the whole thing": "the exponential element . . . arises from the rate of change in [the preference] being proportional to the absolute average degree of [the preference] $(\partial y \propto \bar{y})$" (Fisher 1999, p. 308). In other words, the more the preference has evolved, the greater its current rate of evolution. These letters make it clear that he is thinking about "genetic values"—that is, expected phenotypes (pp. 306, 308). He does not mention genetic correlation of genotypes. The exponential spread of phenotypes for signals and responses, such as Fisher describes, can occur in two ways: (1) by sexual selection, which requires mating by signaler and receiver and proceeds by means of genetic correlation, or (2) by frequency-dependent selection, which does not require mating between signaler and receiver and thus applies to the evolution of all signals. Signals that promote mating are affected by both processes, all others only by the second. Peter O'Donald's (1980) simulations of sexual selection did not involve mating between signaler and receiver, so the dynamics depended entirely on frequency-dependent selection. The simulations clearly illustrate exponential increase in the frequency of alleles for both signal and preference when rare. Eventually, natural selection for both decreases as the alleles approach fixation in the population and the process comes to a standstill.

Fisher made a great contribution by clarifying that preferences for ornaments (and, by extension, responses to any signals) should evolve, just as do other features of behavior. And he was the first to indicate that the evolution of communication had the potential for acceleration. Nevertheless, the first clear explanation of the mechanism of sexual selection, by genetic correlation when signals promote mating with receivers, came in the publications by Russell Lande (1981) and Mark Kirkpatrick (1982). Nevertheless, in recent decades, Fisherian selection has become the usual term for sexual selection by means of genetic correlation, often specifically when male ornaments and female preferences provide no advantages to males or females other than increased access of males to mates. The second mechanism for accelerating evolution of communication, as a result of frequency-dependent selection, has received almost no attention recently.

Early descriptions of sexual selection include . . .

> Darwin, C. 1859. *On the origin of species by means of natural selection.* London: John Murray.
>
> Darwin, C. 1871. *The descent of man, and selection in relation to sex.* London: John Murray.
>
> Huxley, J. S. 1938. Darwin's theory of sexual selection and the data subsumed by it, in the light of recent research. *American Naturalist* 72: 416–433.

For an overview of the history of investigating sexual selection (although not distinguishing it from frequency-dependent selection) see . . .

> Cronin, H. 1991. *The ant and the peacock: Altruism and sexual selection from Darwin to today.* Cambridge: Cambridge University Press.

For a summary of O'Donald's models . . .

> O'Donald, P. 1980. *Genetic models of sexual selection.* Cambridge: Cambridge University Press.
>
> O'Donald, P. 1990. Fisher's contributions to the theory of sexual selection as the basis of recent research. *Theoretical Population Biology* 38: 285–300.

For our current understanding of sexual selection with genetic correlation . . .

> Lande, R. 1981. Models of speciation by sexual selection on polygenic traits. *Proceedings of the National Academy of Sciences of the United States of America* 78: 3721–3725.

Kirkpatrick, M. 1982. Sexual selection and the evolution of female choice. *Evolution* 36: 1–12.

Heisler, I. L. 1984. A quantitative genetic model for the origin of mating preferences. *Evolution* 38: 1283–1295.

Pomiankowski, A. N. 1988. The evolution of female mate preferences for male genetic quality. *Oxford Surveys in Evolutionary Biology* 5: 136–184.

Andersson, M. B. 1994. *Sexual selection*. Princeton, NJ: Princeton University Press.

Kokko, H., M. D. Jennions, and R. Brooks. 2006. Unifying and testing models of sexual selection. *Annual Review of Ecology, Evolution, and Systematics* 37: 43–66.

Kokko, H. 2001. Fisherian and "good genes" benefits of mate choice: How (not) to distinguish between them. *Ecology Letters* 4: 322–326.

For the sexual selection of arbitrary traits as a null hypothesis . . .

Wiley, R. H. 2000. Sexual selection and mate choice: Trade-offs for males and females. In M. Apollonio, M. Festa-Bianchet, and D. Mainardi, eds., *Vertebrate mating systems*, 8–46. Singapore: World Scientific Publishing.

Prum, R. O. 2010. The Lande–Kirkpatrick mechanism is the null model of evolution by intersexual selection: Implications for meaning, honesty, and design in intersexual signals. *Evolution* 64: 3085–3100.

Prum, R. O. 2012. Aesthetic evolution by mate choice: Darwin's really dangerous idea. *Philosophical Transactions of the Royal Society B* 367: 2253–2265.

It is important to emphasize that evolution by genetic correlation is not itself a null hypothesis, but evolution of arbitrary traits by genetic correlation is.

Indirect mate choice, which also evolves by genetic correlation, is described in . . .

Wiley, R. H., and J. Poston. 1996. Perspective: Indirect mate choice, competition for mates, and coevolution of the sexes. *Evolution* 50: 1371–1381.

Wiley, R. H. 1991. Lekking in birds and mammals: Behavioral and evolutionary mechanisms. *Advances in the Study of Behavior* 20: 201–291.

Poston, J. P. 1997. Dominance, access to colonies, and queues for mating opportunities by male Boat-tailed Grackles. *Behavioral Ecology and Sociobiology* 41: 89–98.

Poston, J. P. 1997. Mate choice and competition for mates in the Boat-tailed Grackle. *Animal Behaviour* 54: 525–534.

Poston, J., R. H. Wiley, and D. Westneat. 1999. Male rank, female breeding synchrony, and patterns of paternity in the Boat-tailed Grackle. *Behavioural Ecology* 10: 444–451.

Kirkpatrick, M. 1987. Sexual selection by female choice in polygynous animals. *Annual Review of Ecology and Systematics* 18: 43–70. This review explicitly describes a mating "preference" as any mechanism that narrows the set of potential mates, although it does not present the gamut of possibilities for indirect mate choice.

Net costs of preferences preclude sexual selection. This conclusion is an instance of the general conclusion that communication only evolves when it has advantages on average for both signaler and receiver. A female must on average benefit from responses to a male's signals, either directly as a result of her increased fecundity or survival or indirectly as a result of the male's "good genes" . . .

Pomiankowski, A. 1987. The costs of choice in sexual selection. *Journal of Theoretical Biology* 128: 195–218.

Pomiankowski, A. N. 1988. The evolution of female mating preferences for male genetic quality. *Oxford Surveys in Evolutionary Biology* 5: 136–184.

More evidence that efficiency in signaling influences the direction of evolution by sexual selection . . .

Tazzyman, S. J., Y. Iwasa, and A. Pomiankowski. 2013. Signaling efficacy drives the evolution of larger sexual ornaments by sexual selection. *Evolution* 68: 216–229.

14. cooperation by communication

For specificity and multiplicity in recognition of other individuals, see . . .

Wiley, R. H. 2013. Specificity and multiplicity in the recognition of individuals: Implications for the evolution of social behaviour. *Biological Reviews* 88: 179–195.

Renee Godard's studies of recognition of neighbors by territorial Hooded Warblers are discussed in Chapters 1 and 5 (see references in the notes for Chapter 1). Of numerous such studies of other species, some pioneering contributions include . . .

Falls, J. B., and R. J. Brooks. 1975. Individual recognition by song in White-throated Sparrows. II. Effects of location. *Canadian Journal of Zoology* 53: 1412–1420.

Falls, J. B. 1982. Individual recognition by sounds in birds. In D. E. Kroodsma and E. H. Miller, eds., *Acoustic communication in birds,* vol. 2, *Song learning and its consequences,* 237–278. New York: Academic Press.

Stoddard, P. K., et al. 1991. Recognition of individual neighbors by song in the Song Sparrow, a species with song repertoires. *Behavioral Ecology and Sociobiology* 29: 211–215.

Stoddard, P. K., et al. 1992. Memory does not constrain individual recognition in a bird with song repertoires. *Behaviour* 122: 274–287.

Stoddard, P. K. 1996. Vocal recognition of neighbors by territorial passerines. In D. E. Kroodsma and E. H. Miller, eds., *Ecology and evolution of acoustic communication in birds,* 356–374. Ithaca, NY: Cornell University Press.

Bee, M. A., and H. C. Gerhardt. 2002. Individual voice recognition in a territorial frog *(Rana catesbeiana). Proceedings of the Royal Society B: Biological Science* 269: 1443–1448.

Recognition of individuals by nonhuman primates is important for a variety of complex social interactions . . .

Cheney, D. L., and R. M. Seyfarth. 1980. Vocal recognition in free-ranging Vervet Monkeys. *Animal Behaviour* 28: 362–367.

Cheney, D. L., and R. M. Seyfarth. 1990. *How monkeys see the world: Inside the mind of another species.* Chicago: University of Chicago Press.

Cheney, D. L., and R. M. Seyfarth. 2007. *Baboon metaphysics: The evolution of a social mind.* Chicago: University of Chicago Press.

Bergman, T. J., et al. 2003. Hierarchical classification by rank and kinship in baboons. *Science* 302: 1234–1236.

Bergman, T. J. 2010. Experimental evidence for limited vocal recognition in a wild primate: Implications for the social complexity hypothesis. *Proceedings of the Royal Society B: Biological Sciences* 277: 3045–3053.

Cheney, D. L., et al. 2010. Contingent cooperation between wild female baboons. *Proceedings of the National Academy of Sciences of the United States of America* 107: 9562–9566.

Cheney, D. L. 2011. Extent and limits of cooperation in animals. *Proceedings of the National Academy of Sciences of the United States of America* 108 (supplement 2): 10902–10909.

The pivotal moment in rethinking group selection came with . . .

Williams, G. C. 1966. *Adaptation and natural selection: A critique of some current evolutionary thought.* Princeton, NJ: Princeton University Press.

George Price's equation partitions the variance in offspring produced by individuals among genes and their interactions within and between individuals and groups. Introductions include . . .

Frank, S. A. 1995. George Price's contributions to evolutionary genetics. *Journal of Theoretical Biology* 175: 373–388.

Frank, S. A. 2012. Natural selection. IV. The Price equation. *Journal of Evolutionary Biology* 25: 1002–1019.

Marshall, J. A. R. 2011. Ultimate causes and the evolution of altruism. *Behavioral Ecology and Sociobiology* 65: 503–512.

A new era in thinking about the evolution of cooperation began with Robert Trivers's focus on reciprocity . . .

Trivers, R. L. 1971. The evolution of reciprocal altruism. *Quarterly Review of Biology* 46: 35–57.

Trivers, R. 2006. Reciprocal altruism: 30 years later. In P. M. Kappeler and C. P. Van Schaik, eds., *Cooperation in Primates and Humans: Mechanisms and Evolution,* 67–83. Berlin: Springer-Verlag.

The Prisoner's Dilemma of Game Theory and the tactic of Tit-for-Tat reciprocity became the focus for the evolution of cooperation a decade later . . .

Axelrod, R., and W. D. Hamilton. 1981. The evolution of cooperation. *Science* 211: 1390–1396.

Axelrod, R., and D. Dion. 1988. The further evolution of cooperation. *Science* 242: 1385–1390.

By-product mutualism in cooperative groups is addressed in . . .

Mesterton-Gibbons, M., and L. A. Dugatkin. 1992. Cooperation among unrelated individuals—evolutionary factors. *Quarterly Review of Biology* 67: 267–281.

Grinnell, J., C. Packer, and A. E. Pusey. 1995. Cooperation in male lions: Kinship, reciprocity or mutualism? *Animal Behaviour* 49: 95–105.

Clutton-Brock, T. H. 2002. Breeding together: Kin selection and mutualism in cooperative vertebrates. *Science* 296: 69–72.

Gilby, I., C. 2012. Cooperation among non-kin: reciprocity, markets, and mutualism. In J. C. Mitani et al., eds., 2012, *The evolution of primate societies,* 514–530. Chicago: University of Chicago Press.

It is worth remembering that even in cases such as food-sharing following coordinated hunting (as exemplified by white pelicans, lions, wolves, chimpanzees, and humans) there is the possibility for individuals to exploit each other. Individuals in group hunts might differ in their capabilities, but they might also differ in their inclinations to incur costs in relation to expected rewards. Cheaters might allow others to do a disproportionate amount of the work or to take disproportionate risks. As a result, these forms of apparent by-product mutualism converge with Prisoner's Dilemma or Snowdrift games. For a discussion of such issues, see . . .

Archetti, M., and I. Scheuring. 2011. Coexistence of cooperation and defection in public goods games. *Evolution* 65: 1140–1148.

The possibility of green-beard effects has evoked dozens of articles. Some of the latest tend to confirm to the first arguments that cheating is likely to invade three-gene versions of a green-beard effect (a signal for a tag such as a green beard, a corresponding receptor, and a tendency to behave altruistically) . . .

Dawkins, R. 1976. *The selfish gene.* Oxford: Oxford University Press.
Dawkins, R. 1982. *The extended phenotype.* Oxford: Oxford University Press.
Tanimoto, J. 2007. Does a tag system effectively support emerging cooperation? *Journal of Theoretical Biology* 247: 756–764.
Laird, R. A. 2011. Green-beard effect predicts the evolution of traitorousness in the two-tag prisoner's dilemma. *Journal of Theoretical Biology* 288: 84–91.

Microorganisms form cooperative aggregations that suggest evolution of green-beard effects or by-product mutualism. These organisms (bacteria and protozoans) often express molecules on their surfaces that, in threatening conditions, attach otherwise independent cells together to form a mat or foam of thousands of cells. In bacteria and yeast these aggregates offer advantages to all the constituent cells by resisting penetration by antibacterial substances as well as flotation and resistance to drying. In the case of slime molds (aggregations of a soil-dwelling amoeba), aggregations of cells form when food is scarce and develop a structure consisting of a base, a stiff stalk, and a fruiting body that produces spores borne by the wind to a new location. Some of the cells in an aggregation thus have a chance to reproduce, while others just provide the support. The molecules for attachment bind to each other directly, so the signal and the receptor for recognition are the same molecule. Furthermore, because the attachment itself provides the benefit of aggregation, this molecule seems to produce the cooperative behavior also. It is nevertheless important to consider the possibility that the strength or the conditions of attachment might vary with allelic differences in the molecules. It must be true that cells in different locations in an aggregate, in the spore-producing cap of a slime mold as opposed to the stalk or on the top or bottom or in the center or periphery of a bacterial biofilm, differ in their benefits from aggregation. If so, genes that favor selfish behavior within aggregates might evolve even if they weakened the aggregation overall. The problem of the evolution of cooperation is always contingency in reciprocation or contribution . . .

Queller, D. C., et al. 2003. Single-gene greenbeard effects in the social amoeba *Dictyostelium discoideum. Science* 299: 105–106.

Fiegna, F., and G. J. Velicer. 2003. Competitive fates of bacterial social parasites: Persistence and self-induced extinction of *Myxococcus xanthus* cheaters. *Proceedings of the Royal Society of London B: Biological Sciences* 270: 1527–1534.

Smukalla, S., et al. 2008. Flo1 is a variable green beard gene that drives biofilm-like cooperation in budding yeast. *Cell* 135: 726–737.

Gore, J., H. Youk, and A. Van Oudenarden. 2009. Snowdrift game dynamics and facultative cheating in yeast. *Nature* 459: 253–256.

Zhang, Q. G., et al. 2009. Coevolution between cooperators and cheats in a microbial system. *Evolution* 63: 2248–2256.

Strassmann, J. E., O. M. Gilbert, and D. C. Queller. 2011. Kin discrimination and cooperation in microbes. *Annual Review of Microbiology* 65: 349–367.

Jiricny, N., et al. 2010. Fitness correlates with the extent of cheating in a bacterium. *Journal of Evolutionary Biology* 23: 738–747.

Dumas, Z., and R. Kümmerli. 2012. Cost of cooperation rules selection for cheats in bacterial metapopulations. *Journal of Evolutionary Biology* 25: 473–484.

Possibilities for the evolution of cooperation have proliferated in recent decades. All focus on mechanisms that make interactions between cooperators more likely and those between cooperators and defectors or cheaters less so. Sedentary organisms that interact with only a few neighbors meet this condition. By chance, restricted local neighborhoods can include different numbers of cooperators. If those that cooperate more also produce more progeny that randomly colonize vacancies, co-operation can spread. The exact conditions make a difference to the final result, which can include kaleidoscopic distributions of cooperators and cheaters.

Nowak, M. A., and R. M. May. 1992. Evolutionary games and spatial chaos. *Nature* 246: 15–18.

Doebeli, M., and C. Hauert. 2005. Models of cooperation based on the Prisoner's Dilemma and the Snowdrift game. *Ecology Letters* 8: 748–766.

Ohtsuki, H., et al. 2006. A simple rule for the evolution of cooperation on graphs and social networks. *Nature* 441: 502–505.

Nowak, M. A. 2006. Five rules for the evolution of cooperation. *Science* 314: 1560–1563.

The consequences of spatial restrictions on interactions were anticipated by D. S. Wilson's discussion of structured demes . . .

Wilson, D. S. 1977. Structured demes and evolution of group-advantageous traits. *American Naturalist* 111: 157–185.

Even more attention has centered on possibilities for cooperation when individuals are capable of complex cognition, such as recognizing part-ners with high specificity and multiplicity, keeping scores of other indi-viduals' behavior, and coordinating punishment of cheaters as well as cooperation. There is a bifurcation in the motivations of investigators. Some seek all possible complexities in human cooperation, others the sufficient conditions to explain cooperation in nonhuman animals . . .

Mesterton-Gibbons, M., and L. A. Dugatkin. 1992. Cooperation among unrelated individuals—evolutionary factors. *Quarterly Review of Biology* 67: 267–281.

Nowak, M. A., and K. Sigmund, K. 1998. Evolution of indirect reciprocity by image scoring. *Nature* 393: 573–577.

Nowak, M. A., and K. Sigmund. 2005. Evolution of indirect reciprocity. *Nature* 437: 1291–1298.

Nowak, M. A. 2006. Five rules for the evolution of cooperation. *Science* 314: 1560–1563.

In the ramifying literature on the evolution of cooperation, noise is seldom considered. Random mistakes tend to inhibit the evolution of cooperation, but in the right conditions can promote leniency in co-operation in order to avoid losing opportunities for future benefits . . .

Wu, J. Z., and R. Axelrod. 1995. How to cope with noise in the iterated prisoners-dilemma. *Journal of Conflict Resolution* 39: 183–189.

Vukov, J., G. Szabo, and A. Szolnoki. 2006. Cooperation in the noisy case: Prisoner's dilemma game on two types of regular random graphs. *Physical Review E* 73; doi: 10.1103/PhysRevE.73.067103.

Delton, A. W., et al. 2011. Evolution of direct reciprocity under uncertainty can explain human generosity in one-shot encounters. *Proceedings of the National Academy of Sciences of the United States of America* 108: 13335–13340. The errors investigated here are signal-detection errors as a result of mistakes in judging whether a new partner will be encountered repeatedly or just once.

Kin selection as an explanation for the evolution of cooperation has its own multitudinous literature. W. D. Hamilton's papers, including his original presentations of kin selection in 1964, have been collected . . .

Hamilton, W. D. 1998. *The narrow roads of gene land: The collected papers of W. D. Hamilton*, vol. 1, *Evolution of social behaviour*. Oxford: Oxford University Press.

The anecdote about J. B. S. Haldane, one of the great contributors to our modern understanding of genetics and evolution, was apparently related by John Maynard Smith, one of Haldane's students, during an interview about Maynard Smith's own career . . .

Lewis, R. 1974. Accidental career. *New Scientist* 63: 322–325. The anecdote is on page 325. The episode, a quip in a pub, presumably occurred before Haldane moved to India in 1954 (he died in 1964, the year of Hamilton's publication), but after Maynard Smith arrived in Haldane's lab in London in 1951—and thus possibly as much as a decade before Hamilton had worked out his own theory. Yet it is possible that he made the quip after hearing of Hamilton's result.

Kin selection reached a wide audience as a result of . . .

Wilson, E. O. 1975. *Sociobiology: The new synthesis*. Cambridge, MA: Harvard University Press.

Trivers, R. L. 1974. Parent-offspring conflict. *American Zoologist* 14: 249–264.

Trivers, R. L., and H. Hare. 1976. Haplodiploidy and evolution of social insects. *Science* 191: 249–263.

A recent challenge to kin selection as an explanation for sociality in insects elicited a spate of controversy. One issue was the use of inclusive fitness as a measure of kin selection . . .

Nowak, M. A., C. E. Tarnita, and E. O. Wilson. 2010. The evolution of eusociality. *Nature* 466: 1057–1062. See also the five rebuttals, plus a reply by Nowak and colleagues, published by *Nature* online; doi: 10.1038/nature09831 through doi: 10.1038/nature09836.

Bourke, A. F. G. 2011. The validity and value of inclusive fitness theory. *Proceedings of the Royal Society B: Biological Sciences* 278: 3313–3320.

Note that my application of kin selection to cooperative breeding makes no mention of inclusive fitness, which like other measurements of the "fitness" of phenotypes creates problems (see Chapter 7). Instead, Hamilton's rule is adapted in this chapter and the next to compare the expected numbers of individuals at the start of the next cohort with alleles associated with two alternatives, either helping or not. The application of kin selection to Stripe-backed Wrens in Chapter 15 has followed the same procedure (see references there).

For excellent introductions to cooperative breeding in birds . . .

Koenig, W. B., and P. B. Stacey, eds. 1990. *Cooperative breeding in birds: Long-term studies of ecology and behavior.* Cambridge: Cambridge University Press.

Koenig, W. B., and J. L. Dickinson, eds. 2004. *Ecology and evolution of cooperative breeding in birds.* Cambridge: Cambridge University Press.

For discussions about cooperative hunting by mammalian carnivores . . .

MacNulty, D. R., et al. 2012. Nonlinear effects of group size on the success of wolves hunting elk. *Behavioral Ecology* 23: 75–82.

Sand, H., et al. 2006. Effects of hunting group size, snow depth and age on the success of wolves hunting moose. *Animal Behaviour* 72: 781–789.

Packer, C., D. Scheel, and A. E. Pusey. 1990. Why lions form groups: Food is not enough. *American Naturalist* 136: 1–19.

Scheel, D., and Packer, C. 1991. Group hunting behavior of lions—a search for cooperation. *Animal Behaviour* 41: 697–709.

Creel, S., and N. M. Creel. 1995. Communal hunting and pack size in African Wild Dogs, *Lycaon pictus. Animal Behaviour* 50: 1325–1339.

E. O. Wilson's criticism of kin selection is featured in his recent book . . .

Wilson, E. O. 2013. *The social conquest of earth.* New York: W. W. Norton.

References for superorganisms are presented in the notes for the next chapter.

15. COMPLEX SOCIETIES

For the investigation of Stripe-backed Wrens in Venezuela . . .

Rabenold, K. N. 1990. Campylorhynchus wrens: The ecology of delayed dispersal and cooperation in the Venezuelan savanna. In W. B. Koenig and P. B. Stacey, eds., *Cooperative breeding in birds: Long-term studies of ecology and behavior,* 157–196. Cambridge: Cambridge University Press.

Rabenold, K. N. 1984. Cooperative enhancement of reproductive success in tropical wren societies. *Ecology* 65: 871–885. This and the preceding reference provide information about the number of young produced by groups of different sizes in an intensively studied population.

Rabenold, K. N. 1985. Cooperation in breeding by nonreproductive wrens: Kinship, reciprocity, and demography. *Behavioral Ecology and Sociobiology* 17: 1–17.

Zack, S., and K. Rabenold. 1989. Assessment, age and proximity in dispersal contests among cooperative wrens: Field experiments. *Animal Behaviour* 38: 235–247.

Rabenold, P. P., et al. 1990. Shared paternity revealed by genetic analysis in cooperatively breeding tropical wrens. *Nature* 348: 538–540.

Piper, W. H., and P. P. Rabenold. 1992. Use of fragment-sharing estimates from DNA fingerprinting to determine relatedness in a tropical wren. *Molecular Ecology* 1: 69–78.

Piper, W. H., and G. Slater. 1993. Polyandry and incest avoidance in the cooperative Stripe-backed Wren of Venezuela. *Behaviour* 124: 227–247.

Piper, W. H. 1994. Courtship, copulation, nesting behavior and brood parasitism in the Venezuelan Stripe-backed Wren. *Condor* 96: 654–671.

Piper, W. H., P. G. Parker, and K. N. Rabenold. 1995. Facultative dispersal by juvenile males in the cooperative Stripe-backed Wren. *Behavioral Ecology* 6: 337–342.

Stevens, E. 1995. Genetic consequences of restricted dispersal and incest avoidance in a cooperatively breeding wren. *Journal of Theoretical Biology* 175: 423–436.

Yaber, M. C., and K. N. Rabenold. 2002. Effects of sociality on short-distance, female-biased dispersal in tropical wrens. *Journal of Ecology* 71: 1042–1055. This reference includes an analysis of pooled observations of the number of young produced by groups of different sizes.

For the general finding that kin selection does not provide a sufficient explanation for cooperative breeding, see the books edited by Walter Koenig and Peter Stacey, and by Koenig and Janis Dickinson, cited in Chapter 15; see also . . .

> Clutton-Brock, T. H. 2002. Breeding together: Kin selection and mutualism in cooperative vertebrates. *Science* 296: 69–72.

Although not a sufficient explanation, kin selection still occurs in nearly all species with cooperative breeding because individuals, with few exceptions, help relatives. Even if helping has other benefits for individuals, kin selection explains why helpers preferentially join groups with relatives. Furthermore, even if kin selection in groups of relatives does not provide a sufficient increment in the spread of alleles for helping, it provides some increment. The consequence is that requirements for other advantages of helping are thereby relaxed. The fundamental requisite for the evolution of helping is that alleles associated with helping must spread faster in a population than alleles associated with not helping (selfishness). A combination of weak contributions to the spread of alleles for helping can add up to enough for the evolution of helping.

For investigation of queuing for favorable opportunities for breeding . . .

> Wiley, R. H., and K. N. Rabenold. 1984. The evolution of cooperative breeding by delayed reciprocity and queuing for favorable social positions. *Evolution* 38: 609–621.
>
> Kokko, H., and R. A. Johnstone. 1999. Social queuing in animal societies: A dynamic model of reproductive skew. *Proceedings of the Royal Society of London. Series B: Biological Sciences* 266: 571.
>
> Johnstone, R. A., and M. A. Cant. 1999. Reproductive skew and the threat of eviction: A new perspective. *Proceedings of the Royal Society of London. Series B: Biological Sciences* 266: 275.

Reports on the Bicolored Wren's less pronounced cooperation in breeding . . .

> Austad, S. N., and K. N. Rabenold. 1985. Reproductive enhancement by helpers and an experimental inquiry into its mechanism in the Bicolored Wren. *Behavioral Ecology and Sociobiology* 17: 19–27.
>
> Haydock, J., P. G. Parker, and K. N. Rabenold. 1996. Extra-pair paternity uncommon in the cooperatively breeding Bicolored Wren. *Behavioral Ecology and Sociobiology* 38: 1–16.

For investigations of the vocalizations of Stripe-backed Wrens, including the extraordinary intragroup calls . . .

Wiley, R. H., and M. S. Wiley. 1977. Recognition of neighbors' duets by Stripe-backed Wrens *Campylorhynchus nuchalis*. *Behaviour* 62: 10–34.

Price, J. J. 1998. Family- and sex-specific vocal traditions in a cooperatively breeding songbird. *Proceedings of the Royal Society B: Biological Sciences* 265: 497–502.

Price, J. J. 1999. Recognition of family-specific calls in Stripe-backed Wrens. *Animal Behaviour* 57: 483–492.

For cooperation with inequality . . .

Grinnell, J., C. Packer, and A. E. Pusey. 1995. Cooperation in male lions: Kinship, reciprocity or mutualism? *Animal Behaviour* 49: 95–105.

Tit-for-Tat reciprocity in territorial warblers has been documented in . . .

Godard, R. 1993. Tit for Tat among neighboring Hooded Warblers. *Behavioral Ecology and Sociobiology* 33: 45–50.

For social insects as superorganisms . . .

Wilson, D. S., and E. Sober. 1989. Reviving the superorganism. *Journal of Theoretical Biology* 136: 337–356.

Seeley, T. D. 1989. The honey bee colony as a superorganism. *American Scientist* 77: 546–553.

Wilson, D. S., and E. O. Wilson. 2007. Rethinking the theoretical foundation of sociobiology. *Quarterly Review of Biology* 82: 327–348.

Hölldobler, B., and E. O. Wilson. 2009. *The superorganism: The beauty, elegance, and strangeness of insect societies.* New York: Norton.

Theory behind the evolution of superorganisms has now been considered by a number of investigators following John Maynard Smith and Eörs Szathmáry's masterful book . . .

Maynard Smith, J., and E. Szathmáry. 1995. *The major transitions in evolution.* Oxford: Oxford University Press.

Reeve, H. K., and B. Hölldobler. 2007. The emergence of a superorganism through intergroup competition. *Proceedings of the National Academy of Sciences of the United States of America* 104: 9736.

Queller, D. C., and J. E. Strassmann. 2009. Beyond society: The evolution of organismality. *Philosophical Transactions of the Royal Society B: Biological Sciences* 364: 3143.

Nowak, M. A., C. E. Tarnita, and T. Antal. 2010. Evolutionary dynamics in structured populations. *Philosophical Transactions of the Royal Society B: Biological Sciences* 365: 19.

Bourke, A. F. G. 2011. *Principles of social evolution.* Oxford: Oxford University Press.

A possible case of genetic modifications accompanying the evolution of cooperation in vertebrates is the skewed sex ratio of the cooperative Seychelles Warbler . . .

Komdeur, J., et al. 1997. Extreme adaptive modification in sex ratio of the Seychelles Warbler's eggs. *Nature* 385: 522–525.

Komdeur, J., M. J. L. Magrath, and S. Krackow. 2002. Pre-ovulation control of hatchling sex ratio in the Seychelles Warbler. *Proceedings of the Royal Society of London. Series B: Biological Sciences* 269: 1067–1072.

Packs of African Wild Dogs are also on the cusp of becoming superorganisms in the same sense that Stripe-backed Wrens are. Pairs and small packs have low chance of reproducing, and the sex ratio at birth is skewed toward males to promote cooperation in packs . . .

Malcolm, J. R., and K. Marten. 1982. Natural selection and the communal rearing of pups in African wild dogs, *Lycaon pictus. Behavioral Ecology and Sociobiology* 10: 1–13.

Courchamp, F., and D. W. Macdonald. 2001. Crucial importance of pack size in the African Wild Dog *Lycaon pictus. Animal Conservation* 4: 169–174.

Not discussed in this chapter is the evolution of reproductive skew (the uneven distribution of reproduction) in social groups. The Stripe-backed Wrens have a high degree of reproductive skew (for the most part only two individuals reproduce in each group). Theory indicates that queuing for advantageous social positions has a strong influence on the evolution of such extreme skew . . .

Kokko, H., and R. A. Johnstone. 1999. Social queuing in animal societies: A dynamic model of reproductive skew. *Proceedings of the Royal Society of London. Series B: Biological Sciences* 266: 571–578.

Ragsdale, J. E. 1999. Reproductive skew theory extended: the effect of resource inheritance on social organization. *Evolutionary Ecology Research* 1: 859–874.

Alberts, S. C., H. E. Watts, and J. Altmann. 2003. Queuing and queue-jumping: Long-term patterns of reproductive skew in male savannah baboons, *Papio cynocephalus. Animal Behaviour* 65: 821–840.

Hager, R., and C. B. Jones, eds. 2009. *Reproductive skew in vertebrates: Proximate and ultimate causes.* Cambridge: Cambridge University Press. Chapters present reviews of theory and observation.

16. MOLECULAR SIGNALS

Although the topics in this chapter are outside my usual scope, it is clear that principles of signal detection have hardly appeared in the literature on molecular biology. The basics of immunology are presented in any up-to-date textbook on medicine or physiology. This is one field that has recognized the relevance of signal detection for molecular signaling. Recent reviews that consider the specificity and errors of immune responses include . . .

Fu, G., et al. 2013. Themis sets the signal threshold for positive and negative selection in T-cell development. *Nature* 504: 441–445.

Råberg, L., D. Sim, and A. F. Read. 2007. Disentangling genetic variation for resistance and tolerance to infectious diseases in animals. *Science* 318: 812–814.

Hogquist, K. A., T. A. Baldwin, and S. C. Jameson. 2005. Central tolerance: Learning self-control in the thymus. *Nature Reviews Immunology* 5: 772–782.

Tseng, S. Y., and M. L. Dustin. 2002. T-cell activation: A multidimensional signaling network. *Current Opinion in Cell Biology* 14: 575–580.

The issue of what constitutes a "signal" for the immune system has provoked debate for the past decade or two . . .

Greenspan, N. S. 2001. Dimensions of antigen recognition of levels of immunological specificity. *Cancer Research* 80: 147–187.

Matzinger, P. 2007. Friendly and dangerous signals: Is the tissue in control? *Nature Immunology* 8: 11–13.

Raz, E. 2007. Organ-specific regulation on innate immunity. *Nature Immunology* 8: 3–4.

A discussion of "sticky" enzymes . . .

Brettner, L. A., and J. Masel. 2012. Protein stickiness, rather than number of functional protein-protein interactions, predicts expression noise and plasticity in yeast. *BMC Systems Biology* 6: 128; doi:10.1186/1752-0509-6 -128. http://www.biomedcentral.com/1752-0509/6/128.

Specificity of interactions in networks of proteins . . .

Nam, H., et al. 2012. Network context and selection in the evolution to enzyme specificity. *Science* 337: 1101–1104.

Deceptive signals from cancer cells . . .

Hendrix, M. J. C., et al. 2003. Vasculogenic mimicry and tumour-cell plasticity: Lessons from melanoma. *Nature Reviews Cancer* 3: 411–421.

And from viruses . . .

Oldstone, M. B. A. 1998. Molecular mimicry and immune-mediated diseases. *FASEB Journal* 12: 1255–1265.

Zhao, Z.-S., et al. 1998. Molecular mimicry by herpes simplex virus-type 1: Autoimmune disease after viral infection. *Science* 279: 1344–1347.

Wucherpfennig, K. W., and J. L. Strominger. 1995. Molecular mimicry in T cell–mediated autoimmunity: Viral peptides activate human T cell clones specific for myelin basic protein. *Cell* 80: 695–705.

G proteins, a large class of receptor molecules, are embedded cell membranes, with receptor domains on the exterior surfaces of cells and response domains inside cells. Recent reports show that associated proteins modify the sensitivity or selectivity of their responses . . .

Siekhaus, D. E., and D. G. Drubin. 2003. Spontaneous receptor-independent heterotrimeric G-protein signalling in an RGS mutant. *Nature Cell Biology* 5: 231–235.

Hao, N., et al. 2007. Systems biology analysis of G protein and MAP kinase signaling in yeast. *Oncogene* 26: 3254–3266.

Getsy-Palmer, D., and L. M. Luttrell. 2011. Refining efficacy: Exploiting functional selectivity for drug discovery. *Advances in Pharmacology* 62: 79–107.

Dixit, G., et al. 2014. Cellular noise suppression by the regulator of G protein signaling Sst2. *Molecular Cell* 55: 85–96.

Pharmacologists seeking new drugs are understandably interested in false alarms (side effects) as well as missed detections (ineffective responses). Drugs are often unnatural molecules that interact with enzymes that normally respond to other signals . . .

Egeblad, L., et al. 2012. Pan-pathway based interaction profiling of FDA-approved nucleoside and nucelobase analogs with enzymes of the human

nucleotide metabolism. *PLoS One* 7: e37724; doi: 10.1371/journal
.pone.0037724.

Welin, M., and P. Nordlund. 2010. Understanding specificity in metabolic
pathways—structural biology of human nucleotide metabolism.
Biochemical and Biophysical Research Communications 396: 157–163.

17. HUMAN COMMUNICATION

Comparisons of human gestures and facial expressions across cultures
and species began with Charles Darwin. Although nearly all studies report some cross-cultural similarities in expression, some focus on the
differences instead. Like language, expressions no doubt result from an
interaction of genes and environment. Human expression has greater canalization, while language has greater plasticity. In both cases investigators have tended to focus attention on the less obvious effect, cultural
universals (canalization) in language and cultural specificity (plasticity)
in expressions.

Darwin, C. 1872. *The expression of the emotions in man and animals.* London:
John Murray. Reprint, Chicago: University of Chicago Press, 1965.

Ekman, P., ed. 1973. *Darwin and facial expression.* New York: Academic
Press.

Biehl, M., et al. 1997. Matsumoto and Ekman's Japanese and Caucasian facial
expressions of emotion (JACFEE): Reliability data and cross-national
differences. *Journal of Nonverbal Behavior* 21: 3–21.

Elfenbein, H. A., et al. 2007. Toward a dialect theory: Cultural differences in
the expression and recognition of posed facial expressions. *Emotion* 7:
131–146.

Jack, R. E., et al. 2012. Facial expressions of emotion are not culturally
universal. *Proceedings of the National Academy of Sciences of the United
States of America* 109: 7241–7244. A report that poses nature and nurture
as a dichotomy rather than an interaction. See also the exchange of letters
following this article.

Engelmann, J. B., and M. Pogosyan. 2013. Emotion perception across cultures:
The role of cognitive mechanisms. *Frontiers in Psychology* 4: 118; doi:
10.3389/fpsyg.2013.00118.

Self-deception as a prerequisite for deception is argued in . . .

Trivers, R. 2014. *The folly of fools: The logic of deceit and self-deception in
human life.* New York: Basic Books.

The relationship between language and the external world has had the attention of philosophers for millennia. The modern view of external referents for words or propositions derives from . . .

Saussure, F. de. 1959. *Course in general linguistics.* Translated by W. Baskin. New York: Philosophical Library. Originally published as *Cours de linguistique generale,* 1916.

Ogden, C. K., and I. A. Richards. 1923. *The meaning of meaning: A study of the influence of language upon thought.* New York: Harcourt Brace.

Peirce, C. S. 1955. *Philosophical writings of Peirce.* Edited by J. Buchler. New York: Dover. A contemporary of Saussure, Peirce also emphasized three elements: the object, the sign, and the interpretant or thought that results.

René Descartes first proposed his equivalence of thinking and personal existence in French in 1637. The equivalent proposition in Latin came seven years later . . .

Descartes, R. 1980. *Discourse on the method of rightly conducting one's reason and of seeking truth in the sciences.* Translated by D. A. Cress. Indianapolis: Hackett. Originally published as *Discours de la méthode . . . ,* 1637.

Descartes, R. 2008. *Principles of philosophy.* Translated by J. Veitch. New York: Barnes & Noble. Originally published as *Principia philosophiae,* 1644.

Following Descartes's treatment of optics, in which he first recognized the correct spherical angle of a rainbow, and then Newton's experiments with prisms, which decomposed white light into a rainbow, the modern realization that human perception of light did not entirely correlate with its physical properties came with Goethe and especially Helmholtz . . .

Goethe, J. W. von. 1971. *Goethe's color theory.* Translated by H. Aach. New York: Van Nostrand Reinhold. Originally published as *Farbenlehre,* 1840.

Helmholtz, H. von. 1962. *Helmholtz's treatise on physiological optics.* Translated by J. P. C. Southall. New York: Dover Publications. Originally published as *Handbuch der physiologischen Optik,* 1867.

My position in the present volume is that brain and behavior are associated, but with some noise. In a previously published essay, I accepted a necessary uncertainty about any conclusion that a mind is completely predicted by its associated brain . . .

Wiley, R. H. 2013. Communication as a transfer of information: Measurement, mechanism and meaning. In U. Stegmann, ed., *Animal signals and communication: Information and influence,* 113–129. Cambridge: Cambridge University Press.

Ludwig Wittgenstein discusses rules and family resemblances in language and acknowledges that the latter preclude logic . . .

Wittgenstein, L. 1968. *Philosophical investigations,* 3rd ed. Translated by G. E. M. Anscombe. Oxford: Blackwell. Especially pertinent paragraphs include 66–67, 108, and 201–207.

Niko Tinbergen's experiments on the behavior of diverse animals led to his concept of releasing mechanisms, neural mechanisms that respond to a selection of the features of an animal's overall stimulation . . .

Tinbergen, N. 1951. *The study of instinct.* Oxford: Oxford University Press.
Tinbergen, N. 1953. *The Herring Gull's world: A study of the social behaviour of birds.* London: Collins.

The importance of innate constraints on learning has been argued by ethologists and linguists . . .

Chomsky, N. 1959. Verbal behavior by B. F. Skinner. *Language* 35: 26–58. Pages 41–44 (section 5) of this paper specifically address predispositions for learning in general and for learning language in particular. His book below expands this argument.
Lorenz, K. 1965. Evolution and modification of behavior. Chicago: University of Chicago Press.
Chomsky, N. 1966. *Cartesian linguistics.* New York: Harper & Row.

A "critical period" in learning language is a contentious issue, despite routine observation that late exposure to a second language results in a persistent "accent." Some of the discussion seems to hinge on the term *critical period*, which suggests rigid boundaries for learning. Students of birdsong have by now rejected this term in favor of the term *sensitive period* because the limitations on learning are susceptible to modification by experience. Learning of birdsong, and presumably language, are further examples of developmental interactions between innate (relatively canalized) and experiential (relatively plastic) mechanisms. It also seems clear that features of language, such as grammar, vocabulary, and pronunciation, differ in limitations to sensitive periods. A persistent "accent" after late learning of a second language pertains most obviously to pronunciation as a result of the difficult acquisition of new phonemes by adults. Casual observation also suggests that sensitive periods for learning language vary among individuals, at least to a degree. Furthermore, there are several possible neural mechanisms that might explain a developmental sensitive period, and these might differ for the various features of language. All of these issues need more investigation. A variety of views is presented in . . .

Birdsong, D. 1999. *Second language acquisition and the critical period hypothesis.* London: Routledge.

Birds learning their songs are subject to predispositions, sensitive periods, and social constraints. One great advantage exploited by these studies is the variety of species of songbirds in the suborder Passeri, all of which learn at least some features of their songs but differ in the interactions between innate and plastic mechanisms that influence this learning . . .

Marler, P. 1970. Birdsong and speech development: Could there be parallels? *American Scientist* 58: 669–673.

Marler, P., and H. W. Slabbekoorn, eds. 2004. *Nature's music: The science of birdsong.* San Diego: Elsevier Academic.

Kroodsma, D. E. 1979. Aspects of learning in the ontogeny of bird song: Where, from who, when, how many, which, and how accurately? In B. Burghardt and M. Bekoff, eds., *Ontogeny of Behavior*, 215–230. New York: Garland.

For environmental influences on the sensitive period for song learning in birds . . .

Kroodsma, D. E., and R. Pickert. 1980. Environmentally dependent sensitive periods for avian vocal learning. *Nature* 288: 477–479.

Marler, P., and S. Peters. 1988. Sensitive periods for song acquisition from tape recordings and live tutors in the Swamp Sparrow, *Melospiza georgiana. Ethology* 77: 76–84.

For predispositions in learning by sparrows . . .

Marler, P., and S. Peters. 1977. Selective vocal learning in a sparrow. *Science* 198: 519–521.

Marler, P., and S. Peters. 1988. The role of song phonology and syntax in vocal learning preferences in the Song Sparrow, *Melospiza melodia. Ethology* 77: 125–149.

Soha, J. A., and P. Marler. 2000. A species-specific acoustic cue for selective song learning in the White-crowned Sparrow. *Animal Behaviour* 60: 297–306.

Soha, J. A., and P. Marler. 2001. Cues for early discrimination of conspecific song in the White-crowned Sparrow *(Zonotrichia leucophrys). Ethology* 107: 813–826.

Young birds exposed to recordings of both degraded and undegraded songs selectively learn the undegraded versions. They reconstruct the songs as they are normally produced without the added noise . . .

Morton, E. S., S. L. Gish, and M. Van Der Voort. 1986. On the learning of degraded and undegraded songs in the Carolina Wren. *Animal Behaviour* 34: 815–820.

Nowicki, S., et al. 1992. Is the tonal quality of birdsong learned? Evidence from Song Sparrows. *Ethology* 90: 225–235.

Peters, S., E. P. Derryberry, and S. Nowicki. 2012. Songbirds learn songs least degraded by environmental transmission. *Biology Letters* 8: 736–739.

Interactions with rivals can influence which songs are learned . . .

Payne, R. B. 1981. Song learning and social interaction in Indigo Buntings. *Animal Behaviour* 29: 688–697.

Kroodsma, D. E., and R. Pickert. 1984. Sensitive phases for song learning: Effects of social interaction and individual variation. *Animal Behaviour* 32: 389–394.

In some species learning is restricted to specific social partners . . .

Hile, A. G., T. K. Plummer, and G. F. Striedter. 2000. Male vocal imitation produces call convergence during pair bonding in Budgerigars, *Melopsittacus undulatus. Animal Behaviour* 59: 1209–1218.

Berg, K. S., et al. 2012. Vertical transmission of learned signatures in a wild parrot. *Proceedings of the Royal Society B: Biological Sciences* 279: 585–591.

Mundinger, P. C. 1970. Vocal imitation and individual recognition of finch calls. *Science* 168: 480–482.

18. TRUTH IN LANGUAGE

Many of the philosophical issues addressed in this and the following chapters are so fundamental that they have received the attention of numerous thinkers. In many cases it is futile to try to cite one or a few sources. Furthermore, discussion of these issues has almost never considered the complications added by the inevitability of noisy communication or noisy perception. Therefore my own comments do not address specific points of previous authors. Instead my suggestion is that discussion must be reopened from the start to account for inescapable noise.

For more on Wittgenstein, see the notes for Chapters 17 and 19.

For the problems presented by radical translation . . .

Quine, W. V. 1960. *Word and object.* Cambridge, MA: MIT Press.

Davidson, D. 1984. *Inquiries into truth and interpretation.* Oxford: Oxford University Press. Expanded edition, 2001.

Dennett's two stages of consciousness . . .

Dennett, D. C. 1997. *Kinds of minds: Toward an understanding of consciousness.* New York: Basic Books.

For empathy as an alternative to observing behavior . . .

Wiley, R. H. 2013. Communication as a transfer of information: Measurement, mechanism and meaning. In U. Stegmann, ed., *Animal signals and communication: Information and influence,* 113–129. Cambridge: Cambridge University Press.

For the use of rising pitch at the ends of statements, as well as questions, see . . .

Ching, M. 1982. The question intonation in assertions. *American Speech* 57: 95–107.

McLemore, C. A. 1991. The pragmatic interpretation of English intonation: Sorority speech. Ph.D. diss., University of Texas–Austin. *Dissertation Abstracts International A: The Humanities and Social Sciences* 52(4): 1311-A.

Warren, P. 2005. Patterns of late rising in New Zealand English: Intonational variation or intonation change? *Language Variation and Change* 17: 209–230.

Fletcher, J., E. Grabe, and Warren, P. 2005. Intonational variation in four dialects of English: The high rising tone. In S.-A. Jun, ed., *Prosodic typology: The phonology of intonation and phrasing,* 390–409. Oxford: Oxford University Press.

In claiming that the inevitability of noise has never been addressed, I do so advisedly. Noise is in fact widely discussed these days, but not the consequences of the *inevitability* of noise. The sociological theorist Niklas Luhmann, for instance, in his application of systems theory to human societies, contends that social systems spontaneously form in response to noise. In his more recent books, however, noise is only infrequently mentioned. The systems theorist who really considers noise is Norbert Wiener in his classic *Cybernetics*. He showed mathematically that feedback counteracts the effects of noise and thus improves the performance of systems. For Wiener, feedback allows a system to search for stability or for information in noise. Rather than a necessary condition for the spontaneous generation of systems, noise is a sufficient condition for the utility of feedback. In the search for information in noise, the contribution of feedback is related to that of redundancy. Luhmann emphasizes that feedback is a form of communication but does not indicate its relationship to redundancy. As Chapter 6 describes, redundant interactions between signaler and receiver can reduce the receiver's errors—that is, noise. Chapter 19 suggests that such communication reduces errors in our use of language and ultimately produces a degree of consilience among human minds. Like feedback, redundant communication thus does not require noise (the possibility of error), but it only serves a purpose in the presence of noise. If the thesis of the present volume is correct, noise is universal. Consequently, so is the utility of feedback and redundancy.

Luhmann, N. 1995. *Social systems*. Translated by J. Bednarz Jr. and D. Baecker. Palo Alto, CA: Stanford University Press. Originally published as *Soziale Systeme: Grundriß einer allgemeinen Theorie*, 1984.

Wiener, N. 1948. *Cybernetics*. New York: Wiley.

The philosopher Robert Kane connects noise to free will, but his brief argument has little relation to the one made in this book. He contends that noise in nervous systems allows us to interpret another person's actions as a product of that person's will. It seems to me, though, that this simple consequence of noise would result in just the opposite conclusion. A person using a noisy nervous system to wrestle with a moral dilemma would eventually act unpredictably. It might not even appear to a woman (Kane's example) that she had willed her decision. In my approach, noise contributes to uncertainty in responses or perceptions, but the emphasis on decision making leaves receivers (at least those capable of higher-order cognition) in control of their own criteria for actions and perceptions. A thinking receiver could adjust the probabilities of false alarms and missed detections, although never entirely eliminate both. An outside observer would see either patterns or unpredictability in another's behavior depending on the observer's own disposition as a receiver of signals . . .

Kane, R. 2002. Free will: New directions for an ancient problem. In R. Kane, ed., *Free will,* 222–248. Oxford: Blackwell.

Kane, R. 2011. Some neglected pathways in the free will labyrinth. In R. Kane, ed., *Oxford handbook of free will,* 406–439. Oxford: Oxford University Press. Repeats the 2002 argument almost verbatim.

For a different approach to free will, try . . .

Wiley, K. 2014. *A taxonomy and metaphysics of mind-uploading.* Los Angeles: Humanity+ Press and Alautun Press.

19. SUBJECTIVITY

Turing's argument about machines and consciousness . . .

Turing, A. 1950. Computing machinery and intelligence. *Mind* 59: 433–460.

Searle, J. R. 1980. Minds, brains, and programs. *Behavioral and Brain Sciences* 3: 417–424.

Searle, J. R. 1997. *The mystery of consciousness.* New York: New York Review of Books.

Searle's Chinese Room rephrases Turing's test to emphasize a point. Turing imagined a judge who could communicate by means of a keyboard with two players—one a human and the other a machine—in order to decide if possible which is which. Searle condenses the situation to a judge in communication with only one player. The judge and the player understand different languages—for instance, Chinese and English, respectively. The judge submits written questions or comments to the player in Chinese. The player then uses a book of rules to transpose each of these sequences into a response in a way that mimics a person conversing in Chinese. As Searle argues, the judge would not be able to distinguish an uncomprehending player with a book of rules from a machine that uses the same rules. Consequently, Turing's test cannot determine whether or not a machine can understand a human language and, therefore, Searle extrapolates, whether or not it is conscious.

Notice that Searle's proposed example of successful rote translation also shows that a judge would not be able to determine whether or not the player following rules understands Chinese. If understanding a language can be mimicked by rote performance, it is not surprising that communication provides no way to distinguish between the two. Furthermore, comprehension in this case seems to require no more than internalizing the rules.

To my mind, Turing's fundamental question is a different one: Does human language follow rules? If it does follow rules, no matter how complex, then an adequately programmed computer could mimic a comprehending human just as well as an uncomprehending one. A judge would have no way to distinguish uncomprehending from comprehending humans nor, we should also acknowledge, any way to distinguish uncomprehending from comprehending machines!

The challenge is specifying the rules. The difficulty increases if the rules can have some unpredictable variation. Machines and humans (including programmers and players) make mistakes in any way that any receiver or signaler can. Programs or books of rules can also incorporate random variation by design. Another complication surfaces when we realize that programs can include enumerations of possibilities in addition to abstract rules. Language might also make use of pat phrases. Linguists sometimes claim that the combinatorial feature of human language permits humans to understand infinite phrases, but brains are finite and there is a critical difference between a large number and infinity. Judging whether or not brains or computers follow rules will not be easy.

Perhaps in the future a robot can be programmed to settle the question. Computers can already play chess and manage simple conversations in phone trees and might soon play soccer like a human. Alternatively, linguists might identify all of the rules and routine phrases (and their random variation) for the natural use of language by a specified set of people. In the meantime, we should recognize, as I have emphasized throughout this book, that noise in perception and in communication limits the precision of all parties in Turing's and Searle's tests—the judge and the players, both human and machine. Some precision in determining the correspondence of different players is no doubt possible, but perfection is not.

Another possibility, perhaps related to Searle's concern with consciousness, is that a computer with sufficiently high levels of abstraction, compartmentalization, and covert internal communication might also develop a sense of subjectivity like a person's. Because this chapter argues that subjectivity is a consequence of unavoidable noise in perception and communication, combined with higher-levels of internal functions, it seems possible that a computer of sufficient complexity might also experience subjectivity. I leave this possibility open. Aside from questions about machines, there are those about brains of animals other than humans.

Dolphins, with brains superficially like humans, are capable of some remarkable language-like skills. For instance, they quickly understand associations of arbitrary signals with objects and also simple syntax based on the order of words, including the distinction all human languages make between the subject and predicate of a statement . . .

Richards, D. G., J. P. Wolz, and L. M. Herman. 1984. Vocal mimicry of computer-generated sounds and vocal labeling of objects by a Bottle-nosed Dolphin, *Tursiops truncatus. Journal of Comparative Psychology* 98: 10–28.

Herman, L. M., D. G. Richards, and J. P. Wolz. 1984. Comprehension of sentences by Bottlenosed Dolphins. *Cognition* 16: 129–219.

Richards, D. G. 1986. Dolphin vocal mimicry and vocal object labeling. In R. J. Schusterman, J. A. Thomas, and F. G. Wood, eds., *Dolphin cognition and behavior,* 273–288. Hillsdale, NJ: Erlbaum.

The discussion of other minds is picked up in Chapter 20. The discussion of empathy is based on my previous thoughts . . .

Wiley, R. H. 2013. Communication as a transfer of information: Measurement, mechanism and meaning. In U. Stegmann, ed., *Animal signals and communication: Information and influence,* 113–129. Cambridge: Cambridge University Press.

The summary of neural connections and development, including David Hubel and Thorsten Wiesel's classic experiments on vision, can be found in many textbooks of comparative physiology or neurobiology.

20. VERIFICATION

For information about tool use by Common Chimpanzees and other animals . . .

Goodall, J. 1986. *The chimpanzees of Gombe: Patterns of behavior.* Cambridge, MA: Harvard University Press.

Sanz, C., J. Call, and C. Boesch, eds. 2014. *Tool use in animals: Cognition and ecology.* Cambridge: Cambridge University Press.

Wittgenstein famously discusses the possibility of private languages in . . .

Wittgenstein, L. 1968. *Philosophical investigations.* Translated by G. E. M. Anscombe. Oxford: Blackwell. Originally published 1953. His discussion in paragraphs 377–381 comes close to my arguments in this chapter. To summarize his insight that private words are inherently unstable without some consilience achieved by communication, he writes, "How do I know that this colour is red?—It would be an answer to say: 'I have learnt English'" (paragraph 381).

For explications of Wittgenstein's thought, with minimal contentiousness . . .

Fogelin, R. J. 1976. *Wittgenstein.* London: Routledge and Kegan Paul.

Garver, N. 1994. *This Complicated form of life: Essays on Wittgenstein.* Peru, IL: Open Court.

There is more to say about private language than hinted in this chapter and Chapter 19. First is the issue of private perceptions. Anyone's pain (or any sensation) is private. This situation might change in the future. If it became possible to specify the state of every neuron in a person's brain, one could in principle know the state of a person's brain when that person was exposed to a stimulus. Yet to confirm that person's perception (pain or other sensation) would require asking the person to tell us. So we would also have to predict this higher-order association of the stimulus. If the state of every neuron in a person's brain could be specified, this prediction also seems possible. Two problems arise. First, it is yet uncertain whether humans ever can reach such complete specification of a brain in real time. Second, noisy perception limits both the predictability of any person's perceptions of sensations and thus any observer's perceptions of the state of another person's brain. Perfection in predicting a person's perceptions is thus an unattainable ideal in either case.

Next is the issue of a private language to describe one's own perceptions. One might thereby describe one's own reality—both perceptions
of the external world and second-order perceptions. In principle, a person's brain might construct such a language *ab initio.* That person would,
as Wittgenstein emphasized, have no way to confirm whether or not this
language produced descriptions of reality corresponding to anyone else's
language. Neither the categorization of sensations by naming nor the
grammar for encoding their relationships would necessarily correspond
to the categories or grammar of any other human's language. For instance, a combined sensation of red and blue light (as specified by
wavelengths) might be labeled by one person in the same way as a sensation of violet light (both called "purple") but by another person as a
distinct sensation of magenta. If any one language a person might construct has no confirmed relationship with external or internal sensations,
then any other language would share the same indeterminacy. Any such
private language would thus be an arbitrary locus in chaos circumscribed
by innate predispositions. Predispositions and the resulting universals
of language presumably can stabilize only rudimentary aspects of human
language. Only during the development and use of a public language,
during communication, can a person seek consilience with others in
naming and syntax. This chapter thus concludes that a private language,
even if possible, is anchored to reality by a rode of precariously short
scope. Because thought requires language, as I have proposed, a person's
subjectivity is on a lee shore. It can be rescued from drafts of doubt only
by communication. Yet there is still a difficulty.

Presumably two or more people might reach consilience that is no less
arbitrary than a private language, no more than an agreed upon position beyond possibility of correction. This outcome, however, is unlikely
for organisms subject to natural selection. This principle sets limits for
the arbitrariness of language in relation to reality and instead yields some
correspondence between signals and reality, both external and internal.
An arbitrary language would not provide any advantage in an organism's dealings with reality. Natural selection of brains would instead produce a relationship between signals and environment with some net
advantage for an organism on average. As argued in Parts II and III, honesty is the norm in the evolution of communication. For perception,
which (as we have also seen) is a form of communication, accuracy is

the analog of honesty. Accuracy is thus the norm in the evolution of perception—at least accuracy with enough precision to provide an advantageous correspondence with reality. Nevertheless, natural selection produces adaptations for promoting the spread of alleles, so it is likely that it would promote brains capable of accurate communication with others about perceptions of reality. Communication could refine perceptions of reality, so people resolutely seeking consilience in language might approach perfection in use of language to describe reality in the external environment. How far this process might go we do not know. As urged in this book, perceptions are subject to noise. Recall that noise comes from sources both external and internal to any observer (a receiver of signals). It includes, for instance, unavoidable (molecular and thus thermal) noise in a receiver's nervous system. Noise then limits the precision of any person's language. Consilience can improve the precision of language, but no amount of averaging can convert error in a small sample (one person) to absolute certainty in a large sample (consilience among many). Communication rescues subjectivity from idiosyncrasies of solipsism, but it does not assure perfection in perceptions of reality.

As for human language as a shared, cultural phenomenon, its essential properties and their evolution have evoked much discussion. I have emphasized two features of language: (1) innate predispositions to focus initial stages of learning; and (2) complexity, including higher-order relationships, in the categorization and association of perceptions (signals) and responses by receivers. It seems to me that any complex communication is bound to involve combinations of elements and predispositions for learning. All evidence is consistent with these two features of language, and we can hope for agreement on some finer points before long. New evidence continues to accumulate at a great rate. Some examples of the current state of investigation include . . .

Fitch, W. T. 2010. *The evolution of language.* Cambridge: Cambridge University Press.

Hauser, M. D., N. Chomsky, and W. T. Fitch. 2002. The faculty of language: What is it, who has it, and how did it evolve? *Science* 298: 1569–1579.

Pinker, S., and R. Jackendoff. 2005. The faculty of language: What's special about it? *Cognition* 95: 201–236.

The evolution of communication, and language in particular, has also been studied by computer simulation. Much of this work has focused on the evolution of stable communication and reliable references for signals. For a recent summary . . .

Scott-Phillips, T. C., and S. Kirby. 2013. Information, influence and inference in language evolution. In U. E. Stegmann, ed., *Animal communication theory: Information and influence,* 421–438. Cambridge: Cambridge University Press.

Mathematical models of communication by Martin Nowak and colleagues include a demonstration that errors limit the optimal number of simple signals. Combinations of signals thus become important. The following study includes error as variation around a mean, not as the interacting errors of signal detection . . .

Nowak, M. A., D. C. Krakauer, and A. Dress. 1999. An error limit for the evolution of language. *Proceedings of the Royal Society B: Biological Sciences* 266: 2131–2136.

Rudimentary communication by means of home sign in deaf children with no exposure to a sign language provides evidence for predispositions in the development of language. Although their gestures are not simply imitations, these children have engaged in basic communication with their families over a prolonged period . . .

Goldin-Meadow, S., et al. 1994. Nouns and verbs in a self-styled gesture system: What's in a name? *Cognitive Psychology* 27: 259–319. This article describes the investigators' procedures in some detail.

Goldin-Meadow, S., C. Mylander, and C. Butcher. 1995. The resilience of combinatorial structure at the word level: Morphology in self-styled gesture systems. *Cognition* 56: 195–262.

Goldin-Meadow, S., and C. Mylander. 1998. Spontaneous sign systems created by deaf children in two cultures. *Nature* 391: 279–281.

Zheng, M., and S. Goldin-Meadow. 2002. Thought before language: How deaf and hearing children express motion events across cultures. *Cognition* 85: 145–175.

Goldin-Meadow, S. 2003. *The resilience of language: What gesture creation in deaf children can tell us about how all children learn language.* New York: Psychology Press.

On the other hand, deaf children without a full sign language provide clear evidence that language is needed for complex thinking (numerical and spatial cognition in these studies) . . .

Spaepen, E., et al. 2011. Number without a language model. *Proceedings of the National Academy of Sciences of the United States of America* 108: 3163–3168.

Pyers, J. E., et al. 2010. Evidence from an emerging sign language reveals that language supports spatial cognition. *Proceedings of the National Academy of Sciences of the United States of America* 107: 12116–12120.

In the large literature on Nicaraguan Sign Language (NSL), recent studies have documented the emergence of properties in common with all human languages. Although it is clear that many properties of mature NSL are not imitations of other sign languages, nor of gestures of speaking Nicaraguans, nor written or spoken Spanish, the early stages of NSL were not well documented.

Senghas, A., S. Kita, and A. Özyürek. 2004. Children creating core properties of language: Evidence from an emerging sign language in Nicaragua. *Science* 305: 1779–1782.

Sandler, W., et al. 2005. The emergence of grammar: Systematic structure in a new language. *Proceedings of the National Academy of Sciences of the United States of America* 102: 2661–2665.

Senghas, R. J., A. Senghas, and J. E. Pyers. 2005. The emergence of Nicaraguan Sign Language: questions of development, acquisition, and evolution. In S. T. Parker, J. Langer, and C. Milbrath, eds., *Biology and knowledge revisited: From neurogenesis to psychogenesis,* 287–306. Mahwah, NJ: Erlbaum.

For social influences on development of song in songbirds . . .

Baptista, L. F., and L. Petrinovich. 1984. Social interaction, sensitive phases and the song template hypothesis in the White-crowned Sparrow. *Animal Behaviour* 32: 172–181.

West, M. J., and A. P. King. 1988. Female visual displays affect the development of male song in the cowbird. *Nature* 334: 244–246.

Byers, B. E., and D. E. Kroodsma. 1992. Development of two song categories by Chestnut-sided Warblers. *Animal Behaviour* 44: 799–810.

Liu, W. C., and D. E. Kroodsma. 1999. Song development by Chipping Sparrows and Field Sparrows. *Animal Behaviour* 57: 1275–1286.

Nordby, J. C., et al. 2000. Social influences during song development in the Song Sparrow: A laboratory experiment simulating field conditions. *Animal Behaviour* 59: 1187–1197.

The positions of John Locke, George Berkeley, Aristotle, Karl Popper, and Michael Polanyi are all explained in their well-known publications.

The paraphrase of Paul comes from the resonant verses, 1 Corinthians 13: 12–13, in the translation organized by James I: "For now we see through a glass, darkly; but then face to face; now I know in part; but then shall I know even as also I am known. And now abideth faith, hope, charity, these three; but the greatest of these is charity." The term translated as "charity" is the Greek *agape,* also often translated as "love"—in the modern sense of brotherly (or charitable) love. The Greek term translated as "glass" has uncertain meaning (the problem of translation surfaces here). Most commentators seem to feel that its most likely referent is a mirror (in ancient times usually polished brass), and it is sometimes translated thus. The polished brass of ancients seems likely to have produced the noisy perception implied by Paul.

ACKNOWLEDGMENTS

It has been a long road and many have paved the way for me. The first steps were assisted by my parents, Marybeth and Richard, who imparted some of their flair for art and music, nature and science, and especially adventure. It was decades before I could merge these strains. All would have come to naught without the foresight of three perspicacious teachers. If Donald Griffin and Keith Porter had not helped with an early decision, I would never have studied with Peter Marler. He encouraged, altogether altruistically, my direction along the path that led to this book. For me his scientific acumen remains unrivaled. He was a model of dedication with humor. After my teachers, my students continued the impetus. My research students have kept me thinking about far-ranging issues and buoyed the journey with their exploits and laughter. Many are mentioned in this book, but the others have been just as lively companions. And the nearly one thousand undergraduates who accompanied me overnight on many trips to remote and frozen corners of North Carolinian wilderness made me think clearly about the most basic, and hence most critical, issues. My colleagues in Chapel Hill, especially Helmut Mueller, helped with numerous challenges, and for a decade a rollicking course taught jointly with Steve Nowicki at our two institutions prodded me along the way.

Sabbatical visits with Peter Slater and John Maynard Smith at the University of Sussex and with Nick Davies at the University of Cambridge provided opportunities to develop ideas. The Department of

Ornithology at Harvard University's Museum of Comparative Zoology has twice been an oasis for recharging my resolve. Yet the critical impetus to surmount the final hurdles came from a symposium in my honor arranged by my former students Marc Naguib and Jordan Price at a meeting of the Animal Behavior Society. They also subsequently edited a special volume of *Behaviour,* which included reports by many of those present at the symposium. All those who contributed on these occasions have my gratitude.

Colleagues who have helped more than could be expected by critically reading or discussing inchoate versions of chapters include Bob Dooling, Michael Green, Albert Harris, Elizabeth Harris, Alan Jones, Laurie McNeil, Marc Naguib, David Pfennig, Walter Piper, Jordan Price, Maria Servedio, Rod Suthers, and James Whitehead. Two anonymous reviewers also had helpful suggestions. Michael Fisher's editorial advice has proven to be critical at several difficult passes. Needless to say, any shortcomings that remain are my own.

The illustrations could never have been completed so quickly or expertly without the cheerful assistance of Brian Nalley and Susan Whitfield. Corey Johnson helped with a critical dissection. Some of the research described herein was supported by the National Science Foundation and by Kenan and Pogue Fellowships from the University of North Carolina at Chapel Hill.

My two children, aside from their joyful company in many zany ventures, made special contributions, Aleta with her sharp ear for words and eye for images, and Keith with his keen nose for edgy science and his comradery during some ornery programming. Yet the one without whom this pilgrimage would never have been so much fun and would never have sailed the oceans and passed the mountains along the way is my wife, Minna. She made sure we got where we wanted to go, from outermost reef to innermost forest, and always in just the right style. She also proposed the suitably ambiguous title for this book, which addresses diverse noise matters but also reveals why noise matters.

INDEX